Titles in This Series

Volume

1 **Markov random fields and their applications,** Ross Kindermann and J. Laurie Snell

2 **Proceedings of the conference on integration, topology, and geometry in linear spaces,** William H. Graves, Editor

3 **The closed graph and P-closed graph properties in general topology,** T. R. Hamlett and L. L. Herrington

4 **Problems of elastic stability and vibrations,** Vadim Komkov, Editor

5 **Rational constructions of modules for simple Lie algebras,** George B. Seligman

6 **Umbral calculus and Hopf algebras,** Robert Morris, Editor

7 **Complex contour integral representation of cardinal spline functions,** Walter Schempp

8 **Ordered fields and real algebraic geometry,** D. W. Dubois and T. Recio, Editors

9 **Papers in algebra, analysis and statistics,** R. Lidl, Editor

10 **Operator algebras and K-theory,** Ronald G. Douglas and Claude Schochet, Editors

11 **Plane ellipticity and related problems,** Robert P. Gilbert, Editor

12 **Symposium on algebraic topology in honor of José Adem,** Samuel Gitler, Editor

13 **Algebraists' homage: Papers in ring theory and related topics,** S. A. Amitsur, D. J. Saltman, and G. B. Seligman, Editors

14 **Lectures on Nielsen fixed point theory,** Boju Jiang

15 **Advanced analytic number theory. Part I: Ramification theoretic methods,** Carlos J. Moreno

16 **Complex representations of $GL(2, K)$ for finite fields K,** Ilya Piatetski-Shapiro

17 **Nonlinear partial differential equations,** Joel A. Smoller, Editor

18 **Fixed points and nonexpansive mappings,** Robert C. Sine, Editor

19 **Proceedings of the Northwestern homotopy theory conference,** Haynes R. Miller and Stewart B. Priddy, Editors

20 **Low dimensional topology,** Samuel J. Lomonaco, Jr., Editor

21 **Topological methods in nonlinear functional analysis,** S. P. Singh, S. Thomeier, and B. Watson, Editors

22 **Factorizations of $b^n \pm 1$, b = 2, 3, 5, 6, 7, 10, 11, 12 up to high powers,** John Brillhart, D. H. Lehmer, J. L. Selfridge, Bryant Tuckerman, and S. S. Wagstaff, Jr.

23 **Chapter 9 of Ramanujan's second notebook—Infinite series identities, transformations, and evaluations,** Bruce C. Berndt and Padmini T. Joshi

24 **Central extensions, Galois groups, and ideal class groups of number fields,** A. Fröhlich

25 **Value distribution theory and its applications,** Chung-Chun Yang, Editor

26 **Conference in modern analysis and probability,** Richard Beals, Anatole Beck, Alexandra Bellow, and Arshag Hajian, Editors

27 **Microlocal analysis,** M. Salah Baouendi, Richard Beals, and Linda Preiss Rothschild, Editors

28 **Fluids and plasmas: geometry and dynamics,** Jerrold E. Marsden, Editor

29 **Automated theorem proving,** W. W. Bledsoe and Donald Loveland, Editors

30 **Mathematical applications of category theory,** J. W. Gray, Editor

31 **Axiomatic set theory,** James E. Baumgartner, Donald A. Martin, and Saharon Shelah, Editors

32 **Proceedings of the conference on Banach algebras and several complex variables,** F. Greenleaf and D. Gulick, Editors

33 **Contributions to group theory,** Kenneth I. Appel, John G. Ratcliffe, and Paul E. Schupp, Editors

34 **Combinatorics and algebra,** Curtis Greene, Editor

Titles in This Series

Volume

35 **Four-manifold theory,** Cameron Gordon and Robion Kirby, Editors

36 **Group actions on manifolds,** Reinhard Schultz, Editor

37 **Conference on algebraic topology in honor of Peter Hilton,** Renzo Piccinini and Denis Sjerve, Editors

38 **Topics in complex analysis,** Dorothy Browne Shaffer, Editor

39 **Errett Bishop: Reflections on him and his research,** Murray Rosenblatt, Editor

40 **Integral bases for affine Lie algebras and their universal enveloping algebras,** David Mitzman

41 **Particle systems, random media and large deviations,** Richard Durrett, Editor

42 **Classical real analysis,** Daniel Waterman, Editor

43 **Group actions on rings,** Susan Montgomery, Editor

44 **Combinatorial methods in topology and algebraic geometry,** John R. Harper and Richard Mandelbaum, Editors

45 **Finite groups–coming of age,** John McKay, Editor

46 **Structure of the standard modules for the affine Lie algebra $A_1^{(1)}$,** James Lepowsky and Mirko Primc

47 **Linear algebra and its role in systems theory,** Richard A. Brualdi, David H. Carlson, Biswa Nath Datta, Charles R. Johnson, and Robert J. Plemmons, Editors

48 **Analytic functions of one complex variable,** Chung-chun Yang and Chi-tai Chuang, Editors

49 **Complex differential geometry and nonlinear differential equations,** Yum-Tong Siu, Editor

50 **Random matrices and their applications,** Joel E. Cohen, Harry Kesten, and Charles M. Newman, Editors

51 **Nonlinear problems in geometry,** Dennis M. DeTurck, Editor

52 **Geometry of normed linear spaces,** R. G. Bartle, N. T. Peck, A. L. Peressini, and J. J. Uhl, Editors

53 **The Selberg trace formula and related topics,** Dennis A. Hejhal, Peter Sarnak, and Audrey Anne Terras, Editors

54 **Differential analysis and infinite dimensional spaces,** Kondagunta Sundaresan and Srinivasa Swaminathan, Editors

55 **Applications of algebraic K-theory to algebraic geometry and number theory,** Spencer J. Bloch, R. Keith Dennis, Eric M. Friedlander, and Michael R. Stein, Editors

56 **Multiparameter bifurcation theory,** Martin Golubitsky and John Guckenheimer, Editors

57 **Combinatorics and ordered sets,** Ivan Rival, Editor

58.I **The Lefschetz centennial conference. Proceedings on algebraic geometry,** D. Sundararaman, Editor

58.II **The Lefschetz centennial conference. Proceedings on algebraic topology,** S. Gitler, Editor

58.III **The Lefschetz centennial conference. Proceedings on differential equations,** A. Verjovsky, Editor

59 **Function estimates,** J. S. Marron, Editor

60 **Nonstrictly hyperbolic conservation laws,** Barbara Lee Keyfitz and Herbert C. Kranzer, Editors

61 **Residues and traces of differential forms via Hochschild homology,** Joseph Lipman

62 **Operator algebras and mathematical physics,** Palle E. T. Jorgensen and Paul S. Muhly, Editors

63 **Integral geometry,** Robert L. Bryant, Victor Guillemin, Sigurdur Helgason, and R. O. Wells, Jr., Editors

64 **The legacy of Sonya Kovalevskaya,** Linda Keen, Editor

65 **Logic and combinatorics,** Stephen G. Simpson, Editor

66 **Free group rings,** Narian Gupta

67 **Current trends in arithmetical algebraic geometry,** Kenneth A. Ribet, Editor

Titles in This Series

Volume

68 **Differential geometry: The interface between pure and applied mathematics,** Mladen Luksic, Clyde Martin, and William Shadwick, Editors

69 **Methods and applications of mathematical logic,** Walter A. Carnielli and Luiz Paulo de Alcantara, Editors

70 **Index theory of elliptic operators, foliations, and operator algebras,** Jerome Kaminker, Kenneth C. Millett, and Claude Schochet, Editors

71 **Mathematics and general relativity,** James A. Isenberg, Editor

72 **Fixed point theory and its applications,** R. F. Brown, Editor

73 **Geometry of random motion,** Rick Durrett and Mark A. Pinsky, Editors

74 **Geometry of group representations,** William M. Goldman and Andy R. Magid, Editors

75 **The finite calculus associated with Bessel functions,** Frank M. Cholewinski

76 **The structure of finite algebras,** David C. Hobby and Ralph Mckenzie

77 **Number theory and its applications in China,** Wang Yuan, Yang Chung-chun, and Pan Chengbiao, Editors

78 **Braids,** Joan S. Birman and Anatoly Libgober, Editors

79 **Regular differential forms,** Ernst Kunz and Rolf Waldi

80 **Statistical inference from stochastic processes,** N. U. Prabhu, Editor

81 **Hamiltonian dynamical systems,** Kenneth R. Meyer and Donald G. Saari, Editors

82 **Classical groups and related topics,** Alexander J. Hahn, Donald G. James, and Zhe-xian Wan, Editors

83 **Algebraic K-theory and algebraic number theory,** Michael R. Stein and R. Keith Dennis, Editors

84 **Partition problems in topology,** Stevo Todorcevic

85 **Banach space theory,** Bor-Luh Lin, Editor

86 **Representation theory and number theory in connection with the local Langlands conjecture,** J. Ritter, Editor

87 **Abelian group theory,** Laszlo Fuchs, Rüdiger Göbel, and Phillip Schultz, Editors

88 **Invariant theory,** R. Fossum, W. Haboush, M. Hochster, and V. Lakshmibai, Editors

89 **Graphs and algorithms,** R. Bruce Richter, Editor

90 **Singularities,** Richard Randell, Editor

91 **Commutative harmonic analysis,** David Colella, Editor

92 **Categories in computer science and logic,** John W. Gray and Andre Scedrov, Editors

93 **Representation theory, group rings, and coding theory,** M. Isaacs, A. Lichtman, D. Passman, S. Sehgal, N. J. A. Sloane, and H. Zassenhaus, Editors

94 **Measure and measurable dynamics,** R. Daniel Mauldin, R. M. Shortt, and Cesar E. Silva, Editors

95 **Infinite algebraic extensions of finite fields,** Joel V. Brawley and George E. Schnibben

96 **Algebraic topology,** Mark Mahowald and Stewart Priddy, Editors

97 **Dynamics and control of multibody systems,** J. E. Marsden, P. S. Krishnaprasad, and J. C. Simo, Editors

98 **Every planar map is four colorable,** Kenneth Appel and Wolfgang Haken

99 **The connection between infinite dimensional and finite dimensional dynamical systems,** Basil Nicolaenko, Ciprian Foias, and Roger Temam, Editors

100 **Current progress in hyperbolic systems: Riemann problems and computations,** W. Brent Lindquist, Editor

101 **Recent developments in geometry,** S.-Y. Cheng, H. Choi, and Robert E. Greene, Editors

102 **Primes associated to an ideal,** Stephen McAdam

103 **Coloring theories,** Steve Fisk

Titles in This Series

Volume

104 Accessible categories: The foundations of categorical model theory, Michael Makkai and Robert Paré

105 Geometric and topological invariants of elliptic operators, Jerome Kaminker, Editor

106 Logic and computation, Wilfried Sieg, Editor

107 Harmonic analysis and partial differential equations, Mario Milman and Tomas Schonbek, Editors

108 Mathematics of nonlinear science, Melvyn S. Berger, Editor

109 Combinatorial group theory, Benjamin Fine, Anthony Gaglione, and Francis C. Y. Tang, Editors

110 Lie algebras and related topics, Georgia Benkart and J. Marshall Osborn, Editors

111 Finite geometries and combinatorial designs, Earl S. Kramer and Spyros S. Magliveras, Editors

Finite Geometries and Combinatorial Designs

Dale M. Mesner

CONTEMPORARY MATHEMATICS

111

Finite Geometries and Combinatorial Designs

Proceedings of the AMS Special Session in
Finite Geometries and Combinatorial Designs
held October 29–November 1, 1987

Earl S. Kramer and
Spyros S. Magliveras, Editors

AMERICAN MATHEMATICAL SOCIETY • PROVIDENCE, RHODE ISLAND

The AMS Special Session in Finite Geometries and Combinatorial Designs was held at the University of Nebraska, Lincoln, Nebraska, October 29–November 1, 1987.

1980 *Mathematics Subject Classification* (1985 *Revision*). Primary 05BXX, 51EXX.

Library of Congress Cataloging-in-Publication Data

Finite geometries and combinatorial designs: proceedings of the AMS Special Session in Finite Geometries and Combinatorial Designs held October 29–November 1, 1987 [Lincoln, Nebraska]/Earl S. Kramer and Spyros S. Magliveras, editors.
 p. cm.—(Contemporary mathematics, ISSN 0271-4132; v. 111)
 Includes bibliographical references.
 ISBN 0-8218-5118-7
 1. Finite geometries—Congresses. 2. Combinatorial designs and configurations—Congresses. I. Kramer, Earl S. (Earl Sidney), 1940–. II. Magliveras, Spyros S. (Spyros Simos), 1938–. III. Title. IV. Series: Contemporary mathematics (American Mathematical Society); v. 111.
QA167.2.A47 1987
515'.13—dc20

90-45302
CIP

DEDICATED TO

DALE M. MESNER

Contents

Preface   xiii

List of Contributors   xv

List of Participants   xvii

A Note on Hamilton Cycles in Block-Intersection Graphs
B. ALSPACH, K. HEINRICH, AND B. MOHAR   1

All Dicyclic Groups of Order at Least Twelve Have Symmetric Sequencings
B. A. ANDERSON   5

Concerning Pairwise Balanced Designs With Prime Power Block Sizes
F. E. BENNETT   23

On the p-Rank of Incidence Matrices and a Question of E.S. Lander
A. A. BRUEN AND U. OTT   39

On the Dempwolff Plane
M. J. DE RESMINI   47

Difference Sets in 2-Groups
J. F. DILLON   65

On the Existence of Room Squares with Subsquares
J. H. DINITZ AND D. R. STINSON   73

A Bound for Blocking Sets in Finite Projective Planes
D. A. DRAKE   93

Sets With More Than One Representation as an Algebraic Curve of Degree Three
J. W. P. HIRSCHFELD AND J. A. THAS   99

Flocks and Partial Flocks of Quadric Sets
N. L. JOHNSON   111

The Finite Flag-Transitive Linear Spaces with an Exceptional
Automorphism Group
P. B. Kleidman   117

Constructing 6-(14,7,4) Designs
D. L. Kreher and S. P. Radziszowski   137

On the Classification of Finite C_n-Geometries with Thick Lines
A. Pasini   153

Cyclic Codes and Cyclic Configurations
V. Pless   171

Designs and Approximation
J. J. Seidel   179

Flocks, Maximal Exterior Sets, and Inversive Planes
J. A. Thas   187

Self-Orthogonal Designs
V. D. Tonchev   219

Nonembeddable Quasi-Residual Designs
T. van Trung   237

Finite Planes And Clique Partitions
W. D. Wallis   279

Oval Designs in Quadrics
M. A. Wertheimer   287

On the Order of a Finite Projective Plane and its Collineation Group
C. Y. Ho   299

Some Geometric Aspects of Root Finding in $GF(q^m)$
P. C. van Oorschot and S. A. Vanstone   303

Automorphism Groups as Linear Groups
J. Siemons   309

Preface

A special session on *Finite Geometries and Combinatorial Designs* was held at a regional conference of the American Mathematical Society in Lincoln, Nebraska during October 29–November 1, 1987. The organizers of the session were Dale Mesner and the editors of these proceedings. We were pleasantly surprised at the large number of participants from several countries who came to Lincoln for the three-day conference. We were even more pleased at the very high quality of results presented at the conference and submitted to the proceedings. We wish to thank conference participants, contributors to this volume, and referees. The editors take the liberty of dedicating this volume to Dale M. Mesner in appreciation of his contributions to combinatorics.

The editors express their thanks to the University of Nebraska, especially the Department of Computer Science and the Department of Mathematics, for funding and other support. Very special gratitude is due to Leanne Magliveras for her expert typesetting of the manuscripts.

Spyros S. Magliveras
Earl S. Kramer
Lincoln, Nebraska
November, 1989

List of Contributors

ALSPACH, Brian	Simon Fraser University, Burnaby BC
ANDERSON, Bruce A.	Arizona State University, Tempe, AZ
BENNETT, Frank E.	Mount St. Vincent University, Halifax, Nova Scotia
BRUEN, Aiden A.	University of Western Ontario, London, Canada
de RESMINI, Marialuisa J.	Università di Roma "La Sapienza", Italy
DILLON, John F.	New Carrolton, MD
DINITZ, Jeff H.	University of Vermont, Burlington, VT
DRAKE, David A.	University of Florida, Gainesville, FL
HEINRICH, Katherine	Simon Fraser University, Burnaby BC
HIRSCHFELD, James W.P.	University of Sussex, Brighton UK
HO, Chat Y.	University of Florida, Gainesville, FL
JOHNSON, Norman L.	University of Iowa, Iowa City, IA
KLEIDMAN, Peter B.	Trinity College, Cambridge, UK
KREHER, Don L.	Rochester Institute of Technology, Rochester, NY
MOHAR, Bojan	University of Ljubljana, Yugoslavia
OTT, U.	University of Braunschweig, West Germany
PASINI, Antonio	Università di Siena, Italy
PLESS, Vera S.	University of Illinois, Chicago, IL
RADZISZOWSKI, Stanislaw P.	Rochester Institute of Technology, Rochester, N.Y.
SEIDEL, J. J.	Technische Hogeschool Eindhoven, The Netherlands
SIEMONS, Johannes	University of East Anglia, Norwich, UK
STINSON, Doug R.	University of Manitoba Winnepeg, Manitoba
THAS, Jef A.	State University of Ghent, Belgium
TONCHEV, Vladimir, D.	Bulgarian Academy of Sciences, Sofia, Bulgaria
van OORSCHOT, P. C.	University of Waterloo, Ontario, Canada
van TRUNG, Tran	University of Heidelberg, West Germany
VANSTONE, Scott A.	University of Waterloo, Ontario, Canada
WERTHEIMER, Michael A.	Department of Defense, Fort George G. Meade, MD
WALLIS, Walter D.	Southern Illinois University, Carbondale, IL

Conference Participants

List of Participants

ADAMS, Michael
ANDERSON, Bruce A.   Arizona State University, Tempe, AZ
ARCHDEACON, Dan   University of Vermont, Burlington, VT
ATKINSON, Michael D.   Carleton University, Ottawa, Canada
BAKER, Ronald D.   University of Delaware, Newark, DE
BANNAI, Eiichi   Ohio State University, Columbus, OH
BATTEN, Lynn M.   University of Winnipeg, Manitoba, Canada
BENNETT, Frank   Mount St. Vincent University, Halifax, Nova Scotia, Canada
BHATTACHARYA, Prabir   University of Nebraska Lincoln, NE
BEUTELSPACHER, A.   Siemens - München, West Germany
BRIDGES, W   University of Wyoming, Laramie, WY
BROWN, Julia M.   York University, North York, Ontario, Canada
BRUALDI, Richard A.   University of Wisconsin, Madison, WI
BRUEN, Aiden A.   University of Western Ontario, London, Ontario, Canada
CALDERBANK, R.   Bell Laboratories, Murray Hill, NJ
CARRAMINARA, Rodrigo   University of Iowa, Iowa City, IA
CHEROWITZO, Bill E.   University of Colorado at Denver, CO
CHOUINARD, Leo G.   University of Nebraska, Lincoln, NE
COLBOURN, Charles J.   University of Waterloo, Ontario, Canada
CORDERO-BRANA, Minerva   University of Iowa, Iowa City, IA
CRITTENDEN, Andre   Colorado State University, Fort Collins, CO
de RESMINI, Marialuisa J.   Università di Roma "La Sapienza", Italy
DILLON, John F.   New Carrolton, MD
DINITZ, Jeff H.   University of Vermont, Burlington, VT
DOOB, Michael   University of Manitoba, Winnipeg, Canada
DOYEN, Jean   Université Libre de Bruxelles, Belgium
DRAKE, David A.   University of Florida, Gainesville, FL
EALY, Clifton E., Jr.   Northern Michigan University, Marquette,MI
ELLARD, Cecil Andrew   Miami University, Oxford, OH
FIGUEROA, Raul   University of Iowa, Iowa City, IA
FURINO, Steven   University of Waterloo, Ontario, Canada
GRAMS, Gerhard
GAWGIRDWIBOON, Suwimon   University of Nebraska, Lincoln, NE
HARTMAN, Alan   IBM - Israel
HEINRICH, Katherine   Simon Fraser University, Burnaby, BC, Canada
HIRSCHFELD, James W.P.   University of Sussex, Brighton, UK
HO, Chat Y.   University of Florida, Gainesville, FL
HOBART, Sylvia   University of Wyoming, Laramie, WY
HUGHES, Dan R.   Queen Mary College, London, UK
JOHNSON, Norman L.   University of Iowa, Iowa City, IA
KLARNER, David A.   University of Nebraska, Lincoln, NE

KLEIDMAN, Peter B.	Trinity College, Cambridge, UK
KRAMER, Earl E.	University of Nebraska, Lincoln, NE
KREHER, Don L.	Rochester Institute of Technology, Rochester, NY
LAMKEN, Esther	Institute for Defense Analyses, Princeton, NJ
LA ROSA, Myrna	University of Iowa, Iowa City, IA
LIEBLER, Bob	Colorado State University, Fort Collins, CO
MAGLIVERAS, Spyros	University of Nebraska, Lincoln, NE
MATHON, Rudi	University of Toronto, Ontario, Canada
MCLEAN, Jeffrey Thomas	College of St. Thomas, St. Paul, MN
MEMON, Nasir	University of Nebraska, Lincoln, NE
MENDELSOHN, Eric	University of Toronto, Ontario, Canada
MENDELSOHN, Nathan S.	University of Manitoba, Winnipeg, Canada
MESNER, Dale	University of Nebraska, Lincoln, NE
OSTROM, T. G.	Washington State University, Pullman, WA
PADMANABHAN, R.	University of Manitoba, Winnipeg, Canada
PASINI, Antonio	Università di Siena, Italy
PAYNE, Stan E.	University of Colorado at Denver, CO
PLESS, Vera S.	University of Illinois, Chicago, IL
POMAREDA, Rolando	Universidad de Chile, Santiago, Chile
RAY-CHAUDHURI, Dijen K.	Ohio State University, Columbus, OH
RODGERS, Chris	Auburn University, Alabama
ROSA, Alex	McMaster University, Hamilton, Ontario, Canada
SCHELLENBERG, Paul J.	University of Waterloo, Ontario, Canada
SEIDEL, Jaap J.	Technische Hogeschool Eindhoven, The Netherlands
SHULT, Ernie	Kansas State University, Manhattan, KS
SIEMONS, Johannes	University of East Anglia, Norwich, UK
SPENCE, Ted	University of Glasgow, Scotland, UK
STINSON, Doug R.	University of Manitoba, Winnipeg, Canada
TAM, Kok C.	University of Western Ontario, London, Ontario, Canada
TEIRLINCK, Luc	Auburn University, Alabama
THAS, Jef A.	State University of Ghent, Belgium
VALDES, Silvia	University of Iowa, Iowa City, IA
VANSTONE, Scott A.	University of Waterloo, Ontario, Canada
van TRUNG, Tran	University of Heidelberg, West Germany
WALLIS, Walter D.	Southern Illinois University, Carbondale, IL
WARD, John J.	
WERTHEIMER, Michael A.	Department of Defense, Fort George G. Meade, MD
WISEMAN, James A.	Massachusetts Institute of Technology, Cambridge, MA
WILSON, Stephen	Northern Arizona University, Flagstaff, Arizona
WU, Qiu-Rong	University of Nebraska, Lincoln, NE

Contemporary Mathematics
Volume 111, 1990

A Note on Hamilton Cycles in
Block-Intersection Graphs

BRIAN ALSPACH, KATHERINE HEINRICH, AND BOJAN MOHAR

ABSTRACT. We show that the block-intersection graph of a pairwise balanced design has a Hamilton cycle provided that the cardinality of a largest block in the design is no more than twice the cardinality of a smallest block.

A set $V = \{1, 2, \ldots, n\}$ together with a collection $\mathbf{B} = \{B_1, \ldots, B_m\}$ of subsets of V is called an *incidence structure*. The elements of $\mathbf{B}$ are called *blocks* and the *size* of a block is the number of elements in it. Let $\mathbf{B}$ be an incidence structure. The *block intersection graph* of $\mathbf{B}$, denoted $\mathbf{G}(\mathbf{B})$, has the blocks of $\mathbf{B}$ for its vertices and two vertices are adjacent if and only if the blocks have non-empty intersection.

EXAMPLES: (i) Let $\mathbf{G}$ be a graph. Letting the vertices of $\mathbf{G}$ be the elements of V and the edges of $\mathbf{G}$ be the blocks, $\mathbf{G}$ determines an incidence structure $\mathbf{B}$. The block intersection graph of $\mathbf{B}$ is then the ordinary line graph of $\mathbf{G}$. C. Thomassen has conjectured that every 4-connected line graph has a Hamilton cycle.

(ii) Let $\mathbf{B}$ be a Steiner triple system. At a regional meeting of the American Mathematical Society in March 1987, R. L. Graham asked if the block-intersection graph of a Steiner triple system has a Hamilton cycle. This question is answered in the affirmative in [1] and is a corollary of the main result in this paper.

Let $\mathbf{B}$ be an incidence structure on a set V. Define $\mathbf{M}(\mathbf{B})$ to be the multigraph with vertex-set V such that for each pair of distinct vertices a and b, there is an edge between a and b of color B_i if $\{a, b\}$ is a subset of B_i for $B_i \in \mathbf{B}$.

1980 *Mathematics Subject Classification* (1985 *Revision*). Primary 05C38, 05C45; Secondary 05B05, 05B07.

This research was partially supported by the Natural Sciences and Engineering Research Council of Canada under Grants A-4792 and A-7829.

This paper is in final form and no version of it will be submitted for publication elsewhere.

A *trail* in a multigraph is a walk in which no edge appears twice. A trail in $\mathbf{M(B)}$ is said to be *block-dominating* if the vertices of the trail intersect every block of $\mathbf{B}$.

LEMMA. *Let* $\mathbf{B}$ *be an incidence structure. The block intersection graph of* $\mathbf{B}$ *is hamiltonian if and only if* $\mathbf{M(B)}$ *has a block-dominating closed trail* W *with no two edges of* W *having the same color.*

PROOF. Let $H = B_1, B_2, B_3, ..., B_m, B_1$ be a Hamilton cycle in $\mathbf{G(B)}$. Let a_i be an element of $B_i \cap B_{i+1}$, $i = 1, 2, ..., m$, where B_{m+1} is taken to be B_1. Consider the sequence $a_1, a_2, ..., a_m, a_1$ of elements of V. Replace each maximal length constant subsequence of consecutive elements by the single element making up the subsequence. For example, if $a_1 = a_2 = ... = a_j$, then replace this subsequence by a_1. The resulting closed trail is certainly block-dominating and each of its edges has a different color. (A single point is viewed as a closed trail.)

Let $W = a_1, e_1, a_2, e_2, ..., a_k, e_k, a_1$ be a closed block-dominating trail in $\mathbf{M(B)}$ such that each of its edges has a different color. Let B_i be a block containing the edge e_i. We construct a Hamilton cycle in $G(B)$ as follows. Let $C_1, C_2, ..., C_t$ be all the blocks of $\mathbf{B}$, other than $B_1, ..., B_k$, that contain the element a_1. Then $C_1, C_2, ..., C_t, B_1$ is a path in $\mathbf{G(B)}$. Now let $C_{t+1}, ..., C_{t+r}$ be all the blocks of $\mathbf{B}$, other than $B_2, ..., B_k$ and those already appearing in the path, that contain the element a_2. Then $C_1, C_2, ..., C_t, B_1, C_{t+1}, ..., C_{t+r}, B_2$ is a path in $\mathbf{G(B)}$ because B_1 also contains a_2. We continue in this way and produce a Hamilton cycle because W is block-dominating. This completes the proof.

A *pairwise balanced design* is an incidence structure $\mathbf{B}$ in which each unordered pair of distinct elements of V occurs in precisely λ blocks of $\mathbf{B}$.

THEOREM. *Let* B *be a pairwise balanced design with* $\lambda = 1$, M *the maximum size of a block of* $\mathbf{B}$ *and* m *the minimum size of a block of* $\mathbf{B}$. *If* $M \leq 2m$, *then the block intersection graph* $\mathbf{G(B)}$ *of* $\mathbf{B}$ *has a Hamilton cycle.*

PROOF. We construct a block-dominating closed trail in $\mathbf{M(B)}$. In fact, it will be a cycle in $\mathbf{M(B)}$. If $\mathbf{B}$ is the trivial design (that is, has only one block), then $\mathbf{G(B)}$ is K_1 which has a trivial Hamilton cycle. Thus, we assume that $\mathbf{B}$ is a non-trivial pairwise balanced design.

Since $\mathbf{B}$ is non-trivial, there are three points a, b, c in V not all in the same block. Then $C = a, b, c, a$ is a 3-cycle in $\mathbf{M(B)}$ all of whose edges have a different color. If C is block-dominating, we are finished. Otherwise, we extend C to be a longer cycle with edges of different colors. Eventually, we must reach a block-dominating cycle (for example, if the cycle contains all the vertices of $\mathbf{M(B)}$).

Let C be a cycle of $\mathbf{M(B)}$ all of whose edges have different colors and let B be a block of $\mathbf{B}$ that does not intersect C. For simplicity, assume the consecutive vertices of C are 1, 2, ..., k and $B = \{x_1, ..., x_b\}$.

The edge $e_i = i, i+1$ of C is said to be *destructive* if the block B_i containing it intersects B in some element x_j. If there is an edge $i+1, x_r$ (or i, x_r) whose color is not used on C, then C may be extended to a longer cycle by replacing the edge $i, i+1$ on C with the 3-path $i, x_j, x_r, i+1$.

If there are two consecutive destructive edges $e_{i-1} = i-1, i$ and $e_i = i, i+1$, then C may be extended as follows. Let x_j be a vertex of B so that the color of $i-1, x_j$ is the same as that of $i-1, i$ and let x_r be a vertex of B so that the color of $i+1, x_r$ is the same as that of $i, i+1$. Notice that x_j and x_r must be distinct vertices since C has no two edges of the same color. Then replace the two consecutive edges $i-1, i, i+1$ of C with the 3-path $i-1, x_j, x_r, i+1$.

Assume that C cannot be extended. Thus, we may assume that the destructive edges of C form a matching and that if $i, i+1$ is a destructive edge of C, then the colors of the edges i, x_j and $i+1, x_j$ are already used on C for every x_j belonging to B. Let p be the number of destructive edges on C. If the sizes of the corresponding blocks are $b_1, b_2, \ldots, b_p$, they destroy at most

$$(b_1 - 1) + \ldots + (b_p - 1) = \sum_{i=1}^{p} b_i - p$$

edges between $V(C)$ and B. Hence, there are at least

$$bk + p - \sum_{i=1}^{p} b_i$$

edges between $V(C)$ and B whose colors do not appear on C.

The destructive edges of C have $2p$ endvertices so that all the edges between $V(C)$ and B whose colors do not appear on C appear at the remaining $k-2p$ vertices. The number of edges at these vertices is $b(k-2p)$. Hence,

$$(k-2p)b \geq u \geq bk + p - \sum_{i=1}^{p} b_i \geq bk + p - pM,$$

where u denotes the number of edges between $V(C)$ and B whose colors do not already appear on C. The above implies that

$$pM - p - 2pb \geq 0,$$

which in turn implies that

$$M - 1 - 2b \geq 0.$$

The latter is a contradiction and completes the proof.

A *Steiner 2-design* is an incidence structure $\mathbf{B}$ in which each block has the same cardinality and $\lambda = 1$.

COROLLARY. (Horák and Rosa [1]). *The block intersection graph of a Steiner 2-design has a Hamilton cycle.*

REFERENCES

1. P. Horák and A. Rosa, *Decomposing Steiner triple systems into small configurations,* Ars Combinatoria, to appear.

DEPARTMENT OF MATHEMATICS AND STATISTICS, SIMON FRASER UNIVERSITY, BURNABY, BRITISH COLUMBIA, V5A 1S6 CANADA
Current address: B. Mohar, Department of Mathematics, University of Ljubljana, Jadranska 19, 61111 Ljubljana, Yugoslavia

Contemporary Mathematics
Volume **111**, 1990

All Dicyclic Groups of Order at Least Twelve Have Symmetric Sequencings

B. A. ANDERSON

ABSTRACT. If $n \geq 6$ is twice an odd integer, it is shown that the dicyclic group Q_{2n} of order 4n has a symmetric sequencing. It follows, in conjunction with earlier results, that if $n \geq 3$, Q_{2n} has a symmetric sequencing. One consequence is that if $n \geq 3$, the complete graph K_{4n+2} has a left even starter induced 1-factorization whose symmetry group contains the dicyclic group Q_{2n}. It appears that 1-factorizations of this type may be of use in the search for constructions of perfect 1-factorizations.

1. Introduction. It will perhaps be helpful to define several words appearing in the title of this paper and give a short review of previous work.

DEFINITION 1. Suppose G is a finite group of order n with identity e. A *sequencing* of G is an ordering $e, s_2, \ldots, s_n$ of all the elements of G such that the partial products $e, es_2, es_2s_3, \ldots, es_2 \cdots s_n$ are distinct and hence also comprise all of G. If G has order $2n$ and has a unique element z of order 2, a sequencing $e, s_2, \ldots, s_{2n}$ of G will be called a *symmetric sequencing* iff $s_{n+1} = z$ and for $1 \leq i \leq n-1$, $s_{n+1+i} = (s_{n+1-i})^{-1}$.

If z is the unique element of order 2 in G, then z is in the center of G. Thus symmetric sequencings

$$S : e, s_2, \ldots, s_n, z, s_n^{-1}, \ldots, s_3^{-1}, s_2^{-1}$$

have the associated partial product sequence

$$T : e, t_2, \ldots, t_n, t_n z, t_{n-1} z, \ldots, t_2 z, z.$$

DEFINITION 2. Suppose $n \geq 2$ is a positive integer. The *dicyclic group* Q_{2n} is the group of order 4n defined by

$$Q_{2n} = \{a^i b^j : 0 \leq i \leq 2n-1, 0 \leq j \leq 1, a^{2n} = e, b^2 = a^n, ba = a^{2n-1}b\}.$$

1980 *Mathematics Subject Classification* (1985 *Revision*). Primary 05B30, 20D60; Secondary 05C25.

This paper is in final form and no version of it will be submitted for publication elsewhere.

It was known in 1961 [5] that Q_4 has no sequencing. In 1970, a combination of theory and computer testing led to the claim in [4] that Q_6 was not sequenceable. Nevertheless, symmetric sequencings of Q_6 were produced in [2]. This was done as follows.

DEFINITION 3. Suppose $n \geq 2$ is a positive integer. The *dihedral group* D_n is the group of order $2n$ defined by

$$D_n = \{a^i b^j : 0 \leq i \leq n-1, 0 \leq j \leq 1, a^n = e, b^2 = e, ba = a^{n-1}b\}.$$

It is known that Q_{2n} has a unique element of order 2 and that $Q_{2n}/Z_2 \approx D_n$.

DEFINITION 4. Suppose H is a finite group of order n with identity e. A *2-sequencing* of H is an ordering $e, c_2, \ldots, c_n$ of certain elements of H (not necessarily distinct) such that

    i)   the associated partial products $e, ec_2, ec_2c_3, \ldots, ec_2 \cdots c_n$ are distinct and hence all of H,

    ii) if $y \in H$ and $y \neq y^{-1}$, then

$$\left| \{i : 2 \leq i \leq n \text{ and } (c_i = y \text{ or } c_i = y^{-1})\} \right| = 2,$$

    iii) if $y \in H$ and $y = y^{-1}$, then $\left| \{i : 1 \leq i \leq n \text{ and } c_i = y\} \right| = 1$.

The 2-sequencing is *reflective* iff for $2 \leq i, j \leq n$, if $i + j = n + 2$, then $c_i = c_j$.

References [2, 3] contain the following two results. Together they form the starting point for the current investigation.

THEOREM 1. Q_{2n} *has a symmetric sequencing iff* D_n *has a 2-sequencing.*

THEOREM 2. *If* $n \geq 3$ *and* n *is not twice an odd number, then* D_n *has a 2-sequencing.*

It is relatively easy to handle the n odd case in Theorem 2. The method for attacking the n even case described in [3] depends on the following (Z_n is the additive cyclic group of order n).

DEFINITION 5. If s is a 2-sequencing of Z_n, then s will be called *special* iff there is a $y \neq 0$ such that s and the associated partial sum sequence t begin as follows:

$$s_1, s_2, s_3 = 0, 2y, -y$$
$$t_1, t_2, t_3 = 0, 2y, y.$$

A conjecture stated in [3] is that if $n \geq 3$ is odd, then Z_n has a special 2-sequencing. It was shown that if this is true, then for $n \geq 4$, n even, D_n has a 2-sequencing. Although the conjecture has not been verified, there is a way around it that is exploited here.

DEFINITION 6. Suppose G is a group of order $2n$ with identity e and unique element z of order 2. Define

$$E = \{\{x_1, y_1\}, \{x_2, y_2\}, \ldots, \{x_{n-1}, y_{n-1}\}\},$$

a *left even starter* for G iff

    i) every nonidentity element of G except one, denoted m, occurs as an element of some pair of E,

    ii) every nonidentity element of G except z occurs in

$$\{x_i^{-1}y_i, \ y_i^{-1}x_i : 1 \le i \le n-1\}.$$

If E is a left even starter for G, define $E^* = E \cup \{e, m\}$ and $Q^* = \{\{x, xz\} : x \in G\}$.

Note that if G is abelian, the adjective "left" may be omitted from the previous definition. Now think of the elements of G as labeling the complete graph $K_{|G|}$. It is clear that when $m \ne z$, E^* and Q^* are disjoint 1-factors of $K_{|G|}$ and one may consider the associated 2-factor union.

THEOREM 3. [1]. *The group G has a symmetric sequencing iff G has a left even starter E such that $E^* \cup Q^*$ is a Hamiltonian circuit of $K_{|G|}$.*

The constructions to be described all use symmetric sequencings of Z_{2n} to build 2-sequencings of D_{2n}. Thus it will turn out to be crucial to have more detailed information about the left even starter mentioned in Theorem 3. This information, taken from [1], is listed below in the form in which it is used in this paper.

LEMMA 4. *Suppose s is a symmetric sequencing of Z_{2n} and t is the associated partial sum sequence.*

If n is odd, then

    i) $E = \{\{t_2, t_3\}, \dots, \{t_{n-1}, t_n\}, \{t_{n+2}, t_{n+3}\}, \dots, \{t_{2n-1}, t_{2n}\}\}$
    is an even starter for Z_{2n} (note that $t_1 = 0$ and t_{n+1} are not in any pair of E),

    ii) $E+n = \{\{0, t_2\}, \dots, \{t_{n-2}, t_{n-1}\}, \{t_{n+1}, t_{n+2}\}, \dots, \{t_{2n-2}, t_{2n-1}\}\}$
    is a translate of E (in reverse order).

If n is even, then

    iii) $E = \{\{t_2, t_3\}, \dots, \{t_{n-2}, t_{n-1}\}, \{t_{n+1}, t_{n+2}\}, \dots, \{t_{2n-1}, t_{2n}\}\}$ *is an even starter for Z_{2n} (note that $t_1 = 0$ and t_n are not in any pair of E),*

    iv) $E + n = \{\{0, t_2\}, \dots, \{t_{n-1}, t_n\}, \{t_{n+2}, t_{n+3}\}, \dots, \{t_{2n-2}, t_{2n-1}\}\}$
    is a translate E (in reverse order).

2. The constructions. It will be necessary to generalize the idea of a special 2-sequencing.

DEFINITION 7. Suppose s is a 2-sequencing of Z_n and t is the associated partial sum sequence. Define s to be a 2_d-*sequencing* iff there is a j, $2 \le j \le n-1$, and there is a y in Z_n such that

    i) $y \ne 0 \ne 2y$

    ii) $(t_j, t_{j+1}) \in \{(y, 2y), (2y, y)\}$.

There is a special case of this notion that merits a more detailed examination.

DEFINITION 8. Suppose $\Sigma_{2n} : 0, s_2, \ldots, s_{2n}$ is a symmetric sequencing of Z_{2n} and $\Omega_{2n} : 0, t_2, \ldots, t_{2n}$ is the associated partial sum sequence. The statement that Σ_{2n} is a *symmetric d-sequencing* means that there is a j, $2 \leq j \leq 2n - 1$, and there is a y in Z_{2n} such that

   i) $y \neq 0 \neq 2y$
   ii) $(t_j, t_{j+1}) \in \{(y, 2y), (2y, y)\}$.

If $j \leq n - 1$, then (t_j, t_{j+1}) is in the *first* half of Ω_{2n}; otherwise it is in the *second* half. If $(t_j, t_{j+1}) = (y, 2y)$ then Σ_{2n} is *forward*; otherwise it is *backward*. If $\{t_j, t_{j+1}\} \in E$ (see Lemma 4), then Σ_{2n} is *starter*: otherwise it is *translate*.

One is able to speak of things like second forward translate symmetric *d*-sequencings.

The remainder of the paper can now be succinctly outlined. The first goal (Theorem 9) is to show that symmetric d-sequencings of Z_{2n} induce 2-sequencings of D_{2n}. Conclude by verifying (Theorem 15) that all Z_{2n}, $n \geq 2$, have symmetric *d*-sequencings.

REMARK 5. If Σ_{2n} is a symmetric d-sequencing of Z_{2n} then $j \neq 1$, $j \neq n$ and if n is odd, $j \neq 2n - 1$.

The promised constructions of 2-sequencings of D_{2n} coming from symmetric *d*-sequencings of Z_{2n} split into cases according to whether n is even or odd. It will be necessary to have examples of all types of symmetric *d*-sequencings for later use.

Suppose first that n is even. The examples that follow are on Z_{12}. Note that (1) exhibits all "forward" possibilities; first starter, first translate, second starter and second translate. The appropriate pairs are underlined.

(1)
$$\Sigma_{12} : 0, 3, \quad 8, 11, 10, 5, 6, 7, 2, 1, 4, 9$$
$$\Omega_{12} : 0, 3, \underline{11, 10}, \underline{8}, 1, \underline{7, 2, 4}, 5, 9, 6$$

Similarly, (2) gives examples of all backward possibilities for n even.

(2)
$$\Sigma_{12} : 0, 3, \quad 8, 5, 10, 11, 6, 1, \quad 2, 7, 4, 9$$
$$\Omega_{12} : 0, 3, 11, \underline{4, \quad 2, \quad 1}, 7, \underline{8, 1 \underline{0}, 5}, 9, 6$$

Suppose now that n is odd. The examples given are all on Z_{14}. The first one (3) has a first backward starter, a first forward translate, a second backward translate and a second forward starter.

(3)
$$\Sigma_{14} : 0, 10, 9, \quad 8, 12, 11, 1, 7, 13, 3, 2, \quad 6, 5, 4$$
$$\Omega_{14} : 0, \underline{10, 5}, 13, \underline{11, \quad 8}, 9, \underline{2, \quad 1}, 4, \underline{6, 12}, 3, 7$$

The next example (4) contains a first forward starter and a second backward starter.

(4)
$$\Sigma_{14}: 0, 4, 11, 12, 13, 5, 6, 7,\ 8, 9, 1, 2,\ 3, 10$$
$$\Omega_{14}: 0, 4,\ 1, \underline{13, 12},\ 3, 9, 2, \underline{10, 5}, 6, 8, 11,\ 7$$

A second forward translate appears in (5).

(5)
$$\Sigma_{14}: 0, 1, 10, 9, 3, 8, 2,\ 7, 12, 6, 11, 5, 4, 13$$
$$\Omega_{14}: 0, 1, 11, 6, 9, 3, 5, 12, 10, 2, 13, \underline{4, 8,}\ 7$$

Finally, (6) contains a first backward translate.

(6)
$$\Sigma_{14}: 0, 12, 6, 11, 5,\ 4, 13, 7, 1, 10, 9,\ 3, 8, 2$$
$$\Omega_{14}: 0, 12, 4,\ 1, \underline{6, 10},\ 9, 2, 3, 13, 8, 11, 5, 7$$

LEMMA 6. *The following results hold for any dihedral group and for any integers i and j.*

i) $a^j b = ba^{-j}$,

ii) $(a^j b)(a^j b) = e$,

iii) $a^i x = a^j b$ implies $x = a^{j-i}b$, $(a^j b)x = a^i$ implies $x = a^{j-i}b$,

iv) $a^i x = a^j$ implies $x = a^{j-i}$,

v) $(a^i b)x = a^j b$ implies $x = a^{i-j}$.

PROOF. The computations are straight forward.

In what follows, it will be helpful to picture D_n in two rows. The first row will be an ordering of the cyclic subgroup $<a> \subset D_n$. This ordering will be induced by an appropriate ordering of the elements of Z_n. The second row will place $a^i b$ below the a^i in row 1. The plan is to build a Hamiltonian path through D_n that can be thought of as the partial product sequence associated with a 2-sequencing of D_n. If the elements of Z_n are ordered properly, then most of the edges in the paths constructed will join "adjacent vertices" as follows.

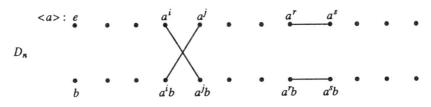

FIGURE 1

If each edge is part of a directed Hamiltonian path that starts at e, then the path induces an orientation of the edges. If an edge is "directed" from vertex c to vertex d, one can define x to be the element of D_n such that $cx = d$. If e is placed first, this gives an ordered collection of elements of D_n whose associated partial product sequence is the given Hamiltonian path.

Note what happens if both edges in the "ex" pattern in Figure 1 are part of a Hamiltonian path. By Lemma 6 (iii), it makes no difference which way these edges are directed.

(7)
$$\text{edge } \{a^i, a^j b\} \text{ yields } x = a^{j-i}b$$
$$\text{edge } \{a^j, a^i b\} \text{ yields } x = a^{i-j}b.$$

If the horizontal lines in Figure 1 are part of a Hamiltonian path, it is clear that the direction of the edges is important. The (a^r, a^s) directed edge yields $x = a^{s-r}$ while the $(a^r b, a^s b)$ directed edge yields $x = a^{r-s}$.

Although the primary concern of this paper is to construct 2-sequencings of D_n, n even, it is most instructive to consider the odd case first. This was done in [3, Theorem 9]. An example will suffice here. Consider the following 2-sequencing of Z_{11}.

$$s: \quad 0, 2, 2, \ 6, 10, 7, 3, \ 10, 7, 3, 6$$

$$t: \quad 0, 2, 4, 10, \ 9, 5, 8, \ 7, 3, 6, 1$$

Construct a Hamiltonian path through D_{11} as in Figure 2.

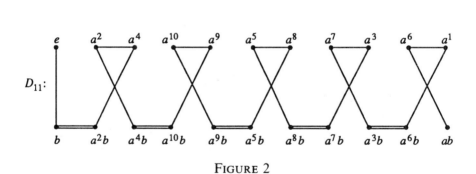

FIGURE 2

The single lines in the t row of Figure 2 give the pairs of a starter [7] on Z_{11}. Each of these lines is "lifted" to " ⤬ " in D_{11}. By (7), directing the "ex" edges gives all $a^i b$, $i \neq 0$, and $a^0 b = b$ comes from the edge (e, b). Thus all elements of the form $a^i b$ occur in the ordering of elements of D_{11} associated with the given Hamiltonian path. The top edges of " ⤬ " will, since they are associated with a starter in Z_{11}, give a collection $\{a^{k_i} : 1 \leq i \leq 5\}$ such that the exponents k_i are the range of a choice function on the pairs $\{\{x, -x\} : x \in Z_{11}\setminus\{0\}\}$. This last collection is the so-called patterned starter for Z_{11} [7].

The double lines in the t row of Figure 2 give the pairs of the translate $F + 1$ of the starter

$$F = \{\{10, 1\}, \{3, 9\}, \{8, 4\}, \{7, 6\}, \{2, 5\}\}.$$

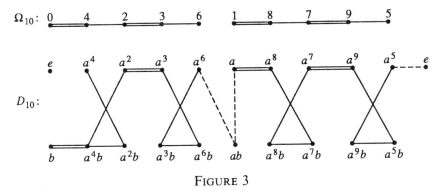

FIGURE 3

These lines are "lifted" to the double lines below them in Figure 2 so these edges in the Hamiltonian path are associated with another collection $\{a^{j_i} : 1 \leq i \leq 5\}$ such that the exponents are the range of a choice function on the patterned starter for Z_{11}.

It is now clear that the Hamiltonian path in Figure 2 comes from a 2-sequencing of D_{11}.

Can this method be made to work on D_n, n even? With suitable modification, it can. An example will illustrate the complications that arise. All constructions of 2-sequencings of D_{2n} will begin with a symmetric d-sequencing of Z_{2n}. This is an analogue of the starter-translate 2-sequencing of Z_{11} above, because it gives an even starter and a translate of it to work with. One problem is that the "ex" parts now omit $a^n b$ as well as b and there is another element a^n of order 2 to include.

Consider the following symmetric d-sequencing of Z_{10}.

$$\Sigma_{10} : 0, 4, 8, 1, 3, 5, 7, 9, 2, 6$$

$$\Omega_{10} : 0, 4, 2, 3, 6, 1, 8, 7, 9, 5$$

Figure 3 is an attempt to mimic the odd case.

There are several things to note about this example. First, since a symmetric sequencing of Z_{2n} always has the property that the last element of the associated partial sum sequence is n, one can start "against the order" of the sequencing to pick up a^n (or $a^n b$) immediately. Since positions n and $n+1$ of Ω_{2n} are related by $t_{n+1} = t_n + n$, use these positions to lift to the elements of $\{b, a^n b, a^n\}$ not already accounted for. This is done by the dotted lines in Figure 3. The single lines in Figure 3 give the pairs of an even starter E on Z_{10}. Each of these lines is lifted to " $\times$ " in D_{10}.

As before, the "ex" parts yield all the elements $a^i b$ not arising via the dotted lines, and the bottom edges of " $\times$ " give a collection $\{a^{k_i} : 1 \leq i \leq 4\}$ such that the exponents k_i are the range of a choice function on the pairs $P = \{\{x, -x\} : x \in Z_{10} \backslash \{0, 5\}\}$.

The double lines in the Ω_{10} row of Figure 3 give the pairs of the translate

$E + 5$. These lines are "lifted" to the edges below them so these edges are associated with another collection $\{\alpha^{j_i} : 1 \le i \le 4\}$ such that the exponents are the range of a choice function on the pairs of P.

Now the edges in Figure 3 do not give a Hamiltonian path through D_{10}. However, if one replaces edge $\{a^4b, a^2b\}$ with edge $\{b, a^2b\}$, a Hamiltonian path is displayed. Furthermore since $4 - 2 = 2 - 0$ and since the direction of the "horizontal" edges is unimportant in the matter of building 2-sequencings, the new path will work.

Since the proof of Theorem 9 is presented largely in terms of diagrams, it is essential that the reader study a modified Figure 3 ($\{a^4b, a^2b\}$ replaced by $\{b, a^2b\}$) until it becomes "clear at a glance" that the Hamiltonian path comes from a 2-sequencing of D_{10}. Purely algebraic definitions of such paths are, in the opinion of the author, unsatisfactory (e.g., see [3, Theorem 12]) as the primary means of exposition.

Three useful facts to keep in mind while looking at the constructions are as follows.

NOTE 7. Suppose Σ_{2n} is a symmetric d-sequencing of Z_{2n} with index j for the special pair (t_j, t_{j+1}) such that

$$(t_j, t_{j+1}) \in \{(y, 2y), (2y, y)\}.$$

The Hamiltonian path constructed through D_{2n} will

   i) finish at a^{2y},

  ii) have $\{a^y b, b\}$ as an edge (think of $\{a^y b, b\}$ as replacing $\{a^y b, a^{2y} b\}$),

 iii) have the property that except for first translate cases, the "lifting" of the even starter pairs will always include the "ex" so as to handle all $a^i b$ except b and $a^n b$.

It will also be helpful to isolate the sections of the constructions that arise in association with the special pair (t_j, t_{j+1}). Note, however, that parts of these diagrams may have to be changed if $j \in \{2, n-1, n+1, 2n-1\}$.

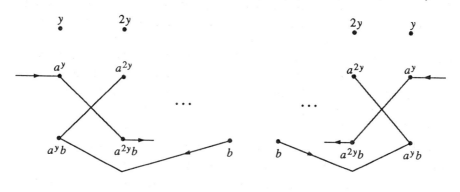

FIGURE 4   FIGURE 5

MODIFICATION 1. Basic crossover.

In the forward starter cases, use Figure 4. The crossover forces the inclusion and also the exclusion of certain horizontal edges in the "..." portion of the diagram in Figure 4 and similar statements hold for Figures 5–7 and 10–11 as well.

In the backward starter cases, use Figure 5. This is exactly what was done in the Figure 3 example mentioned above.

MODIFICATION 2. Basic jump.

In the second forward translate cases, use Figure 6.
In the second backward translate cases, use Figure 7.

NOTE 8. Following observations about symmetric sequencings are easily verified.

i) $1 \le i \le n$ implies $t_{2n+1-i} = t_i + n$

ii) $\{t_i, t_{i+1}\} \in E + n$ iff $\{t_{2n+1-i}, t_{2n+1-(i+1)}\} \in E$ (E and $E + n$ as in Lemma 4),

iii) the two zig-zag edges pictured in Figure 8 yield the same two elements of a proposed 2-sequencing of D_{2n} as do the two edges pictured in Figure 9. The same result holds if the two edges in Figure 8 are replaced by $\{a^{t_i}, a^{t_{i+1}}b\}$ and $\{a^{t_{i+1}+n}, a^{t_i+n}b\}$.

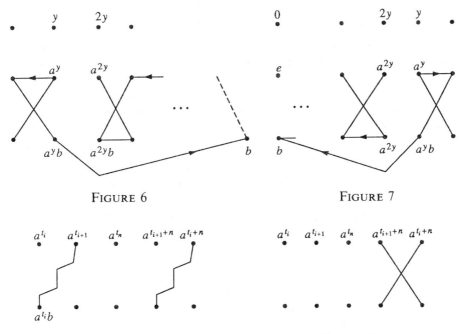

FIGURE 6   FIGURE 7

FIGURE 8   FIGURE 9

MODIFICATION 3. Basic Z-jump construction with $t_i = t_2$.

In the first forward translate cases, use Figure 10. The two zig-zag lines in Figure 10 are the lines of Figure 8 with $t_i = t_2$.

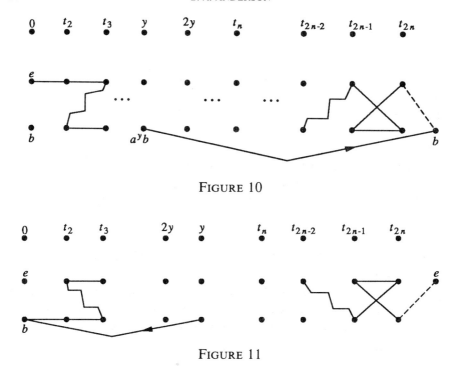

FIGURE 10

FIGURE 11

In the first backward translate cases, use Figure 11.

The portion of the Hamiltonian path not yet considered is the $t_n - t_{n+1}$ section. This is used to handle two of $\{a^n, b, a^n b\}$ and it will become apparent that other parts of the path force certain choices here.

Everything is now in place and the complete constructions for all cases can be summarized. One case will be done in some detail and the remaining cases will be dealt with, in order to save space, by giving a specific example; sometimes in the form of a listing of elements and other times in the form of a diagram. It should be clear how to modify the specific example to solve any other instance of the case under consideration.

Suppose a symmetric d-sequencing of Z_{2n} is given. There are eight cases to consider.

CASE 1. The first forward starter cases. Use the crossover.

A) If n is odd, Figure 12 shows the construction for $j = 4$.

The conventions established in Figure 12 are retained for the remaining cases. The single lines in the t row give the even starter pairs E. The double lines in D_{2n} are the liftings of the even starter translate $E + n$, and the dotted lines give the edges that, when traversed, yield $a^n b$, b and a^n.

It is not hard to satisfy oneself that the construction in Figure 12 still is valid for $j \in \{2, 4, 6, \ldots, n - 1\}$ so long as the crossover is placed at the appropriate point.

B) If n is even, Figure 13 shows the construction, again for $j = 4$. The

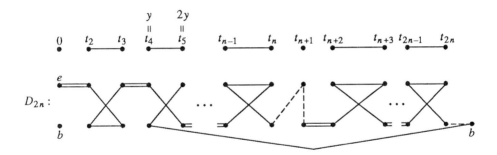

FIGURE 12

difference between the two cases occurs in the edges under t_n and t_{n+1}.

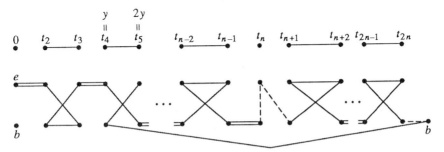

FIGURE 13

CASE 2. The second forward starter cases. Use the crossover.

As in Case 1, the n odd and n even situations are similar. The following is a Hamiltonian path through D_{14} that is based on the $\{6, 12\}$ example of (3) and can be generalized to yield a partial product sequence associated with a 2-sequencing of D_{2n}, given any second forward starter symmetric d-sequencing of Z_{2n}.

$$t: \ e, a^{10}, a^5b, a^{10}b, a^5, a^{13}, a^{11}b, a^{13}b, a^{11}, a^8, a^9b, a^8b, a^9, a^2b,$$
$$a^2, a, a^4b, ab, a^4, a^6, a^{12}b, a^3b, a^7, a^3, a^7b, b, a^6b, a^{12}.$$

CASE 3. The first backward starter cases. Use the crossover.

Here the example uses the pair $\{4,2\}$ of (2).

$$t: \ e, a^6, a^9b, a^6b, a^9, a^5, a^{10}b, a^5b, a^{10}, a^8, a^7b, a^8b, a^7,$$
$$ab, a, a^2, a^4b, a^{11}b, a^3, a^{11}, a^3b, b, a^2b, a^4.$$

CASE 4. The second backward starter cases. Use the crossover.

The $\{10,5\}$ example of (2) is as follows.

$$t: \ e, a^6, a^9b, a^6b, a^9, a^5, a^{10}b, a^8b, a^7, a^8, a^7b, a, ab, a^2b,$$
$$a^4, a^2, a^4b, a^{11}b, a^3, a^{11}, a^3b, b, a^5b, a^{10}.$$

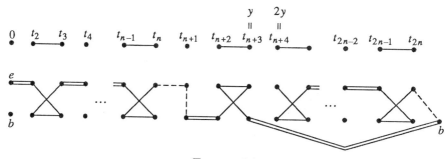

FIGURE 14

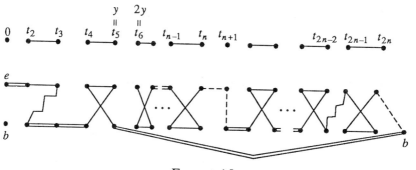

FIGURE 15

CASE 5. The second forward translate cases. Use the jump.
Figure 14 shows the construction for n odd and $j = n + 3$.
CASE 6. The second backward translate cases. Use the jump.
Here is the $\{8,10\}$ example of (2).

$$t: \ e, a^6b, a^9, a^6, a^9b, a^5b, a^{10}, a^5, a^{10}b, b, a^3b, a^{11}, a^3,$$
$$a^{11}b, a^4b, a^2, a^4, a^2b, ab, a, a^7, a^8b, a^7b, a^8.$$

CASE 7. The first forward translate cases. Use the Z-jump with $t_i = t_2$.
Figure 15 shows the construction for n odd and $j = 5$.
CASE 8. The first backward translate cases. Use the Z-jump with $t_i = t_2$.
The $\{6, 10\}$ example of (6) is as follows.

$$t: \ e, a^7b, a^5, a^7, a^5b, a^{11}, a^8b, a^{11}b, a^8, a^{13}, a^3b, a^{13}b, a^3, a^2, a^2b,$$
$$a^9b, a^{10}, a^9, a^{10}b, b, a^{12}b, a^4b, a^{12}, a^4, a, a^6b, ab, a^6.$$

The eight cases just examined constitute a proof of the following:

THEOREM 9. *If* Z_{2n} *has a symmetric d-sequencing, then* D_{2n} *has a 2-sequencing.*

The constructions employed in cases 1–8 are certainly not the only possibilities. The next two examples will indicate some of the conceivable variations.
CASE $2'$. The second forward starter cases. Use the Z-jump with $t_i = t_1$
and the crossover. The $\{7, 2\}$ example of (2) can be used as follows unless

$j = 2n - 1$.
$$t: \quad e, a^3b, a^{11}, a^3, a^{11}b, a^{10}b, a^8, a^{10}, a^8b, ab, a, a^7, a^2b,$$
$$a^4b, a^5, a^4, a^5b, a^9b, a^6b, a^9, a^6, b, a^7b, a^2.$$

CASE 7′. The first forward translate cases. Use the Z-jump with $t_i = t_1$. Here is the $\{11, 10\}$ example of (2).
$$t: \quad e, a^3b, a^{11}, a^3, a^{11}b, b, a^6, a^9, a^6b, a^9b, a^5b, a^4, a^5, a^4b,$$
$$a^2b, a^7, a^2, a^7b, ab, a, a^8, a^{10}b, a^8b, a^{10}.$$

Clearly, now, all that remains is to show that Z_{2n}, $n \geq 2$, always has a symmetric d-sequencing. There is a doubling construction that is very useful in this regard.

DEFINITION 9. Suppose G is a finite group of order $2n$ with identity e and unique element z of order 2. Let $\pi: G \to G/Z_2$ be the natural projection and suppose
$$s: \quad s_1, s_2, \ldots, s_n$$
is a 2-sequencing of G/Z_2. Then
$$S^*: \quad e, S_2, S_3, \ldots, S_n$$
is a *lifting* of s to G iff for each i, $2 \leq i \leq n$, $S_i \in G$ and $\pi(S_i) = s_i$.

When a lifting is present in what follows, capital letters will generally be used in connection with the domain group and lower case letters will pertain to the range group.

DEFINITION 10. Suppose s is a 2_d-sequencing of Z_n with associated partial sum sequence t and special pair $(t_j, t_{j+1}) = (y, 2y)$ [respectively $(2y, y)$]. Suppose further that S^* is a lifting to Z_{2n} with associated partial sum sequence T^*. The statement that S^* is *compatible* with s means that $(T_j, T_{j+1}) = (Y, 2Y)$ [respectively $(2Y, Y)$].

Recall that symmetric d-sequencings are instances of 2_d-sequencings so that Definition 10 applies to that situation as well and n may be even or odd there.

As shown in [2], a 2-sequencing of G/Z_2 can always be lifted to the "first half" of a symmetric sequencing of G. The rules for making sure a lift from Z_n to Z_{2n} (n odd or even) gives a symmetric sequencing are as follows.

POSSIBILITY 1. $v \neq -v$

    i) If v ($-v$ similar) appears twice in the 2-sequencing s, then the two occurrences of $v = \{X, X + N\}$ must be lifted to X and $X + N$.

    ii) If v and $-v$ each appear once in s, then each lift of v forces a unique lift for $-v$. In particular, if v is lifted to X, then $-v$ must be lifted to $-X + N$ and if v is lifted to $X + N$, then $-v$ must be lifted to $-X$.

POSSIBILITY 2. $v = -v$, $v \neq 0$.

In this case v occurs only once in s and can be lifted to either X or $X + N$.

THEOREM 10. *If s is a symmetric d-sequencing of Z_{2n}, $n \geq 2$, then s can be lifted to 2^{n-1} symmetric d-sequencings S of Z_{4n} such that S^* is compatible with s.*

PROOF. By [2, Theorem 4] there are 2^n ways to lift s to a symmetric sequencing of Z_{4n}. The claim is that half these lifts give symmetric d-sequencings S such that S^* is compatible with s. An easy argument can be based on the following two cases.

CASE 1. Suppose the special pair (t_j, t_{j+1}) associated with s is not (t_{2n-1}, t_{2n}). Then the $(y, 2y)$ situation is as described below (the $(2y, y)$ is situation is similiar).

$$s: \left\{ \begin{matrix} 2N \\ 0 \end{matrix} \right\}, s_2, \ldots, \left\{ \begin{matrix} U + 2N \\ U \end{matrix} \right\}, \left\{ \begin{matrix} V + 2N \\ V \end{matrix} \right\}, \ldots, s_{2n}$$

$$t: \left\{ \begin{matrix} 2N \\ 0 \end{matrix} \right\}, t_2, \ldots, \left\{ \begin{matrix} Y + 2N \\ Y \end{matrix} \right\}, \left\{ \begin{matrix} 2Y + 2N \\ 2Y \end{matrix} \right\}, \ldots, t_{2n}.$$

Follow the rules given previously and lift to any symmetric sequencing S of Z_{4n}. There are four possibilities for (T_j, T_{j+1}) of the associated T-row. They are $(Y, 2Y)$, $(Y+2N, 2Y)$, $(Y, 2Y+2N)$ and $(Y+2N, 2Y+2N)$. The first two give S such that S^* is compatible with s. If one of the last two arises, change the lifts associated with the Z_{2n}-inverse elements s_2 and s_{2n}. This adds $2N$ to $T_2, T_3, \ldots, T_{2n-1}$ and replaces $(Y, 2Y+2N)$ with $(Y+2N, 2Y)$ and $(Y+2N, 2Y+2N)$ with $(Y, 2Y)$. It is then clear that the entire set of 2^n lifts can be partitioned into pairs by the two ways of lifting s_2 and s_{2n} and that exactly one lift in each pair has the required properties.

CASE 2. Suppose the special pair (t_j, t_{j+1}) associated with s is (t_{2n-1}, t_{2n}). Since $t_{2n} = n$, $t_{2n-1} \neq 2n = 0$ so $t_{2n-1} \in \{n/2, 3n/2\}$ and n is even. This time pair the lifts that yield symmetric sequencings by the two ways to lift $s_{n+1} = n$. One lift from each such pair will give $T_{2n} = N$ and the other will give $T_{2n} = 3N$. It is easy to see that if $t_{2n-1} = n/2$, then the lifts such that $T_{2n} = N$ are exactly the set of lifts that work, and if $t_{2n-1} = 3n/2$, then the lifts such that $T_{2n} = 3N$ are exactly the set of lifts that work.

There is an analogous result about reflective 2_d-sequencings of Z_n, n odd, that can be verified by arguing as in Case 1 above.

THEOREM 11. *Suppose s is a reflective 2_d-sequencing of Z_n, n odd, with special pair (t_j, t_{j+1}) such that $j \neq n-1$. Then s can be lifted to $2^{(n-3)/2}$ symmetric d-sequencings S of Z_{2n} such that S^* is compatible with s.*

Example 12. A 2_d-sequencing s of Z_{11} that can't be lifted to a symmetric d-sequencing S of Z_{22} such that S^* is compatible with s.

$$s: \quad 0, 7, 8, 1, 4, 5, 10, \ 8, 9, 9, 6$$

$$t: \quad 0, 7, 4, 5, 9, 3, \ 2, 10, 8, 6, 1$$

Note that $(6,1)$ is the only pair (t_j, t_{j+1}) of form $(y, 2y)$ or $(2y, y)$. It will suffice to show that for any lift of s to a symmetric seqencing S of Z_{22}, $T_{11} = 1$. If this holds, then since $T_{10} \in \{6, 17\}$, the result will follow.

By computation,

$$c = \sum_{i=1}^{11} s_i = 67 \equiv 1 \quad (\text{mod } 22).$$

What is $\sum_{i=1}^{11} S_i$ for a valid lift S? By inspection s has two $8s$ and two $9s$. Any valid lift of a "double" will add 11 (mod 22) to c. Also by inspection s has three inverse pairs $\{1, 10\}$, $\{4, 7\}$ and $\{5,6\}$. Any valid lift of an inverse pair adds 0 (mod 22) to c since if $x + y = 11$ then $x + 11 = -y$ (mod 22). Thus, if S is a valid lift,

$$\sum_{i=1}^{11} S_i \equiv (c + 2 \cdot 11 + 3 \cdot 0) \quad (\text{mod } 22) \equiv 1 \quad (\text{mod } 22).$$

Enough information has now been accumulated to allow the generation of 2-sequencings of all D_n, $n \geq 3$, and $n \neq 2^m$, from a single family of 2_d-sequencings of Z_n, n odd, and two constructions; one as in Figure 2 (see [3, Theorem 9]) and the other the doubling construction of Theorems 9-11. The 2^m cases can be handled by the doubling construction applied to this example on Z_4.

$$
\begin{aligned}
s &: \quad 0, 1, 2, 3 \\
t &: \quad 0, 1, 3, 2.
\end{aligned}
$$
(8)

Use the second forward starter construction for D_4 and the first forward translate construction for D_{2m}, $m \geq 3$. It remains to define the family of 2_d-sequencings of Z_n, n odd, just mentioned.

DEFINITION 11. Suppose $n \geq 3$ is odd, PS is the patterned starter for Z_n,

$$PS = \{\{x, -x\} : x \neq 0\}$$
$$\text{and } PS + 1 = \{\{x + 1, -x + 1\} : x \neq 0\}.$$

Let τ_n be the component of the union of PS and $PS + 1$ that contains $\{0, 2\}$ and label

$$\tau_n : t_1, t_2, \ldots \qquad = \tau_n : 0, 2, \ldots \; .$$

It is easy to verify the results of the following:

LEMMA 12. *If $n \geq 3$ is odd and $\tau_n : t_1, t_2, \ldots$ is as above, then*
   i) *$1 \leq i \leq (n - 1)/2$ implies $t_{2i} = 2i$*
   $0 \leq i \leq (n - 1)/2$ implies $t_{2i+1} \equiv -2i$ (mod n),
   ii) *τ_n is a Hamiltonian path through Z_n and there is an associated σ_n ($s_1 = 0$, $s_i = t_i - t_{i-1}$ otherwise) which is a 2-sequencing of Z_n,*

iii) *if* $2 \leq i, j \leq n$ *and* $i + j = n + 2$, *then* $s_i = s_j$ *(so that* σ_n *is reflective)*,

iv) *if* $n = 6m + 1$, $m \geq 1$, *then*

      a) $\{2m + 1, 4m + 2\} \in PS + 1$

      b) $(2m + 1, 4m + 2) = (t_{4m+1}, t_{4m+2})$,

v) *if* $n = 6m + 3$, $m \geq 0$, *then*

      a) $\{2m + 1, 4m + 2\} \in PS$

      b) $(4m + 2, 2m + 1) = (t_{4m+2}, t_{4m+3})$,

vi) *if* $n = 6m + 5$, $m \geq 0$, *then*

      a) $\{2m + 3, 4m + 4\} \in PS + 1$

      b) $(2m + 3, 4m + 4) = (t_{4m+3}, t_{4m+4})$,

vii) σ_n *is a* 2_d-*sequencing (in fact, a starter-translate* 2_d-*sequencing* [3, *Definition 8]) of* Z_n.

Because of the constraint $j \neq n - 1$ in Theorem 11, the case $n = 3$ must be handled separately.

LEMMA 13. *If* $n = 3$ *and* σ_3 *is as in Lemma 12, then there is a first backward starter symmetric d-sequencing S of* Z_6 *such that* S^* *is compatible with* σ_n.

PROOF. S is as follows.

$$S: 0, 2, 5, 3, 1, 4$$
$$T: 0, 2, 1, 4, 5, 3.$$

LEMMA 14. *Suppose* $n \geq 5$ *is odd,* σ_n *and* τ_n *are as in Lemma 12 and S is a symmetric d-sequencing of* Z_{2n} *such that* S^* *is compatible with* σ_n *as promised by Theorem 11.*

   i) *If* $n \equiv 1 \pmod 6$, *then S is a first forward translate symmetric d-sequencing of* Z_{2n}.

   ii) *If* $n \equiv 3 \pmod 6$, *then S can be a first forward starter symmetric d-sequencing (only), a first backward starter symmetric d-sequencing (only) or both on* Z_{2n}.

   iii) *If* $n \equiv 5 \pmod 6$, *then S is a first backward translate symmetric d-sequencing of* Z_{2n}.

PROOF. This follows from Lemma 12.

THEOREM 15. *If* $n \geq 2$, *then* Z_{2n} *has a symmetric d-sequencing.*

PROOF. Use (8), Lemmas 12 and 13 and the doubling constructions. Finally, it is possible to state:

THEOREM 16. *If* $n \geq 3$, *then* D_n *has a 2-sequencing.*

PROOF. This is immediate from Theorems 2, 9 and 15.

COROLLARY 17. *If $n \geq 3$, then Q_{2n} has a symmetric sequencing.*

Another consequence follows from the fact that symmetric sequencings are equivalent to certain left even starters and left even starters on Q_{2n} induce 1-factorizations of K_{4n+2} in a natural way [1].

COROLLARY 18. *If $n \geq 3$, then K_{4n+2} has a 1-factorization whose symmetry group contains Q_{2n}.*

A 1-factorization of K_{2n} with the property that every 2-factor union of distinct 1-factors is a Hamiltonian circuit is said to be *perfect* . Recently Seah and Stinson [6] constructed two examples of perfect 1-factorizations of K_{14} such that in each case the full symmetry group of the 1-factorization is Q_6. Even starter formulations of these examples are given in [2].

This paper shows that for $n \geq 3$, Q_{2n} has left even starters. Can the two examples of [6] be generalized?

REFERENCES

1. B. A. Anderson, *Sequencings and Starters,* Pacific J. Math. **64**(1976), 17–24.
2. ――, *Sequencings of Dicyclic Groups,* Ars Comb. **23**(1987),131–142.
3. ――, *Sequencings of Dicyclic Groups II,* J. of Comb. Math. and Comb. Computing **3**(1988), 5–27.
4. J. Dénes and E. Török, "Groups and Graphs, in Combinatorial Theory and its Applications," North Holland, Amsterdam, 1970, 257–289.
5. B. Gordon, *Sequences in Groups with Distinct Partial Products,* Pacific J. Math **11**(1961), 1309–1313.
6. E. Seah and D. R. Stinson, *Some Perfect On-Factorizations of K_{14} ,* Annals of Discrete Math. **34**(1987), 419–436.
7. W. D. Wallis, A. P. Street and J. S. Wallis, *Combinatorics: Room Squares, Sum-Free Sets, Hadamard Matrices,* Lecture Notes in Math # **292**, Springer-Verlag 1972.

MATHEMATICS DEPARTMENT, ARIZONA STATE UNIVERSITY, TEMPE, ARIZONA 85287

Contemporary Mathematics
Volume **111**, 1990

Concerning Pairwise Balanced Designs With Prime Power Block Sizes

F. E. BENNETT

ABSTRACT. Let K be a set of positive integers. A *pairwise balanced design* (*PBD*) of index unity $B(K, 1; v)$ is a pair $(X, \mathscr{B})$ where X is a v-set (of *points)* and B is a collection of subsets of X (called *blocks*) with sizes from K such that every pair of distinct points of X is contained in exactly one block of B. In this paper, we are concerned with the existence of *PBDs* with specified block sizes which are prime powers exceeding 3. We present a brief survey of some useful known results and provide supplements along with applications to other combinatorial structures.

1. Introduction. Let K be a set of positive integers. A *pairwise balanced design* (*PBD*) of index unity $B(K, 1; v)$ is a pair $(X, \mathscr{B})$ where X is a v-set (of *points*) and $\mathscr{B}$ is a collection of subsets of X (called *blocks*) with sizes from K such that every pair of distinct points of X is contained in exactly one block of $\mathscr{B}$. The number $\mid X \mid = v$ is called the *order* of the PBD.

We shall denote by $B(K)$ the set of all integers v for which there exists a PBD $B(K, 1; v)$. For convenience, we define $B(k_1, k_2, \ldots, k_r)$ to be the set of all integers v such that there is a PBD $B(\{k_1, k_2, \ldots, k_r\}, 1; v)$. A set K is said to be *PBD-closed* if $B(K) = K$.

Pairwise balanced designs are of fundamental importance in combinatorial theory and have been used extensively in the construction of other types of combinatorial designs. Quite often, one is generally interested in constructing $PBDs$ $B(K, 1; v)$ for some specified set K. In this connection, R.M. Wilson's remarkable theory concerning the structure of PBD-closed sets (see [37-39]) often provides us with some form of "asymptotic" results as follows:

THEOREM 1.1 (*Wilson's Theorem*). *Let K be a set of positive integers and*

1980 *Mathematics Subject Classification* (1985 *Revision*). 05B05.

Research supported by the Natural Sciences and Engineering Research Council of Canada under Grant A-5320.

This paper is in final form and no version of it will be submitted for publication elsewhere.

define the two parameters:

$$\alpha(K) = \gcd\{k - 1 : k \in K\}, \ and$$

$$\beta(K) = \gcd\{k(k - 1) : k \in K\}.$$

Then there exists a constant C *(depending on* K*) such that, for all integers* $v > C$, $v \in B(K)$ *if and only if* $v - 1 \equiv 0(\mathrm{mod}\,\alpha(K))$ *and* $v(v - 1) \equiv 0(\mathrm{mod}\,\beta(K))$.

We wish to remark that, for a given set K, Wilson's theory does not really provide any concrete upper bound on the constant C in Theorem 1.1. In this paper we present a brief survey of some useful results, where this problem has been addressed for some specified sets K consisting of prime powers exceeding 3. We also provide supplements with applications to other combinatorial designs. It is safe to say that all of the results contained in this survey have already found applications in the construction of other types of combinatorial structures, such as Mendelsohn designs, Room squares, skew Room squares, perpendicular arrays, orthogonal Steiner systems, orthogonal arrays, Latin squares, and quasigroups with very interesting conjugacy properties. For more details, the interested reader may wish to refer to the references, some of which may not be directly referred to in the body of the paper.

2. Preliminaries. In this section we shall define some terminology and adapt the notations of earlier papers (see, for example, [1]). For more details on PBDs and related designs, the reader is referred to [11, 17, 36].

Definition 2.1. Let K and M be sets of positive integers. A *group divisible design* $(GDD)\,GD(K, 1, M; v)$ is a triple $(X, \mathscr{G}, \mathscr{B})$ where

(i) X is a v-set (of *points*),

(ii) $\mathscr{G}$ is a collection of non-empty subsets of X (called *groups*) with sizes in M and which partition X,

(iii) $\mathscr{B}$ is a collection of subsets of X (called *blocks*), each with size at least two in K,

(iv) no block meets a group in more than one point, and

(v) each pairset $\{x, y\}$ of points not contained in a group is contained in exactly one block.

The *group-type* (or *type*) of GDD $(X, \mathscr{G}, B)$ is the multiset $\{|G| : G \in \mathscr{G}\}$ and we usually use the "exponential" notation for its description: a group-type $1^i 2^j 3^k \dots$ denotes i occurrences of groups of size 1, j occurrences of groups of size 2, and so on.

Definition 2.2. A *transversal design* (TD) $T(k, 1; m)$ is a GDD with km points, k groups of size m and m^2 blocks of size k, where each block meets every group in precisely one point, that is, each block is a transversal of the collection of groups.

Definition 2.3. Let $(X, \mathscr{B})$ be a PBD $B(K, 1; v)$. A *parallel class* in $(X, \mathscr{B})$ is a collection of disjoint blocks of $\mathscr{B}$, the union of which equals X.

$(X, \mathscr{B})$ is called *resolvable* if the blocks of $\mathscr{B}$ can be partitioned into parallel classes. A *GDD* $GD(K, 1, M; v)$ is *resolvable* if its associated *PBD* $B(K \cup M, 1; v)$ is resolvable with M as a parallel class of the resolution.

It is fairly well-known that the existence of a resolvable *TD* $T(k, 1; m)$ (briefly $RT(k, 1; m)$) is equivalent to the existence of a *TD* $T(k+1, 1; m)$ or equivalently $k - 1$ mutually orthogonal Latin squares (MOLS) of order m. In particular, the following two results can be found in [24].

THEOREM 2.4. *For every prime power* q, *there exists a* $T(q + 1, 1; q)$.

THEOREM 2.5. *Let* $m = p_1^{k_1} p_2^{k_2} ... p_r^{k_r}$ *be the factorization of* m *into powers of distinct primes* p_i, *then a* $T(k, 1; m)$ *exists where* $k \leq 1 + \min \{p_i^{k_i}\}$.

The following result can be found in [14, 35].

THEOREM 2.6. *A* $T(5, 1; m)$ *exists for all positive integers* m *with the exception of* $m = 2, 3, 6$, *and possibly excepting* $m = 10$.

We need to establish some more notations. We shall simply write $B(k, 1; v)$ for $B(\{k\}, 1; v)$ and similarly $GD(k, 1, m; v)$ for $GD(\{k\}, 1, \{m\}; v)$. We observe that a *PBD* $B(k, 1; v)$ is essentially a *balanced incomplete block design* (*BIBD*) with parameters v, k and $\lambda = 1$. If $k \notin K$, then $B(K \cup \{k^*\}, 1; v)$ denotes a *PBD* $B(K \cup \{k\}, 1; v)$ which contains a unique block of size k and if $k \in K$, then a $B(K \cup \{k^*\}, 1; v)$ is a *PBD* $B(K, 1; v)$ containing at least one block of size k. We shall sometimes refer to a *GDD* $(X, \mathscr{G}, \mathscr{B})$ as a $K - GDD$ if $|B| \in K$ for every block $B \in \mathscr{B}$.

For some of our recursive constructions of *PBDs* and *GDDs*, we shall make use of Wilson's "Fundamental Construction" (see [36]). We define a *weighting* of a *GDD* $(X, \mathscr{G}, \mathscr{B})$ to be any mapping $w : X \to Z^+ \cup \{0\}$. We present a brief description of Wilson's construction relating to *GDDs* below.

Construction 2.7. (fundamental construction). Suppose that $(X, \mathscr{G}, \mathscr{B})$ is a "master" *GDD* and let $w : X \to Z^+ \cup \{0\}$ be a weighting of the *GDD*. For every $x \in X$, let S_x be $w(x)$ "copies" of x. Suppose that for each block $B \in \mathscr{B}$, a *GDD* $(\cup_{x \in B} S_x, \{S_x : x \in B\}, \mathscr{A}_B)$ is given. Let $X^* = \cup_{x \in X} S_x$, $\mathscr{G}^* = \{\cup_{x \in G} S_x : G \in \mathscr{G}\}$, and $\mathscr{B}^* = \cup_{B \in \mathscr{B}} \mathscr{A}_B$. Then $(X^*, \mathscr{G}^*, \mathscr{B}^*)$ is a *GDD*.

3. Brief survey of known results. In this section, we present a short survey of some useful known results regarding the existence of *PBDs* with prime power block sizes exceeding 3. The interested reader may wish to refer to the appropriate references that are provided for more details and applications.

THEOREM 3.1 (see [17]). *A* $B(4, 1; v)$ *exists if and only if* $v \equiv 1$ *or* $4 \pmod{12}$.

THEOREM 3.2 (see [18]). *A resolvable* $B(4, 1; v)$ *exists if and only if* $v \equiv 4 \pmod{12}$.

THEOREM 3.3 (see [17]). *A* $B(5, 1; v)$ *exists if and only if* $v \equiv 1$ *or* $5 (\mathrm{mod}\, 20)$.

THEOREM 3.4 (see [20, 30]). *A* $B(\{4, 5\}, 1; v)$ *exists if and only if* $v \equiv 0$ *or* $1 (\mathrm{mod}\, 4)$, $v \neq 8, 9, 12$.

THEOREM 3.5 (see [13]). *A* $B(\{4, 7^*\}, 1; v)$ *exists if and only if* $v \equiv 7$ *or* $10 (\mathrm{mod}\, 12)$, $v \neq 10, 19$.

Combining Theorems 3.1 and 3.5 with Theorem 1.1, we easily obtain

THEOREM 3.6. *A* $B(\{4, 7\}, 1; v)$ *exists if and only if* $v \equiv 1 (\mathrm{mod}\, 3)$, $v \neq 10, 19$.

THEOREM 3.7 (see [20]). *Let* $N_4^\infty = \{n \in N : n \geq 4\}$.

Then the following hold:

(a) $B(4, 5, 7, 8, 9, 11) \supseteq M_1$ where $M_1 = N_4^\infty \setminus \{6, 10, 12, 14, 15, 18, 19, 23, 26, 27, 30, 38, 42\}$.

(b) $B(4, 5, 7, 8, 9) \supseteq M_1 \setminus \{51, 86, 90\} = M_2$.

(c) $B(4, 5, 7, 8) \supseteq M_2 \setminus \{62, 66, 74, 78\} = M_3$.

(d) $B(4, 5, 7) \supseteq M_3 \setminus \{39, 50, 54, 63\} = M_4$.

The following theorem is due to R.C. Mullin et al. [26] and, in addition to applications in the construction of frames and skew Room squares, the result has been useful in constructing other combinatorial structures such as resolvable perfect Mendelsohn designs with block size four, nested 4-cycle systems, and a variety of quasigroups (see [2, 3, 34]).

THEOREM 3.8 (see [26]). *A* $B(\{5, 9, 13, 17\}, 1; v)$ *exists for all positive integers* $v \equiv 1 (\mathrm{mod}\, 4)$, *with the possible exception of* $v \in \{29, 33, 49, 57, 93, 129, 133\}$.

The following theorem is due to Mullin and Stinson [27] and it was applied in the construcion of sets of three orthogonal partitioned incomplete Latin squares of type $2^n (n \text{ odd})$, which have the property that two of the squares are mutual transposes and the third is symmetric. Evidently, there are other useful applications.

THEOREM 3.9. (see [27]). *Let* P_5 *denote the set of odd prime powers not less than* 5, *then* $v \in B(P_5)$ *for all odd integers* $v > 3$, *with the possible exception of* $v \in \{15, 33, 39, 51, 75, 87, 93, 183, 195, 219\}$.

The next theorem is also due to Mullin and Stinson [28] where the authors give several applications to the construction of other types of combinatorial structures, including Room squares, skew Room squares, Room cubes, separable orthogonal arrays, and perpendicular arrays.

THEOREM 3.10. (see [28]). *Let* P_7 *denote the set of odd prime powers not less than* 7, *then* $v \in B(P_7)$ *for all odd integers* $v \geq 7$ *with at most* 104 *possible exceptions of which the largest is* 2127.

The result of the next theorem was established independently by Mullin and Stinson [29] and Zhu Lie and Chen Demeng [40], where the result is applied to the construction of orthogonal Steiner triple systems.

THEOREM 3.11. (see [29,40]). *Let* $P_{1,6}$ *denote the set of all prime powers congruent to* 1 *modulo* 6, *then* $v \in B(P_{1,6})$ *for all* $v \equiv 1 \pmod{6}$ *with at most* 31 *possible exceptions of which the largest is* 1921.

The following theorem comes from Bennett, Du and Zhu [4], where the authors apply the result to the construction of perfect Mendelsohn designs with block size 7.

THEOREM 3.12 (see [4]). *Let* Q^* *denote the set of all prime powers congruent to* 0 *or* 1 $\pmod{7}$, *then* $v \in B(Q^*)$ *for all positive integers* $v \equiv 0$ *or* 1 $\pmod{7}$ *with at most* 104 *possible exceptions of which the largest is* 2135.

In [1] the author investigated the existence of PBDs with prime power block sizes exceeding 7, and some applications to the construction of sets of MOLS and Latin squares with pairwise orthogonal conjugates are given. The main result can be summarized in the following theorem.

THEOREM 3.13. (see [1]). *Let* P^* *denote the set of all prime powers exceeding* 7, *then* $v \in B(P^*)$ *for all integers* $v > 2206$ *and most integers below this value.*

4. Supplementary results and applications. In this section, we provide some supplements to those of the preceding section along with applications. We wish to point out that the list of applications for some of the results stated in Section 3 can obviously be extended. For example, in addition to the construction of orthogonal Steiner triple systems provided by Theorem 3.11, the author would like to mention its use in the construction of a class of resolvable perfect Mendelsohn designs described below (see also [2, 5, 25]).

Let v, k, and λ be positive integers. A (v, k, λ)-*Mendelsohn design* (briefly (v, k, λ)-*MD*) is a pair $(X, \mathscr{B})$ where X is a v-set (of *points*) and $\mathscr{B}$ is a collection of cyclically ordered k-subsets of X (called *blocks*) such that every ordered pair of points of X are consecutive in exactly λ of the blocks of $\mathscr{B}$. The pair a_i, a_{i+t} are said to be t-*apart* in a cyclic k-tuple $(a_1, a_2, \ldots, a_k)$ where $i+t$ is taken modulo k. If for all $t = 1, 2, \ldots, k-1$, every ordered pair of points of X are t-apart in exactly λ of the blocks of $\mathscr{B}$, then the (v, k, λ)-MD is called *perfect* and denoted briefly by (v, k, λ)-*PMD*.

A necessary condition for the existence of a (v, k, λ)-MD is $\lambda v(v-1) \equiv 0 \pmod{k}$. This is an immediate consequence of the fact that the number of blocks in such a design is $\lambda v(v-1)/k$.

We shall define the notion of resolvability of a $(v, k, 1)$-MD where, of course, $v(v-1) \equiv 0 \pmod{k}$.

Definition 4.1. If the blocks of a $(v, k, 1)$-MD for which $v \equiv 1 \pmod{k}$ can be partitioned into v sets each containing $(v-1)/k$ blocks which are pairwise disjoint (as sets), we say that the $(v, k, 1)$-MD is resolvable, and each set of $(v - 1)/k$ pairwise disjoint blocks together with the singleton which is the only element not contained in any of its blocks will be called a parallel class.

Definition 4.2. If the blocks of a $(v, k, 1)$-MD for which $v \equiv 0 \pmod k$ can be partitioned into $v - 1$ sets each containing v/k blocks which are pairwise disjoint (as sets), we also say that the $(v, k, 1)$-MD is resolvable and each set of v/k pairwise disjoint blocks will be called a parallel class.

A $(v, k, 1)$-PMD which is resolvable in the sense of either Definition 4.1 or 4.2 will be denoted as a $(v, k, 1)$-$RPMD$.

We have the following lemma from [25, Theorem 2.9].

Lemma 4.3. *Suppose* $v \in B(k_1, k_2, \ldots, k_r)$ *and for each* k_i *there exists a* $(k_i, k, 1)$-PMD. *Then there exists a* $(v, k, 1)$-PMD.

The following result is contained in [5].

Theorem 4.4. *Let* $v = p^r$ *be any prime power and* $k > 2$ *be such that* $k \mid (v - 1)$, *then there exists a* $(v, k, 1)$-$RPMD$.

We shall make use of the following lemma.

Lemma 4.5. *Suppose* $v \in B(k_1, k_2, \ldots, k_r)$ *and for each* k_i *there exists a* $(k_i, k, 1)$-$RPMD$ *in the sense of Definition 4.1. Then there exists a* $(v, k, 1)$-$RPMD$.

Proof. On each block of size k_i in a $B(\{k_1, k_2, \ldots, k_r\}, 1; v)$, we construct a $(k_i, k, 1)$-$RPMD$ in the sense of Definition 4.1. This obviously gives a $(v, k, 1)$-PMD. To show that this resulting design is resolvable (in the sense of Definition 4.1), we observe the following: for every point x of the design, we consider those blocks of the PBD which contain x. Then the parallel class of our $(v, k, 1)$-PMD for which x is the only element not covered by any of its blocks is the union of all the parallel classes of the $(k_i, k, 1)$-$RPMD$s for which x is the only element not covered by their blocks. It is readily checked that this construction provides a $(v, k, 1)$-$RPMD$.

We can now apply the result of Theorem 3.11 to prove the following theorem.

Theorem 4.6. *There exists a* $(v, 6, 1)$-$RPMD$ *for all* $v \equiv 1 \pmod 6$ *with at most* 31 *possible exceptions of which the largest is* 1921.

Proof. Theorem 3.11 guarantees $v \in B(P_{1,6})$ where $P_{1,6}$ denotes the set of all prime powers congruent to 1 modulo 6. For each $q \in P_{1,6}$, Theorem 4.4 gives us a $(q, 6, 1)$-$RPMD$ in the sense of Definition 4.1. The result then follows by applying Lemma 4.5.

In what follows, we shall continue an earlier investigation of the set $B(4, 5, 11)$ and provide some supplements to Theorems 3.1–3.7. The results are then applied to improve the spectra of a variety of Stein quasigroups.

We first observe that Theorem 1.1 guarantees the existence of a constant C such that $v \in B(4, 5, 11)$ for all integers $v > C$. An upper bound of 1198 was given for C in [30] and this was later improved to 298 in [6]. We can actually do a little better here by establishing an upper bound of 154 for C. We need some basic lemmas.

The following two lemmas are taken from [6].

LEMMA 4.7. *If* $k > 1$, *then* $9k + 4 \in B(4, (3k + 1)^*)$.

LEMMA 4.8. *If* $1 \leq m \leq 4k + 1$, *then* $12k + m + 4 \in B(4, 5, m^*)$.

The following lemma is [1, Lemma 2.14] which employs the technique of adding a set of fixed ("infinite") points to a GDD (see also [28] for other generalizations).

LEMMA 4.9. *Let* K *be a set of positive integers and* $s \geq 0$. *Suppose there exists a* K-GDD *of group-type* $T = (m_1, m_2, \ldots, m_n)$.
(a) *If a* PBD $B(K \cup \{s^*\}, 1; m_i + s)$ *exists for* $1 \leq i \leq n$, *then, for each* i, $v + s \in B(K \cup \{(m_i + s)^*\})$ *where* $v = \sum_{1 \leq i \leq n} m_i$.
(b) *If a* PBD $B(K \cup \{s^*\}, 1; m_i + s)$ *exists for* $1 \leq i \leq n - 1$, *then,* $v + s \in B(K \cup \{(m_n + s)^*\})$ *where* $v = \sum_{1 \leq i \leq n} m_i$.

For some of our constructions, we shall require the use of Lemma 4.9 in conjunction with Construction 2.7. We also need some "small" input designs for this purpose.

LEMMA 4.10. *There exist* $\{4\}$-$GDDs$ *of the following group-types:* (a) 3^4, (b) 3^5, (c) $3^4 6^2$, (d) $3^5 6^1$.

PROOF. For (a) and (b), we delete one point from $B(4, 1; v)$ where $v = 13$ and 16, respectively. For the proof of (c), the existence of a $B(\{4, 7\}, 1; 25)$ with precisely two intersecting blocks of size 7 is shown by Drake and Larson [16]. If we delete the point of intersection of the two blocks of size 7, we obtain a $\{4\}$-GDD of group-type $3^4 6^2$. For (d), we delete one point from the unique block of size 7 in a $B(\{4, 7^*\}, 1; 22)$.

LEMMA 4.11. *There exist* $\{4, 5\}$-$GDDs$ *of the following group-types:* (a) $3^4 4^1$, (b) $3^5 4^1$.

PROOF. For the proof of (a), we first construct a $B(\{4, 5\}, 1; 17)$ by adjoining an "infinite" point, say ∞, to a resolvable $B(4, 1; 16)$. We then delete from the resulting design a point $x \neq \infty$ to form a $\{4, 5\}$-GDD of group-type $3^4 4^1$. For the proof of (b), we consider a $B(5, 1; 25)$ and delete one point from a particular block B to get a $\{5\}$-GDD of type 4^6. We then further delete all the points from another block disjoint from B for a $\{4, 5\}$-GDD of type $3^5 4^1$.

LEMMA 4.12. *Suppose a* $T(5, 1; m)$ *exists and* $0 \leq x, y \leq m$ *such that* $x + y \leq m$. *Then there exists a* $\{4, 5\}$-GDD *of group-type* $(3m)^4 (3x + 4y)^1$.

PROOF. In all groups but one of a $T(5, 1; m)$, we give the points weight 3. In the last group, we give x points weight 3 and y points weight 4, and give the remaining points weight 0. We then apply Construction 2.7, using $\{4, 5\}$-$GDDs$ of types 3^4, 3^5, $3^4 4^1$, to obtain the desired $\{4, 5\}$-GDD of group-type $(3m)^4 (3x + 4y)^1$.

LEMMA 4.13. *Suppose a $T(6, 1; m)$ exists and $0 \leq x, y \leq m$. Then there exists a $\{4, 5\}$-GDD of group-type $(3m)^4 (3x)^1 (4y)^1$.*

PROOF. In all groups but two of a $T(6, 1; m)$ we give the points weight 3. In the second last group, we give x points weight 3 and give the remaining points weight 0. In the last group, give y points weight 4 and give the remaining points weight 0. Using $\{4, 5\}$-$GDDs$ of group-types 3^4, 3^5, $3^4 4^1$ and $3^5 4^1$ as input designs, we apply Construction 2.7 for the desired result.

LEMMA 4.14. *Suppose a $T(6, 1; m)$ exists and $0 \leq x, y \leq m$ such that $x + y = m$. Then there exists a $\{4\}$-GDD of group-type $(3m)^4 (3x + 6y)^1 (6m)^1$.*

PROOF. In all groups but two of a $T(6, 1; m)$, give the points weight 3. In the second last group, give x points weight 3 and y points weight 6. In the last group, give all points weight 6. We require $\{4\}$-$GDDs$ of group-types $3^5 6^1$ and $3^4 6^2$, which come from Lemma 4.10, and we apply Construction 2.7 to obtain a $\{4\}$-GDD of group-type $(3m)^4 (3x + 6y)^1 (6m)^1$.

We are now in a position to continue our investigation of $B(4, 5, 11)$. First of all, we have the following result which slightly improves [6, Theorem 4.4].

LEMMA 4.15. *If $v \equiv 3 \pmod{12}$ and $v \neq 3, 15, 27$, then $v \in B(4, 5, 11^*)$.*

PROOF. In Lemma 4.8, we choose $m = 11$ and $k \geq 3$ to obtain $v = 12k + 15 \in B(4, 5, 11^*)$. In [11, 1.16 of VIII] Brouwer gives a construction of a resolvable PBD $B(\{3, 4\}, 1; 28)$ with 11 parallel classes. We can then adjoin a set of 11 "infinite" points to this design, one point being adjoined to each block of one of the parallel classes, to obtain a $B(\{4, 5, 11^*\}, 1; 39)$. This completes the proof of the lemma.

The following result is an improvement of that in [6, Theorem 4.5].

LEMMA 4.16. *If $v \equiv 11 \pmod{12}$ and $v \neq 23$, then $v \in B(4, 5, 11^*)$.*

PROOF. If $n \equiv 0 \pmod{12}$ and $n \neq 12$, then the result of Bermond, Bond and Sotteau in [10, Theorem 5.1] establishes the existence of a PBD $B(\{3, 4\}, 1; ; n)$ in which there is a parallel class of blocks of size 4 and the blocks of size 3 can be partitioned into 10 parallel classes. The introduction of 11 infinite points to this PBD gives $v = n + 11 \in B(4, 5, 11^*)$ and the lemma is proved.

The following lemma is essentially contained in [6, Remark 4.7].

LEMMA 4.17. *If* $v \equiv 7 (\mathrm{mod}\, 12)$, *then* $v \in B(4, 5, 11)$ *for all* $v \geq 55$ *with the possible exceptions of* $v = 67$ *and* 115.

We wish to further remark that Lemma 4.17 can be strengthened with the following:

LEMMA 4.18. *If* $v \equiv 7 (\mathrm{mod}\, 12)$, *then* $v \in B(4, 5, 11^*)$ *for all* $v \geq 79$ *with the possible exception of* $v = 115$.

PROOF. If we take $m = 39$ and $k \geq 10$ in Lemma 4.8, then we get $v = 12k + 43 \in B(4, 5, 39^*) \subseteq B(4, 5, 11^*)$. On the other hand, it is readily checked that $\{79, 91, 103, 127, 139, 151\} \subseteq B(4, 5, 11^*)$ from the constructions in [6, Theorem 4.6].

LEMMA 4.19. $110 \in B(4, 5, 11^*)$.

PROOF. In a $\{4\}$-GDD of group-type $3^5 6^1$, we give all the points weight 5 to obtain a $\{4\}$-GDD of group-type $(15)^5 (30)^1$, using $\{4\}$-GDDs of group-type 5^4 as input design, which we get from a $T(4, 1; 5)$, for example. Using the fact that $20 \in B(4, 5, 5^*)$ and $35 \in B(4, 5, 11^*)$ from Lemma 4.16, we can then apply Lemma 4.9 to adjoin 5 infinite points to the groups of this GDD and obtain $110 \in B(4, 5, 35^*) \subseteq B(4, 5, 11^*)$.

LEMMA 4.20. *If* $v \in \{158, 162, 174\}$, *then* $v \in B(4, 5, 11^*)$.

PROOF. In each of the following cases we shall apply Lemma 4.13 in conjunction with lemma 4.9 by making use of the result in Theorem 3.5 as folllows:

For the case $v = 158$, we apply Lemma 4.13 with $m = 9$, $x = 5$ and $y = 7$ to obtain a $\{4, 5\}$-GDD of type $(27)^4 (15)^1 (28)^1$. The introduction of 7 infinite points to this GDD, by applying Lemma 4.9 with $\{22, 34\} \subseteq B(4, 7^*)$, yields $158 \in B(4, 5, 35^*) \subseteq B(4, 5, 11^*)$.

Similarly, for the case $v = 162$, we apply Lemma 4.13 with $m = 9$, $x = 5$ and $y = 8$ to get $\{4, 5\}$-GDD of type $(27)^4 (15)^1 (32)^1$ and the adjoining of 7 infinite points to this GDD yields $162 \in B(4, 5, 39^*) \subseteq B(4, 5, 11^*)$.

Finally, for $v = 174$, we apply Lemma 4.13 with $m = 9$, $x = 9$, $y = 8$ to get $\{4, 5\}$-GDD of type $(27)^5 (32)^1$ and the introduction of 7 infinite points to this GDD yields $174 \in B(4, 5, 39^*) \subseteq B(4, 5, 11^*)$.

LEMMA 4.21. *If* $v \in \{178, 190, 202, 250, 262, 286\}$, *then* $v \in B(4, 5, 11)$.

PROOF. We shall apply Lemmas 4.13 and 4.14 in conjunction with Lemma 4.9 and the constructions are similar to those given in Lemma 4.20. We have the following:

For $v = 178$, we apply Lemma 4.14 with $m = 8$, $x = 7$ and $y = 1$ to obtain a $\{4\}$-GDD of type $(24)^4 (27)^1 (48)^1$. We then adjoin 7 infinite points to this GDD to get $178 \in B(4, 55^*) \subseteq B(4, 5, 11)$, using $55 \in B(5, 11)$.

For $v = 190$, we take $m = 8$, $x = 3$, $y = 5$ in Lemma 4.14 to get a $\{4\}$-GDD of type $(24)^4(39)^1(48)^1$. The introduction of 7 infinite points to this GDD yields $190 \in B(4, 55^*) \subseteq B(4, 5, 11)$.

For $v = 202$, we apply Lemma 4.14 with $m = 9$, $x = 5$ and $y = 4$ to obtain a $\{4\}$-GDD of type $(27)^4(39)^1(54)^1$. We then adjoin one infinite point to the groups of this GDD to obtain $202 \in B(4, 28, 40, 55) \subseteq B(4, 5, 11)$.

For $v \in \{250, 262, 286\}$, we apply Lemma 4.13 with $m = 16$, $x = 9$ and $y \in \{6, 9, 15\}$. We observe that $\{31, 34, 43, 55, 67\} \subseteq B(4, 7^*)$ from Theorem 3.5 and $55 \in B(5, 11)$. In each case, we can adjoin 7 infinite points to the resulting GDD to get $v \in B(4, 5, 55^*) \subseteq B(4, 5, 11)$. This completes the proof of the lemma.

We are now in a position to prove

THEOREM 4.22. *For all integers* $v \geq 4$, $v \in B(4, 5, 11)$ *holds with the exception of* $v \in \{6, 7, 8, 9, 10, 12, 14, 15, 18, 19, 22, 23, 26, 27, 30, 31, 34\}$, *and the possible exception of* $v \in \{38, 42, 43, 46, 50, 54, 58, 62, 66, 67, 70, 74, 78, 82, 90, 94, 98, 102, 106, 114, 115, 118, 126, 130, 142, 154\}$.

PROOF. First of all, it is an easy matter to check that the values of v listed as exceptions are not possible for $v \in B(4, 5, 11)$ (see, for example, [16]). On the other hand, if $v \equiv 0$ or $1 \pmod 4$, the result follows from Theorem 3.4. If $v \equiv 3 \pmod 4$, the result follows by combining Lemmas 4.15, 4.16 and 4.17. For convenience, we shall break the case $v \equiv 2 \pmod 4$ into the cases $v \equiv 2$ or $6 \pmod{12}$ and $v \equiv 10 \pmod{12}$ and make use of some results contained in [6]. For $v \equiv 2$ or $6 \pmod{12}$, the result is contained in lemmas 4.19, 4.20 and [6, Theorems 4.8 and 4.9]. If $v \equiv 10 \pmod{12}$, then it is shown in the proof of [6, Theorem 5.4] that $v \in B(4, 5, 11)$ for all $v \geq 166$ except for the values of $v \in \{178, 190, 202, 250, 262, 286\}$. However, these exceptional cases can now be handled by applying Lemma 4.21. This completes the proof of the theorem.

Before proceeding, we wish to remark that the set $B(4, 5, 11^*)$ is also of special interest. Since both $\binom{4}{2}$ and $\binom{5}{2}$ are even and $\binom{11}{2}$ is odd, the number of pairsets of points covered in a PBD $B(\{4, 5, 11^*\}, 1; v)$ must be odd. Consequently, a necessary condition for the existence of such a PBD is $v \equiv 2$ or $3 \pmod 4$. By combining the results of Lemmas 4.15, 4.16, and 4.17, we readily obtain the following:

LEMMA 4.23. *If* $v \equiv 3 \pmod 4$ *and* $v \neq 3, 7, 15, 19, 23, 27, 31, 43, 55, 67, 115$, *then* $v \in B(4, 5, 11^*)$.

For the case $v \equiv 2 \pmod 4$, we shall make use of some constructions from [6].

LEMMA 4.24.

(a) *If* $1 \leq s \leq k$, *then* $48k + 12s + 26 \in B(4, 5, 11^*)$.

(b) *If* $1 \leq s \leq k - 1$, *then* $48k + 12s + 14 \in B(4, 5, 11^*)$.

PROOF. For the proof of (a), see [6, Lemma 3.9]. For the proof of (b), see [6, Lemma 3.10].

LEMMA 4.25. *If* $v \equiv 2(mod\ 12)$ *and* $v \neq 2, 14, 26, 38, 50, 62, 74, 98$, *then* $v \in B(4, 5, 11^*)$.

PROOF. By applying Lemma 4.24, we obtain $v \in B(4, 5, 11^*)$ for all the stated values of v except $v = 110$ and 158. However, $\{110, 158\} \subseteq B(4, 5, 11^*)$ from Lemmas 4.19 and 4.20, and this completes the proof.

LEMMA 4.26. *If* $k \geq 2$ *and* $1 \leq s \leq k$, *then* $48k + 12s + 30 \in B(4, 5, 11^*)$.

PROOF. See [6, Lemma 3.11]

LEMMA 4.27. *If* $v \equiv 6(mod\ 12)$ *and* $v \neq 6, 18, 30, 42, 54, 66, 78, 90$, $102, 114, 126$, *then* $v \in B(4, 5, 11^*)$.

PROOF. Lemma 4.26 provides $v \in B(4, 5, 11^*)$ for all the stated values of v except $v = 162, 174$ and 222. But $\{162, 174\} \subseteq B(4, 5, 11^*)$ from Lemma 4.20. For the case $v = 222$, we apply Lemma 4.13 with $m = 13$, $x = 9$ and $y = 8$ to obtain a $\{4, 5\}$-GDD of type $(39)^4(27)^1(32)^1$. The introduction of 7 infinite points to this GDD yields $222 \in B(4, 5, 39^*) \subseteq B(4, 5, 11^*)$. This completes the proof of the lemma.

LEMMA 4.28. *If* $1 \leq s \leq k - 1$, *then we have* $48k + 12s + 10 \in B(4, 5, (12k + 7)^*)$.

PROOF. See [6, Lemma 3.7].

LEMMA 4.29. *If* $v \equiv 10\ (mod\ 12)$ *and* $v \geq 262$, *then* $v \in B(4, 5, 11^*)$.

PROOF. From Lemma 4.18, $12k + 7 \in B(4, 5, 11^*)$ for all $k \geq 6$, $k \neq 9$. Consequently, Lemma 4.28 guarantees the result for all $v \geq 310$, with the exception of $v = 454, 466, 478$, and 490. For $v \in \{454, 466, 478, 490\}$, we can apply Lemma 4.13 with $m = 32$, $x \in \{5, 9, 13, 17\}$ and $y = 12$. Observing that $\{55, 103, 3x + 7\} \subseteq B(4, 7^*)$ and $103 \in B(4, 5, 11^*)$ from Lemma 4.18, we can adjoin 7 infinite points to the resulting GDD in each case and obtain $v \in B(4, 5, 103^*) \subseteq B(4, 5, 11^*)$. The proof that $298 \in B(4, 5, 11^*)$ follows from the construction used in [6, Theorem 5.4]. For $v \in \{262, 274, 286\}$, we can apply Lemma 4.14 with $m = 12$ and $3x + 6y \in \{39, 51, 63\}$, and then adjoin 7 infinite points to the resulting GDD to get $v \in B(4, 79^*) \subseteq B(4, 5, 11^*)$.

Combining the results of Lemmas 4.23, 4.25, 4.27 and 4.29, we have essentially proved the following theorem:

THEOREM 4.30. *The necessary condition* $v \equiv 2$ *or* $3(mod\ 4)$ *for the existence of a PBD* $B(\{4, 5, 11^*\}, 1; v)$ *is sufficient for all* $v \geq 251$.

Theorems 4.22 and 4.30 provide useful supplements to the results in Section 3. For example, the combination of Theorems 3.7(d) and 4.22 gives us the following result, which has been applied in the construction of various types of Latin squares (see [8, 9]).

THEOREM 4.31. *If* $v \geq 4$ *and* $v \notin \{6, 8, 9, 10, 12, 14, 15, 18, 19, 23,$ $26, 27, 30, 38, 42, 50, 54, 62, 66, 74, 78, 90\}$, *then* $v \in B(4, 5, 7, 11)$.

In what follows, we shall apply the results of Theorem 4.22 to further enlarge the spectra of a variety of Stein quasigroups. A quasigroup satisfying the identity $x(xy) = yx$ is called a *Stein quasigroup*. Stein quasigroups are necessarily idempotent and self-orthogonal, that is, the Latin squares defined by such quasigroups are idempotent and self-orthogonal. In [32], Stein originally investigated this identity with the hope of constructing counterexamples to the Euler conjecture. Several authors (see, for example, [5, 21, 30–33]) have given the identity a considerable amount of attention since that time and long after the disproof of the Euler conjecture. Let us denote by $J(x(xy) = yx)$ the spectrum of the identity $x(xy) = yx$. Pelling and Rogers [30, 31]showed that if $n \in \{2, 3, 6, 7, 8, 10, 12, 14\}$, then $n \notin J(x(xy) = yx)$ and moreover, $n \in J(x(xy) = yx)$for all $n > 1042$. This bound was later improved by the author and N.S. Mendelsohn in [6] on the strength of the following:

LEMMA 4.32. $B(4, 5, 9, 11, 19, 31) \subseteq J(x(xy) = yx)$.

LEMMA 4.33 (see [6, Theorems 4.3–4.6, 4.8–4.10]). *If we define* $E =$ $\{6, 7, 8, 10, 12, 14, 15, 18, 22, 23, 26, 27, 30, 34, 35, 38, 39, 42, 43,$ $46, 50, 54, 62, 66, 70, 74, 78, 82, 90, 98, 102, 106, 110, 114, 126,$ $130, 142, 158, 162, 174, 178, 190\}$, *then* $v \in B(4, 5, 9, 11, 19, 31)$ *for all* $v \geq 4$ *where* $v \notin E$.

THEOREM 4.34 (see [6]). $v \in J(x(xy) = yx)$ *holds for all positive integers* v *except* $v \in \{2, 3, 6, 7, 8, 10, 12, 14\}$ *and possibly excepting* $v \in E \setminus$ $\{6, 7, 8, 10, 12, 14\}$, *where* E *is defined in Lemma 4.33*.

Before combining the result of Theorem 4.22 with that of Lemma 4.33, we wish to observe further the following additional constructions.

LEMMA 4.35. *If* $v \in \{106, 130, 142\}$, *then* $v \in B(4, 5, 19, 31)$.

PROOF. For $v = 106$, we have already constructed a $\{4\}$-GDD of grouptype $(15)^5(30)^1$ in the proof of Lemma 4.19. We can then adjoin an infinite point to this GDD and obtain $106 \in B(4, 16, 31^*) \subseteq B(4, 31^*)$. For $v = 130$, we can apply Lemma 4.12 with $m = 8$, $x = 1$ and $y = 6$ to get a $\{4,5\}$-GDD of type $(24)^4(27)^1$. We then adjoin 7 infinite points to this GDD, using the fact that $\{31, 34\} \subseteq B(4, 7^*)$, to obtain $130 \in B(4, 5, 31^*)$. Finally for $v = 142$, we apply Lemma 4.13 with $m = 9$, $x = 5$ and $y = 3$ to get a $\{4, 5\}$-GDD of type $(27)^4(15)^1(12)^1$. Using the fact that $\{22, 34\} \subseteq B(4, 7^*)$, we then adjoin 7 infinite points to this GDD to obtain $142 \in B(4, 5, 19^*)$.

Combining the results of Lemmas 4.33 and 4.35 with Theorem 4.22, we readily obtain the following.

THEOREM 4.36. *For all integers* $v \geq 4$, $v \in B(4, 5, 9, 11, 19, 31)$ *with the exception of* $v \in \{6, 7, 8, 10, 12, 14, 15, 18, 22, 23, 26, 27, 30, 34\}$ *and with the possible exception of* $v \in \{38, 42, 43, 46, 50, 54, 62, 66, 70, 74, 78, 82, 90, 98, 102, 114, 126\}$.

D.G. Rogers [private communication] has recently informed the author that $v = 18$ is now a definite exception in Theorem 4.34. Consequently, we can now provide the following improvement as a consequence of Lemma 4.32 and Theorem 4.36.

THEOREM 4.37. $v \in J(x(xy) = yx)$ *holds for all positive integers* v *except* $v \in \{2, 3, 6, 7, 8, 10, 12, 14, 18\}$ *and possibly excepting* $v \in \{15, 22, 23, 26, 27, 30, 34, 38, 42, 43, 46, 50, 54, 62, 66, 70, 74, 78, 82, 90, 98, 102, 114, 126\}$.

The results of Theorems 4.22 and 4.36 can also be used to enlarge the spectrum of various types of Stein systems which were earlier investigated (see, for example, [6, 30]). In particular, we have the following consequence of Theorem 4.22.

THEOREM 4.38. *For all integers* $n \geq 4$, *there exists a Stein system of order* n *such that every 2-element generated subsystem is of order* 4 *or* 5 *or* 11, *except for* $n \in \{6, 7, 8, 9, 10, 12, 14, 18\}$ *and possibly excepting* $n \in \{15, 19, 22, 23, 26, 27, 30, 31, 34, 38, 42, 43, 46, 50, 54, 58, 62, 66, 67, 70, 74, 78, 82, 90, 94, 98, 102, 106, 114, 115, 118, 126, 130, 142, 154\}$.

As we did in [6], let us write $R^*(v)$ whenever there exists a Stein system of order v which contains both a subsystem of order 4 and a subsystem of order 5. It is fairly obvious that $v \geq 16$ is necessary for $R^*(v)$ to hold. We shall conclude this section with the following improvement of [6, Theorem 5.5] and the reader is referred to [7, 23] for further applications.

THEOREM 4.39. $R^*(v)$ *holds for all* $v \geq 16$ *with the exception of* $v = 18$ *and the possible exception of* $v \in \{16, 19, 22, 23, 25, 26, 27, 30, 31, 34, 38, 42, 43, 46, 50, 54, 55, 58, 62, 66, 70, 74, 78, 82, 90, 94, 98, 102, 114, 126\}$.

PROOF. We can now remove the possible exceptions $v \in \{28, 35, 39, 110, 130, 142, 158, 162, 174, 178, 190\}$ listed in [6, Theorem 5.5] as follows. First of all, it is possible to construct a *PBD* $B(\{4, 5\}, 1; 28)$ containing both block sizes 4 and 5 (see, for example, [11]). Secondly, if $v \in \{35, 39, 110, 130, 142, 158, 162, 174, 178, 190\}$, then it is an easy matter to check that the constructions provided in this paper will guarantee the existence of a *PBD* $B(\{4, 5, 11, 19, 31\}, 1; v)$ which contains both block sizes 4 and 5. In each case, a Stein system of order v can then be constructed with the desired properties.

Note added in proof: Some improvements to results contained in this paper have been found since it was first submitted for publication:

1) In Theorem 3.12, it can now be shown that $v \in B(Q^*)$ for all $v \geq 1415$ where $v \equiv 0$ or $1 \pmod 7$, with at most 84 possible exceptions below this value.

2) In Theorem 4.30, the necessary condition $v \equiv 2$ or $3 \pmod 4$ for the existence of a $B(\{4, 5, 11^*\}, 1; v)$ is sufficient for all $v \geq 227$.

3) The possible exceptions $v = 70$ and 82 listed in Theorems 4.36 and 4.37 can now be removed.

REFERENCES

1. F. E. Bennett, *Pairwise balanced designs with prime power block sizes exceeding 7*, Annals of Discrete Math. **34** (1987), 43–64.

2. ____, *Conjugate orthogonal Latin squares and Mendelsohn designs*, Ars Combinatoria **19** (1985), 51–62.

3. ____, *The spectra of a variety of quasigroups and related combinatorial designs*, Annals of Discrete Math. **77** (1989), 29–50.

4. F. E. Bennett, Du Beiliang and L. Zhu, *On the existence of* $(v, 7, 1)$*-perfect Mendelsohn designs*, Discrete Math., to appear.

5. F. E. Bennett, E. Mendelsohn and N. S. Mendelsohn, *Resolvable perfect cyclic designs*, J. Combinatorial Theory (A) **29** (1980), 142–150.

6. F. E. Bennett and N. S. Mendelsohn, *On the spectrum of Stein quasigroups*, Bull. Austral. Math. Soc. **21** (1980), 47–63.

7. ____, *Conjugate orthogonal Latin square graphs*, Congressus Numerantium **23** (1979), 179–192.

8. F. E. Bennett, Lisheng Wu and L. Zhu, *Conjugate orthogonal Latin squares with equal-sized holes*, Annals of Discrete Math. **34** (1987), 65–80.

9. F. E. Bennett, Lisheng Wu and L. Zhu, *On the existence of COLS with equal-sized holes*, Ars Combin. **26** B (1988), 5–36.

10. J. C. Bermond, J. Bond and D. Sotteau, *Regular packings and coverings*, Annals of Discrete Math. **34** (1987), 81–100.

11. Th. Beth, D. Jungnickel and H. Lenz, "Design Theory", Bibliographisches Institut, Zurich, 1985.

12. R. K. Brayton, D. Coppersmith and A. J. Hoffman, *Self-orthogonal Latin Squares of all orders* $n \neq 2$, 3 *or* 6, Bull. Amer. Math. Soc. **80** (1974), 116–118.

13. A. E. Brouwer, *Optimal packings of* K_4 *'s into a* K_n, J. Combinatorial Theory (A) **26** (1979), 278–297.

14. ____, *The number of mutually orthogonal Latin squares - a table up to order* 10, 000, Research Report ZW123/79, Mathematisch Centrum, Amsterdam, 1979.

15. J. Dénes and A. D. Keedwell, " Latin Squares and Their Applications", Academic Press, New York and London, 1974.

16. D. A. Drake and J. A. Larson, *Pairwise balanced designs whose line sizes do not divide six*, J. Combinatorial Theory (A) **34** (1983), 266–300.

17. H. Hanani, *Balanced incomplete block designs and related designs*, Discrete Math. **11** (1975), 255–369.

18 . H. Hanani, D. K. Ray-Chaudhuri and R. M. Wilson, *On resolvable designs*, Discrete Math. **3** (1972), 343–357.

19. J. F. Lawless, *Pairwise balanced designs and the construction of certain combinatorial systems*, Proc. of the Second Louisiana Conf. on Combinatorics, Graph Theory and Computing, Baton Rouge (1971), 353–366.

20. H. Lenz, *Some remarks on pairwise balanced designs*, Mitt. Math. Sem. Giessen No. **165** (1984), 49–62.
21. C. C. Lindner, *Construction of quasigroups satisfying the identity* $x(xy) = yx$, Canad. Math. Bull. **14** (1971), 57–59.
22. ____, *Quasigroup identities and orthogonal arrays*, in: *Surveys in Combinatorics* (edited by E.K. Lloyd) London Math. Soc. Lecture Notes Ser. **82**, Cambridge Univ. Press, 1983, 77–105.
23. C. C. Lindner, E. Mendelsohn, N. S. Mendelsohn and B. Wolk, *Orthogonal Latin square graphs*, J. Graph Theory **3** (1979), 325–338.
24. H. F. MacNeish, *Euler squares*, Ann. of Math. **23** (1922), 221–227.
25. N. S. Mendelsohn, *Perfect cyclic designs*, Discrete Math. **20** (1977), 63–68.
26. R. C. Mullin, P. J. Schellenberg, S. A. Vanstone and W. D. Wallis, *On the existence of frames*, Discrete Math. **37** (1981), 79–104.
27. R. C. Mullin and D. R. Stinson, *Holey SOLSSOMs*, Utilitas Math. **25** (1984), 159–169.
28. ____, *Pairwise balanced designs with odd block sizes exceeding* 5, Discrete Math., to appear.
29. ____, *Pairwise balanced designs with block sizes* $6t + 1$, Graphs and Combinatorics **3** (1987), 365–377.
30. M. J. Pelling and D. G. Rogers, *Stein quasigroups I: combinatorial aspects*, Bull. Austral. Math. Soc. **18** (1978), 221–236.
31. ____, *Stein Quasigroups II: algebraic aspects*, Bull. Austral. Math. Soc. **20** (1979), 321–334.
32. S. K. Stein, *On the foundations of quasigroups*, Trans. Amer. Math. Soc. **85** (1957), 228–256.
33. ____, *Homogeneous quasigroups*, Pacific J. Math. **14** (1964), 1091–1102.
34. D. R. Stinson, *On the spectrum of nested* 4-*cycle systems*, Utilitas Math. **33** (1988), 47–50.
35. D. T. Todorov, *Three mutually orthogonal Latin squares of order* 14, Ars Combinatoria **20** (1985), 45–48.
36. R. M. Wilson, *Constructions and uses of pairwise balanced designs*, Mathematical Centre Tracts **55** (1974), 18–41.
37. ____, *An existence theory for pairwise balanced designs I*, J. Combinatorial Theory (A) **13** (1972)), 220–245.
38. ____, *An existence theory for pairwise balanced designs II*, J. Combinatorial Theory (A) **13** (1972), 246–273.
39. ____, *An existence theory for pairwise balanced designs III*, J. Combinatorial Theory (A) **18** (1975), 71–79.
40. Zhu Lie and Chen Demeng, *Orthogonal Steiner triple systems of order* $6m + 1$, preprint.

DEPARTMENT OF MATHEMATICS, MOUNT SAINT VINCENT UNIVERSITY, HALIFAX, NOVA SCOTIA B3M 2J6, CANADA

Contemporary Mathematics
Volume **111**, 1990

On the p-Rank of Incidence Matrices and a Question of E.S. Lander

A. A. BRUEN AND U. OTT

Summary. For p a prime we examine the p-rank x for an incidence matrix M of a finite linear space X. (This question is not without interest. For example, let us specialize for the moment to the case when X is a finite projective plane of order n where p divides n. Assume also that p^2 does not divide n. Then it follows that $x = \frac{1}{2}(n^2 + n + 2)$. From this it is well-known that one can fairly easily derive the Bruck-Ryser nonexistence criterion for the projective plane X.)

In this note we introduce a new technique which also seems promising for future developments. Our main result is a very general relationship involving the dimensions of a certain tensor product space and the space of "circuits" in X. From this we obtain a lower bound for x. In particular if X is a finite projective plane of order n then it is shown that $x \geq n\sqrt{n} + 1$. This represents a substantial improvement on the lower bound of $3n - 2$ obtained in the monograph of E. S. Lander [2, p. 68]. As remarked, the bound is obtained from very general principles. However, perhaps surprisingly, it is quite close to being "best possible" as is shown by the dimension x of an incidence matrix associated with $PG(2, q)$, $q = 2^t$, $p = 2$.

1. Background, notation. Formal definitions are given below. Let M be an incidence matrix for a finite linear space X on v points with b blocks. The rows of M then correspond to the lines of X. Let T be a set of lines in X with the following property: each line ℓ of T contains a point not lying on any other line of T.

Then the rows of X corresponding to the lines of T are linearly independent, working over any field K. Therefore the K-rank of M is at least $|T|$. Specialize for the moment to the case when X is a finite projective plane of order n. Assume also that T is a dual blocking set. Then from [1]

1980 *Mathematics Subject Classification* (1985 *Revision*). 05B20.

This paper is in final form and no version of it will be submitted for publication elsewhere.

$|T| \leq n\sqrt{n}+1$ with equality when and only when T is (the set of tangents to) a unital. On the other hand notice that if $X = PG(2, q)$, $q = 2^t$, $K = GF(2)$ then the 2-rank of M is about $q^{1.6} = n^{1.6}$. From these remarks it would appear that a reasonable (but fairly good) lower bound to aim for in connection with the p-rank of M would be of the order of $n\sqrt{n}$.

One of the main results implies this. However our results are valid not just for projective planes but for arbitrary finite linear spaces.

For general background we refer to [2], [3], [4]. Although our main interest here is in projective and affine spaces we initially work with more general structures, the linear spaces. Recall that a (finite) linear space X consists of a finite set Ω_0 of points and a collection Ω_1 of subsets of points called lines satisfying the following axiom:

AXIOM. Any two points A, B in X are joined by (contained in) a unique line y of X.

We say that A is *incident* with y and write AIy if point A lies on line y. Then the incidenct pair (A, y) is called a *flag* of X. We denote by Λ the set of all flags.

We denote the total number of lines, points of X by b, v respectively. Let us label the points of X as $P_1, P_2, \ldots, P_v$ and the lines of X as $\ell_1, \ell_2, \ldots, \ell_b$. We then form a so-called *incidence matrix* $M = (m_{ij})$ of size $b \times v$ where

$$m_{ij} = \begin{cases} 1 & \text{if } \ell_i IP_j \\ 0 & \text{otherwise} \end{cases}.$$

Different labellings of the points and lines give rise to different incidence matrices M. However, working over some field K, the rank of M over K (i.e. the K-rank of M) does not depend on the labelling. Of course the rank *will* depend on our choice of K. We are mainly interested in the case when $K = GF(p)$ the field of order p, p a prime. In this case the rank of M over K is denoted by $\text{Rank}_p M$.

We will use the K-vector spaces V_0, V_1, V admitting Ω_0, Ω_1 and Λ as a basis, respectively.

One of the key ideas is to exploit the K-linear map φ from the tensor product (over K) of V_0 and V_1 to V

$$\varphi : V_0 \otimes V_1 \to V,$$

which is given by its action on a basis, namely

$$\varphi(A \otimes a) = A \wedge a = \begin{cases} (A, a) & \text{if } AIa \\ 0 & \text{otherwise} \end{cases}.$$

Before examining this map φ we introduce some further notation. For each line a, let

$$[a] = \sum_{A I a} A = \sum_{A \wedge a \neq 0} A,$$

and put

$$U_0 = < [a] \, | \, a \in \Omega_1 > \subseteq V_0.$$

Dually, for each point A let

$$[A] = \sum_{A I a} a = \sum_{A \wedge a \neq 0} a,$$

and put

$$U_1 = < [A] \, | \, A \in \Omega_0 > \subseteq V_1.$$

We utilize the vector

$$\overline{[a]} = \sum_{A \nmid a} A = \sum_{A \wedge a = 0} A,$$

which we call the *affine plane of* a. Note that $\overline{[a]}$ is a vector in V_0. Let

$$W_0 = < \overline{[a]} | a \in \Omega_1 > \subseteq V_0,$$
$$W_1 = < [A] - [B] | A, B \in \Omega_0 > \subset U_1 \subset V_1.$$

Finally, for each line a and point A we denote by $|a|$ and $|A|$ the number of points on a and lines through A, respectively.

2.1 LEMMA. *Put* $U = U_0 \otimes U_1$ *with*

$$x_i = \dim(U_i), \qquad i = 0, 1.$$

Then
(a) $x_0 = x_1 = \mathrm{Rank}_p(M) = x$, *say.*
(b) $\dim(U) = x^2$.

PROOF. By definition we have

$$\dim(U) = \dim(U_0 \otimes U_1) = \dim(U_0)\dim(U_1) = x_0 x_1.$$

Thus it suffices to prove (a). With respect to the bases Ω_0 of V_0 and Ω_1 of V_1 each generator $[a]$ of U_0 and generator $[A]$ of U_1 is represented by a row or column of an incidence matrix M. Since the row rank of M equals the column rank of M, we obtain the desired result.

In order to study the image of W under the action of our linear map φ, we introduce special "circuits". Let A_1, A_2, A_3 denote an ordered triangle of points. There is a unique line a_i joining A_j and A_k where $i \neq j \neq k \neq i$. Then the *triangular circuit* $\Delta = \Delta(A_1, A_2, A_3)$ associated with the given triangle is an element of the space V, defined by the rule

$$\Delta = (A_1, a_3) - (A_2, a_3) + (A_2, a_1) - (A_3, a_1) + (A_3, a_2) - (A_1, a_2).$$

2.2 LEMMA. *Assume that $\overline{[a]}$ is in the space U_0 for every line a. Then every triangular circuit is an element of the image $\varphi(W_0 \otimes W_1) \subset \varphi(U)$.*

PROOF. Let $\Delta = \Delta(A_1, A_2, A_3)$ be any triangular circuit. Put

$$E = \overline{[a_1]} \wedge ([A_2] - [A_3]) + \overline{[a_2]} \wedge ([A_3] - [A_1]) + \overline{[a_3]} \wedge ([A_1] - [A_2]).$$

We conclude that $\Delta = E$ by checking which flag (A, a) appears in the linear combination corresponding to E with respect to the base of the flags in the flag space V : it is convenient to examine the cases where a point is or is not incident with a side of the given triangle A_1, A_2, A_3. By assumption $\overline{[a_i]}$ is an element of U_0, so that E is an element of $\varphi(U)$. The result follows.

2.3 COROLLARY. *Suppose that $\overline{[a]}$ is in U_0 for every line a. Then the dimension of $\varphi(U_0 \otimes W_1)$ is greater than or equal to the maximum number of linearly independent triangular circuits in the flag space V.*

We also obtain:

2.4 THEOREM. *Assume that $\overline{[g]}$ is in U_0 for every line g. For any line a of the linear space we have*

$$\dim \varphi(U_0 \otimes W_1) \geq (v - |a|)(|a| - 1).$$

PROOF. We fix a point P on the line a. Let Q be any point not on a, and $R \neq P$ a further point on a.

We claim that the set of all triangular circuits $\Delta(P, R, Q)$ for different choices of Q, R is linearly independent. This follows from the fact that the flag (Q, q), where q is the line joining Q and R, has coefficient 1 in the triangular circuit $\Delta(P, R, Q)$ but coefficient 0 in all the other triangular circuits. There are $v - |a|$ choices for Q and $|a| - 1$ choices for R. Using Corollary 2.3 we get the result.

We are therefore led to an examination of the question of when $\overline{[a]}$ is in U_0, i.e. the question of when W_0 is contained in U_0.

2.5 LEMMA. *Assume that there exists an integer $\theta \not\equiv 0 \pmod{p}$ such that $|A| \equiv \theta \pmod{p}$ for all points A. Then we have*
(a) *for any line a, $\overline{[a]}$ is in U_0,*
(b) *$\dim(W_1) = x - 1$,*
(c) *$\dim \varphi(U_0 \otimes W_1) \leq x(x - 1)$.*

PROOF. We compute the element

$$\overline{[a]} = \sum_{A \text{ in } \Omega_0} A - ([a])$$

$$= \frac{1}{\theta} \sum_{A \text{ in } \Omega_0} |A| A - ([a])$$

$$= \frac{1}{\theta} \sum_{y \text{ in } \Omega_1} [y] - ([a]),$$

which, by definition, is in U_0. This proves (a).

For (b) we only need to show that W_1 is a proper subspace of U_1. This can be seen as follows.

With respect to the base Ω_1 of V_1 each generator $[A]$ of U_1 is represented as a column of the incidence matrix M. Using the usual dot product, each generator $[A] - [B]$, and so each element of W_1, is orthogonal, over $GF(p)$, to the all one column j. However no column of M is orthogonal to j, since $|A| \not\equiv 0 \pmod{p}$. Thus (b) is proved.

Statement (c) follows from (b), and the proof of 2.5 is complete. Using 2.4 and 2.5 we obtain the following result.

2.6 THEOREM. *Let X be any finite linear space on v points and let M be an incidence matrix for X. Assume that the number of lines through a point is, modulo a prime p, a nonzero constant. Then*

$$\text{Rank}_p M (\text{Rank}_p M - 1) \geq (v - |a|)(|a| - 1),$$

for any line a of X.

2.7 LEMMA. *Assume that there are integers R, $K \not\equiv 0 \pmod{p}$ such that $|A| \equiv R$ and $|a| \equiv K$ for all points A and lines a. Assume also that $R \equiv b \pmod{p}$.*

Then

$$\dim W_0 \leq x - 1.$$

PROOF. By lemma 2.5 we know that W_0 is a subspace of U_0. In 2.5 we obtained the formula

$$\overline{[a]} = \frac{1}{R} \sum_{y \text{ in } \Omega_1} [y] - ([a]).$$

This yields also

$$\overline{[a]} = \frac{1}{R} \sum_{y \text{ in } \Omega_1} ([y] - [a]) + \frac{b}{R}[a] - [a]$$

$$= \frac{1}{R} \sum_{y \text{ in } \Omega_1} ([y] - [a]).$$

Thus W_0 is in fact a subspace of $L = \langle [y_1] - [y_2] | y_1, y_2 \text{ in } \Omega_1 \rangle \subseteq U_0$. It remains to show that L is a proper subspace of U_0. This can be shown by a proof dual to that of 2.5, part (b). The lemma is proved.

The proof of the next result is similar to that of 2.6.

2.8 THEOREM. *Let X be a finite linear space on v points, b lines and let M be its incidence matrix. Assume that the number of lines (points) of X through a point (on a line) is a nonzero constant $R(K)$ modulo p, p a prime. Assume also that $b \equiv R (\mathrm{mod}\, p)$.*

Then for any line a of X we have

$$(\mathrm{Rank}_p M - 1)^2 \geq (v - |a|)(|a| - 1).$$

REMARK. Actually we have shown that $(\mathrm{Rank}_p M - 1)^2$ is bigger or equal to the maximum number of linearly independent triangular circuits in the flag space V.

2.9 COROLLARY. *Let Π be a projective plane of order n and let p be a prime divisor of n.*

Then

$$\mathrm{Rank}_p M \geq n\sqrt{n} + 1,$$

where M is an incidence matrix of Π.

We should remark that this bound on the p-rank of a projective plane is significantly better than the only known one, $\mathrm{rank}_p(M) \geq 3n - 2$ (see Lander [2]).

Furthermore (see [3]) the 2-rank of a desarguesian plane of order $n = 2^e$ is $3^e + 1$ which is approximately $n^{1.6}$. This shows that the bound in 2.9 is fairly close to being "best possible".

Let D be a design with the parameters b, v, r, k, λ having their customary meaning. As usual we define the *line* of D through any two different points A and B of D to be the intersection of all blocks through A and B. Then the points and lines of D give rise to a finite linear space X. Let M be its incidence matrix and let x denote the p-rank of M. We then obtain the following result.

2.10 THEOREM. *Assume that the number of lines through any point of D is a nonzero constant modulo p. Suppose that any line which is not contained in a block meets the block in at most one point.*

Then

$$x(x - 1) \geq (v - k)(k - 1),$$

where k is the block size of D.

PROOF. We use 2.3, 2.5 and we modify the proof of 2.4 by choosing a block b instead of the line a there.

The counting of triangular circuits in the proof of theorem 2.10 yields also a further application of theorem 2.8 and the subsequent remark there.

2.11 COROLLARY. *Let $\mathscr{P}$ be a projective space of dimension $n \geq 2$ over a finite field with $q = p^e$ elements and let M be the incidence matrix of the linear space consisting of points and lines of $\mathscr{P}$.*
Then

$$(\operatorname{rank}_p M - 1)^2 \geq q^n \left[\frac{q^n - 1}{q - 1} - 1 \right].$$

REMARKS. 1. It is also of interest to examine the kernel of the mapping φ discussed in Section 2 for the case of classical planes.

2. It is easy to see how some of the methods of this paper apply not just to linear spaces but to generalized n-gons, partial geometries, etc.

3. Some of the discussion goes through for arbitrary circuits instead of triangular circuits.

REFERENCES

1. A. A. Bruen and J. A. Thas, *Blocking Sets*, Geometriae Dedicata **6**, 1977, 193–203.
2. E. S. Lander, *Symmetric Designs: An Algebraic Approach*, L.M.S. Lecture Notes Series **74**, Cambridge University Press, 1983.
3. F. J. MacWilliams and N. J. A. Sloane, *The Theory of Error-Correcting Codes*, North Holland, 1978.
4. U. Ott, *Some Remarks On Representation Theory in Finite Geometry*, in Geometries and Groups, Eds. M. Aigner and D. Jungnickel, Springer Lecture Notes No. **893**, 1981, 68–110.

DEPARTMENT OF MATHEMATICS, UNIVERSITY OF WESTERN ONTARIO, LONDON, ONTARIO, N6A 5B7 CANADA

INSTITUTE OF GEOMETRY, UNIVERSITY OF BRAUNSCHWEIG, 3300 BRAUNSCHWEIG, WEST GERMANY

Contemporary Mathematics
Volume 111, 1990

On the Dempwolff Plane

MARIALUISA J. DE RESMINI

ABSTRACT. Some combinatorial properties are investigated of the Dempwolff plane Π. It is shown that Π admits generating quadrangles and any such quadrangle has at most one point on the line at infinity. It is proved that Π contains maximal Fano subplanes, i.e. Fano subplanes which are not contained in any Baer subplane, and Fano subplanes which are contained in one, two or three Baer subplanes. The Fano subplanes are also classified according to their behaviour w.r.t. the line at infinity. Finally, complete arcs are investigated in Π. In particular, it is shown that Π contains complete 14-arcs. Some results on hyperovals are provided too.

1. Introduction. The Dempwolff plane [4] is a translation plane of order 16 and all the translation planes of such an order have been classified [5]. Moreover, their full collineation groups are known (see [16] and references there), hence the properties implied by them are known too. On the other hand, to the author's best knowledge, up to now there are just some results on the combinatorial structures of these planes [2, 8, 10].

In this paper we provide some information on the combinatorial properties of the Dempwolff plane, say Π. First of all, we briefly summarize the most important features of Π. For more detail the reader is referred to [4, 5, 6, 14, 15, 16].

The plane Π is of Lenz-Barlotti class IV.a.1 [3, 4]. It admits SL(2,4) as a collineation group [4, 14, 15]. This is remarkable as Π is the only translation plane of order q^2 which admits a collineation group isomorphic to SL(2,q) and which is not of dimension two over its kernel. That is, the kernel of Π is GF(2) [5, 14, 15]. Another exceptional feature of the Dempwolff plane is that the lengths of the orbits on the line at infinity are 1, 1, 15 w.r.t. its full collineation group [4, 15]. Finally, Π may be derived from the semifield plane of order 16 with kernel GF(2) [15, 16].

We shall regard Π as a projective plane but keep in mind that, being a translation plane, it has a special line which will obviously be referred to as

---

1980 *Mathematics Subject Classification* (1985 *Revision*). Primary 51E15, 05B25.
This paper is in final form and no version of it will be submitted for publication elsewhere.

the line at infinity, $\ell\infty$.

We show that Π is singly generated by quadrangles and a generating quadrangle can have one vertex at most on $\ell\infty$. A quadrangle generates a Fano subplane whenever it does not generate Π and some of these quadrangles can be characterized (Section 3). No Baer subplane of Π was found which is tangent to the line at infinity (Section 4).

The Fano subplanes of Π can be classified from two different standpoints (Section 5). On the one hand, we can look at their behaviour w.r.t. the line at infinity. Such an approach yields three inequivalent types of Fano subplanes and this classification can obviously be refined by looking in which orbits on $\ell\infty$ the points at infinity lie of the considered Fano subplane. On the other hand, we can classify the subplanes of order two according to their being or not contained in (extendable to) Baer subplanes. In the latter case, a subplane of order two will be called maximal and Π contains maximal Fano subplanes. It is well known that in the Desarguesian plane PG(2,16) every Fano subplane is contained in a unique Baer subplane. This is not true in Π which, besides the maximal Fano subplanes, contains subplanes of order two which extend to one, two or three Baer subplanes (Section 5).

Finally, we investigate complete arcs in Π. The smallest possible size for such an arc turns out to be ten (Section 6). It is well known that in the Desarguesian plane of order 16, PG(2,16), the maximum possible size for a complete arc other than a hyperoval is 13 [12] and the existence of such arcs was shown in [11]. On the other hand, a non-Desarguesian plane of order 16 can contain complete arcs of sizes 14, 15, and 16. Examples are known of complete 14-arcs in the Lorimer plane and in its dual plane [10], the latter being not a translation plane, of complete 16-arcs in the Hall plane [19]. In Section 7 we show that Π contains complete 14-arcs. We observe that the semifield plane with kern GF(2) from which Π can be derived seems to contain no complete k-arc other than a hyperoval with $k > 13$ [8].

In Section 8 some examples are given of hyperovals in Π pointing out the existence of quadruples of hyperovals with some special intersection properties. Notice that the hyperovals in all the translation planes of order 16 are thoroughly investigated and completely classified in [12].

For background on projective planes and, in particular, on translation planes the reader is referred to [3, 12, 13, 18, 20].

2. Notation and terminology. The description of Π provided in [5] enables us to write down all its lines as the sets of their points. This way of representing Π is used here. We are not writing down all the lines of Π but we are now giving enough information to do it.

The points of Π are denoted by $A0\,A1\,A2\ldots A16$, $B1\,B2\ldots B16$, Cj, Dj, Fj, Hj, Kj, Lj, Mj, Nj, Pj, Rj, Sj, Tj, Wj, Xj, Zj, $j = 1, 2, \ldots, 16$. The lines of Π are denoted by the corresponding lower case letters and same subscripts. The line at infinity is the line $a0$: $A0\,A1\,A2\ldots$

$A16$. Each of the remaining lines aj contains all the points with the same letter and all the subscripts, $a1$ the B's, $a2$ the C's, ..., $a16$ the Z's, their common point at infinity being $A0$, e.g. $a3$: $A0\,D1\,D2...D16$. The line bj, $j = 1, 2, ..., 16$, contains all the points with the subscript j and its point at infinity is $A1$, e.g. $b7$: $A1\,B7\,C7...Z7$. The points at infinity of the remaining lines are as follows. $A2$ of the lines cj's, $A3$ of the dj's, ..., $A16$ of the zj's. To give the finite points of all these lines we use

TABLE I

$c1$:	1	2	3	4	5	6	7	8	9	10	11	12	13	14	15	16
$c3$:	3	6	1	7	11	2	4	13	16	12	5	10	8	15	14	9
$c4$:	4	10	7	1	8	12	3	5	14	2	13	6	11	9	16	15
$c5$:	5	16	11	8	1	9	13	4	6	15	3	14	7	12	10	2
$c7$:	7	12	4	3	13	10	1	11	15	6	8	2	5	16	9	14
$c8$:	8	15	13	5	4	14	11	1	12	16	7	9	3	6	2	10
$c9$:	9	11	16	14	6	5	15	12	1	13	2	8	10	4	7	3
$c13$:	13	14	8	11	7	15	5	3	10	9	4	16	1	2	6	12
$d1$:	1	10	8	2	14	16	15	13	7	4	6	5	3	12	9	11
$d3$:	3	12	13	6	15	9	14	8	4	7	2	11	1	10	16	5
$d4$:	4	2	5	10	9	15	16	11	3	1	12	8	7	6	14	13
$d5$:	5	15	4	16	12	2	10	7	13	8	9	1	11	14	6	3
$d7$:	7	6	11	12	16	14	9	5	1	3	10	13	4	2	15	8
$d8$:	8	16	1	15	6	10	2	3	11	5	14	4	13	9	12	7
$d9$:	9	13	12	11	4	3	7	10	15	14	5	6	16	8	1	2
$d13$:	13	9	3	14	2	12	6	1	5	11	15	7	8	16	10	4
$f1$:	1	4	13	10	12	11	9	3	15	2	16	14	8	5	7	6
$f3$:	3	7	8	12	10	5	16	1	14	6	9	15	13	11	4	2
$f4$:	4	1	11	2	6	13	14	7	16	10	15	9	5	8	3	12
$f5$:	5	8	7	15	14	3	6	11	10	16	2	12	4	1	13	9
$f7$:	7	3	5	6	2	8	15	4	9	12	14	16	11	13	1	10
$f8$:	8	5	3	16	9	7	12	13	2	15	10	6	1	4	11	14
$f9$:	9	14	10	13	8	2	1	16	7	11	3	4	12	6	15	5
$f13$:	13	11	1	9	16	4	10	8	6	14	12	2	3	7	5	15
$h1$:	1	11	15	6	7	12	13	10	2	16	9	4	5	3	14	8
$h3$:	3	5	14	2	4	10	8	12	6	9	16	7	11	1	15	13
$h4$:	4	13	16	12	3	6	11	2	10	15	14	1	8	7	9	5
$h5$:	5	3	10	9	13	14	7	15	16	2	6	8	1	11	12	4
$h7$:	7	8	9	10	1	2	5	6	12	14	15	3	13	4	16	11
$h8$:	8	7	2	14	11	9	3	16	15	10	12	5	4	13	6	1
$h9$:	9	2	7	5	15	8	10	13	11	3	1	14	6	16	4	12
$h13$:	13	4	6	15	5	16	1	9	14	12	10	11	7	8	2	3

$k1$:	1	14	9	12	15	4	8	11	16	5	7	6	2	13	10	3
$k3$:	3	15	16	10	14	7	13	5	9	11	4	2	6	8	12	1
$k4$:	4	9	14	6	16	1	5	13	15	8	3	12	10	11	2	7
$k5$:	5	12	6	14	10	8	4	3	2	1	13	9	16	7	15	11
$k7$:	7	16	15	2	9	3	11	8	14	13	1	10	12	5	6	4
$k8$:	8	6	12	9	2	5	1	7	10	4	11	14	15	3	16	13
$k9$:	9	4	1	8	7	14	12	2	3	6	15	5	11	10	13	16
$k13$:	13	2	10	16	6	11	3	4	12	7	5	15	14	1	9	8
$\ell1$:	1	16	10	11	3	8	14	5	13	6	12	7	15	4	2	9
$\ell3$:	3	9	12	5	1	13	15	11	8	2	10	4	14	7	6	16
$\ell4$:	4	15	2	13	7	5	9	8	11	12	6	3	16	1	10	14
$\ell5$:	5	2	15	3	11	4	12	1	7	9	14	13	10	8	16	6
$\ell7$:	7	14	6	8	4	11	16	13	5	10	2	1	9	3	12	15
$\ell8$:	8	10	16	7	13	1	6	4	3	14	9	11	2	5	15	12
$\ell9$:	9	3	13	2	16	12	4	6	10	5	8	15	7	14	11	1
$\ell13$:	13	12	9	4	8	3	2	7	1	15	16	5	6	11	14	10
$m1$:	1	5	11	14	13	3	10	2	8	12	4	15	9	6	16	7
$m3$:	3	11	5	15	8	1	12	6	13	10	7	14	16	2	9	4
$m4$:	4	8	13	9	11	7	2	10	5	6	1	16	14	12	15	3
$m5$:	5	1	3	12	7	11	15	16	4	14	8	10	6	9	2	13
$m7$:	7	13	8	16	5	4	6	12	11	2	3	9	15	10	14	1
$m8$:	8	4	7	6	3	13	16	15	1	9	5	2	12	14	10	11
$m9$:	9	6	2	4	10	16	13	11	12	8	14	7	1	5	3	15
$m13$:	13	7	4	2	1	8	9	14	3	16	11	6	10	15	12	5
$n1$:	1	6	5	16	4	9	2	15	14	11	8	3	10	7	13	12
$n3$:	3	2	11	9	7	16	6	14	15	5	13	1	12	4	8	10
$n4$:	4	12	8	15	1	14	10	16	9	13	5	7	2	3	11	6
$n5$:	5	9	1	2	8	6	16	10	12	3	4	11	15	13	7	14
$n7$:	7	10	13	14	3	15	12	9	16	8	11	4	6	1	5	2
$n8$:	8	14	4	10	5	12	15	2	6	7	1	13	16	11	3	9
$n9$:	9	5	6	3	14	1	11	7	4	2	12	16	13	15	10	8
$n13$:	13	15	7	12	11	10	14	6	2	4	3	8	9	5	1	16
$p1$:	1	12	2	5	6	7	16	9	10	14	3	13	11	15	8	4
$p3$:	3	10	6	11	2	4	9	16	12	15	1	8	5	14	13	7
$p4$:	4	6	10	8	12	3	15	14	2	9	7	11	13	16	5	1
$p5$:	5	14	16	1	9	13	2	6	15	12	11	7	3	10	4	8
$p7$:	7	2	12	13	10	1	14	15	6	16	4	5	8	9	11	3
$p8$:	8	9	15	4	14	11	10	12	16	6	13	3	7	2	1	5
$p9$:	9	8	11	6	5	15	3	1	13	4	16	10	2	7	12	14
$p13$:	13	16	14	7	15	5	12	10	9	2	8	1	4	6	3	11

$r1$:	1	13	7	3	9	5	4	16	6	8	15	11	14	2	12	10
$r3$:	3	8	4	1	16	11	7	9	2	13	14	5	15	6	10	12
$r4$:	4	11	3	7	14	8	1	15	12	5	16	13	9	10	6	2
$r5$:	5	7	13	11	6	1	8	2	9	4	10	3	12	16	14	15
$r7$:	7	5	1	4	15	13	3	14	10	11	9	8	16	12	2	6
$r8$:	8	3	11	13	12	4	5	10	14	1	2	7	6	15	9	16
$r9$:	9	10	15	16	1	6	14	3	5	12	7	2	4	11	8	13
$r13$:	13	1	5	8	10	7	11	12	15	3	6	4	2	14	16	9
$s1$:	1	9	4	7	10	14	3	6	11	15	2	16	12	8	5	13
$s3$:	3	16	7	4	12	15	1	2	5	14	6	9	10	13	11	8
$s4$:	4	14	1	3	2	9	7	12	13	16	10	15	6	5	8	11
$s5$:	5	6	8	13	15	12	11	9	3	10	16	2	14	4	1	7
$s7$:	7	15	3	1	6	16	4	10	8	9	12	14	2	11	13	5
$s8$:	8	12	5	11	16	6	13	14	7	2	15	10	9	1	4	3
$s9$:	9	1	14	15	13	4	16	5	2	7	11	3	8	12	6	10
$s13$:	13	10	11	5	9	2	8	15	4	6	14	12	16	3	7	1
$t1$:	1	8	16	13	2	10	12	14	4	3	5	9	7	11	6	15
$t3$:	3	13	9	8	6	12	10	15	7	1	11	16	4	5	2	14
$t4$:	4	5	15	11	10	2	6	9	1	7	8	14	3	13	12	16
$t5$:	5	4	2	7	16	15	14	12	8	11	1	6	13	3	9	10
$t7$:	7	11	14	5	12	6	2	16	3	4	13	15	1	8	10	9
$t8$:	8	1	10	3	15	16	9	6	5	13	4	12	11	7	14	2
$t9$:	9	12	3	10	11	13	8	4	14	16	6	1	15	2	5	7
$t13$:	13	3	12	1	14	9	16	2	11	8	7	10	5	4	15	6
$w1$:	1	15	6	9	8	13	5	12	3	7	14	10	4	16	11	2
$w3$:	3	14	2	16	13	8	11	10	1	4	15	12	7	9	5	6
$w4$:	4	16	12	14	5	11	8	6	7	3	9	2	1	15	13	10
$w5$:	5	10	9	6	4	7	1	14	11	13	12	15	8	2	3	16
$w7$:	7	9	10	15	11	5	13	2	4	1	16	6	3	14	8	12
$w8$:	8	2	14	12	1	3	4	9	13	11	6	16	5	10	7	15
$w9$:	9	7	5	1	12	10	6	8	16	15	4	13	14	3	2	11
$w13$:	13	6	15	10	3	1	7	16	8	5	2	9	11	12	4	14
$x1$:	1	3	14	8	11	15	6	7	12	13	10	2	16	9	4	5
$x3$:	3	1	15	13	5	14	2	4	10	8	12	6	9	16	7	11
$x4$:	4	7	9	5	13	16	12	3	6	11	2	10	15	14	1	8
$x5$:	5	11	12	4	3	10	9	13	14	7	15	16	2	6	8	1
$x7$:	7	4	16	11	8	9	10	1	2	5	6	12	14	15	3	13
$x8$:	8	13	6	1	7	2	14	11	9	3	16	15	10	12	5	4
$x9$:	9	16	4	12	2	7	5	15	8	10	13	11	3	1	14	6
$x13$:	13	8	2	3	4	6	15	5	16	1	9	14	12	10	11	7

$z1:$	1	7	12	15	16	2	11	4	5	9	13	8	6	10	3	14
$z3:$	3	4	10	14	9	6	5	7	11	16	8	13	2	12	1	15
$z4:$	4	3	6	16	15	10	13	1	8	14	11	5	12	2	7	9
$z5:$	5	13	14	10	2	16	3	8	1	6	7	4	9	15	11	12
$z7:$	7	1	2	9	14	12	8	3	13	15	5	11	10	6	4	16
$z8:$	8	11	9	2	10	15	7	5	4	12	3	1	14	16	13	6
$z9:$	9	15	8	7	3	11	2	14	6	1	10	12	5	13	16	4
$z13:$	13	5	16	6	12	14	4	11	7	10	1	3	15	9	8	2

the existence of an involution, which is a translation, that keeps the letters fixed, both of points and lines, and pairs off the subscripts of points and lines as follows: (1,2), (3,6),(4,10), (5,16), (7,12), (8,15),(9,11), (13,14). Thus in Table I we write down the subscripts of the points off $l\infty$ of half of the lines, and the used letters are to be inserted in alphabetic order. For instance, from Table I it follows that the points on $f3$ are

$f3: A4\ B3\ C7\ D8\ F12\ H10\ K5\ L16\ M1\ N14\ P6\ R9\ S15\ T13\ W11\ X4\ Z2$

and we can immediately write its conjugate line, namely

$f6: A4\ B6\ C12\ D15\ F7\ H4\ K16\ L5\ M2\ N13\ P3\ R11\ S8\ T14\ W9\ X10\ Z1.$

Notice that with our notation the two orbits of length one on the line at infinity are $\{A0\}$ and $\{A1\}$.

3. Quadrangles. Of course, quadrangles in Π can be distinguished according to the number, zero, one or two, of vertices they have on $l\infty$. However, such a classification usually does not allow us to foresee what the quadrangle generates.

PROPOSITION 3.1. *If a quadrangle has two of its diagonal points on $l\infty$, then the third diagonal point lies on $l\infty$ too and the quadrangle generates a Fano subplane.*

PROOF. It is a consequence of Π being a translation plane whose translation group is an elementary abelian 2-group (see also [8]).

COROLLARY 3.2. *A quadrangle two of whose vertices lie on $l\infty$ generates a Fano subplane.*

Next, we list some quadrangles which generate Fano subplanes. Notice that this result follows from Proposition 3.1 for the first five quadrangles, from Corollary 3.2 for those containing two points on $a0$ and is achieved by checking for the remaining ones.

$B1\ B2\ C1\ C2,\quad B1\ B6\ C2\ C3,\quad B1\ B3\ M4\ M7,\quad B1\ B2\ C3\ C6,$

$\quad C2\ C5\ M7\ M14,\quad A5\ A7\ B1\ C3,\quad A3\ A5\ B7\ L10,\quad A0\ A3\ D8\ N9,$

$\quad B2\ F4\ F10\ P10,\quad C2\ C5\ K9\ T9,\quad C2\ C5\ M7\ B7,\quad C2\ C5\ M7\ B10,$

$\quad C2\ C5\ M7\ D11,\quad C2\ C5\ M7\ F6,\quad C2\ C5\ M7\ F8,\quad A1\ B4\ D7\ F3,$

$\qquad\qquad\qquad A0\ B10\ C4\ P1,\quad A2\ B2\ F1\ F4.$

The incidences of some Fano subplanes are given in Section 5 where such subplanes are classified.

PROPOSITION 3.3. Π *is singly generated by quadrangles. Furthermore, a generating quadrangle of* Π *can have at most one vertex on the line at infinity.*

PROOF. The second part of the statement follows from Corollary 3.2. As to the first part, it is easy to check that any of the following quadrangles generates Π.

$A0\ B1\ C3\ D4$,	$A0\ B1\ C2\ Z3$,	$A0\ B1\ Z2\ D3$,	$A0\ B1\ Z1\ D2$,
$A0\ B1\ M1\ C2$,	$A1\ B2\ P4\ Z6$,	$A2\ L3\ M7\ P12$,	$A7\ F14\ M16\ Z12$,
$B1\ C3\ D4\ F5$,	$B1\ B5\ C3\ C7$,	$B1\ B3\ M4\ M12$,	$B1\ Z2\ C3\ F5$,
$L9\ L13\ M3\ M10$,	$B1\ C2\ D6\ F10$,	$D8\ H5\ L7\ X9$,	$C3\ C7\ N14\ W6$,
$C2\ C5\ H3\ L13$,	$D3\ D4\ N4\ S7$,	$M3\ M7\ F7\ K10$,	$C2\ C5\ M7\ K9$,
$C2\ C5\ L15\ S4$,	$C2\ C5\ M7\ S15$,	$C2\ C5\ H9\ M9$,	$C2\ C5\ M7\ B5$,
$C2\ C5\ M7\ D12$,	$C2\ C5\ M7\ M3$,	$C2\ C5\ M7\ F7$.	

We observe that to have an idea of the distribution of the generating quadrangles, we fixed the triangle $C2\ C5\ M7$ and let the fourth vertex of the quadrangle range over different lines. When the fourth vertex ran on $a1$ and $a4$, there were precisely two points, $B7$, $B10$ and $F6$, $F8$, respectively, for which the quadrangle generated a Fano subplane, whereas when it ranged over $a3$ and $a8$ there was just one point, $D11$ and $M14$, respectively, giving a generating quadrangle for a Fano subplane. The so obtained subplanes of order two behave in different ways w.r.t. the line at infinity. Since the full collineation group of Π is not transitive on triangles, one can get different results by starting with other triangles.

4. Baer subplanes. It is well known that Π contains Baer subplanes being a derived plane. Moreover, all the Baer subplanes of a derivation set have five points on the line at infinity. On the other hand, no Baer subplane was found which is tangent to $a0$. The same situation occurs in the semifield plane with kern GF(2) from which Π may be derived [8].

Baer subplanes in Π can be obtained either by extending suitable Fano subplanes (see Section 5) or by looking for those Baer subplanes which belong to a derivation set. Some examples provided by the former construction will be given in Section 5. All the so obtained Baer subplanes contain the point $A0$ and this in agreement with all of them sharing the line at infinity $a0$ [1].

5. Fano subplanes. First of all, we classify the Fano subplanes of Π according to their behaviour with respect to the line at infinity. This yields three inequivalent types of subplanes of order two. The subplanes of type (i) have three points on $a0$, those of type (ii) are tangent to $a0$, and, finally, the line at infinity is an exterior line for the subplanes of type (iii). Obviously, subplanes belonging to different types are projectively inequivalent.

On the other hand, subplanes of the same type do not need to be projectively equivalent. Next, for each of the three considered types we refine the classification by looking at the behaviour of the points on $a0$ and/or the points at infinity of the lines of the subplane with respect to the three orbits $\{A0\}$, $\{A1\}$, $\{A2, A3, \ldots, A16\} = \Omega$ on $l\infty$ of the full collineation group of Π. Thus, for the subplanes of type (i) we have the following subtypes (we list the points at infinity of the subplanes):

(i)(a) $A0\ A1\ Aj$, $Aj \in \Omega$;
(i)(b) $A0\ Ai\ Aj$, $Ai, Aj \in \Omega$;
(i)(c) $A1\ Ai\ Aj$, $Ai, Aj \in \Omega$;
(i)(d) $Ai\ Aj\ Au$, $Ai, Aj, Au \in \Omega$;

For a subplane of type (ii) there are three possibilities for its point at infinity: A0, Al, the point is in Ω. Finally, for the subplanes of type (iii) we can look at the points at infinity of their lines. This yields four obvious cases. On the other hand, all the Fano subplanes of this type we found have one line either on $A0$ or on $A1$.

Next, we list some examples.

Type (i). All the subplanes have three points on $\ell\infty = a0$ and the subtypes (a)–(d) are clear from the incidences.

(1)	(2)	(3)	(4)
$a0:A0\ A1\ A2$	$a0\ :A5\ A7\ A15$	$a0\ :A15\ A0\ A2$	$a0\ :A15\ A3\ A5$
$a1:A0\ B1\ B2$	$h1\ :A5\ B1\ K12$	$a1\ :A0\ B1\ B6$	$d7\ :A3\ B7\ P3$
$a2:A0\ C1\ C2$	$h5\ :A5\ C3\ D10$	$a2\ :A0\ C2\ C3$	$d5\ :A3\ K2\ L10$
$b1:A1\ B1\ C1$	$\ell1\ :A7\ B1\ D10$	$c1\ :A2\ B1\ C2$	$h7\ :A5\ B7\ K2$
$b2:A1\ B2\ C2$	$\ell9\ :A7\ C3\ K12$	$c6\ :A2\ B6\ C3$	$h9\ :A5\ P3\ L10$
$c1:A2\ B1\ C2$	$x1\ :A15\ B1\ C3$	$x1\ :A15\ B1\ C3$	$x7\ :A15\ B7\ L10$
$c2:A2\ B2\ C1$	$x11:A15\ D10\ K12$	$x6\ :A15\ B6\ C2$	$x8\ :A15\ K2\ P3$

(5)	(6)	(7)	(8)
$a0:A15\ A0\ A9$	$a0\ :A1\ A3\ A4$	$a0\ :A0\ A9\ A11$	$a0\ :A12\ A1\ A3$
$a1:A0\ B1\ B2$	$b1\ :A1\ B1\ C1$	$a2\ :A0\ C2\ C5$	$b1\ :A1\ B1\ T1$
$a2:A0\ C3\ C6$	$b2\ :A1\ F2\ P2$	$a8\ :A0\ M7\ M14$	$b2\ :A1\ F2\ R2$
$n1:A9\ B1\ C6$	$d1\ :A3\ B1\ F2$	$n3\ :A9\ C2\ M14$	$d1\ :A3\ B1\ F2$
$n2:A9\ B2\ C3$	$d10:A3\ C1\ P2$	$n9\ :A9\ C5\ M7$	$d3\ :A3\ R2\ T1$
$x1:A15\ B1\ C3$	$f1\ :A4\ B1\ P2$	$r7\ :A11\ C5\ M14$	$s1\ :A12\ B1\ R2$
$x2:A15\ B2\ C6$	$f4\ :A4\ C1\ F2$	$r14:A11\ C2\ M7$	$s12:A12\ F2\ T1$

	(9)		(10)		(11)		(12)
a0:A15 A0 A16		a0 :A0 A3 A9		a0 :A0 A1 A2		a0 :A0 A1 A7	
a1:A0 B1 B3		a3 :A0 D2 D8		a13:A0 T8 T15		a1 :A0 B3 B6	
a8:A0 M4 M7		a9 :A0 N7 N9		a14:A0 W8 W15		a15:A0 X3 X6	
x1:A15 B1 M7		d1 :A3 D8 N7		b8 :A1 T8 W8		b3 :A1 B3 X3	
x3:A15 B3 M4		d15:A3 D2 N9		b15:A1 T15 W15		b6 :A1 B6 X6	
z1:A16 B1 M4		n16:A9 D2 N7		c3 :A2 T8 W15		ℓ3 :A7 B3 X6	
z3:A16 B3 M7		n4 :A9 D8 N9		c6 :A2 T15 W8		ℓ6 :A7 B6 X3	

Type (ii). The Fano subplanes of this type are tangent to $a0$. Again the subtypes are clear from the incidences.

	(1)		(2)		(3)
a1 : A0 B4 B10		c1 : A2 F4 P10		a2 : C2 C5 C9	
a2 : A0 C2 C4		c2 : A2 B2 F10		b9 : C9 K9 T9	
a10 : A0 P1 P4		c4 : A2 C10 F1		m16 : A8 C2 K9	
b4 : B4 C4 P4		b10 : C10 F10 P10		m1 : A8 C5 T9	
d4 : B4 C2 P1		a4 : F1 F4 F10		m6 : A8 C9 M3	
c10 : C4 P1 B10		d2 : B2 F1 P10		t15 : C2 M3 T9	
f10 : C2 P4 B10		f2 : B2 C10 F4		w10 : C5 K9 M3	

	(4)		(5)		(6)
a2 : C2 C5 C7		a2 : C1 C2 C5		b4 : A1 B4 L4	
b7 : B7 C7 M7		r14 : C2 M7 N8		b7 : A1 D7 L7	
r14 : A11 C2 M7		n9 : A9 C5 M7		b3 : A1 F3 L3	
r7 : A11 B7 C5		n3 : A9 C2 D11		c4 : B4 D7 L3	
r5 : A11 C7 K1		n6 : A9 C1 N8		s4 : B4 F3 L7	
n9 : C5 K1 M7		f4 : C1 M7 D11		r1 : D7 F3 L4	
p7 : B7 C2 K1		m1 : C5 N8 D11		a7 : L3 L4 L7	

Type (iii). For these subplanes $a0$ is an exterior line.

a2 : C2 C5 C13	a2 : C2 C5 C13	a2 : C2 C3 C5	b3 : B3 D3 R3
r14 : C2 M7 P6	r14 : C2 M7 D16	r14 : C2 M7 S10	x3 : B3 M4 Z11
n9 : C5 M7 W15	n9 : C5 M7 L11	n9 : C5 M7 H14	t9 : D3 M4 X5
f10 : B10 C2 W15	t15 : C2 F6 L11	s11 : C2 F8 H14	c5 : M4 T7 R3
w10 : B10 C5 P6	z13 : C5 D16 F6	d15 : C5 F8 S10	w3 : B3 T7 X5
s10 : B10 M7 C13	s10 : C13 F6 M7	x1 : C3 F8 M7	z10 : D3 T7 Z11
z5 : C13 P6 W15	h4 : C13 D16 L11	t13 : C3 H14 S10	r14 : R3 X5 Z11

In any projective plane Π^* a subplane Π_0^* is maximal if it is not contained in any proper subplane of Π^*. In other words, if for any point y on a line of Π_0^* but not in Π_0^*, $\Pi_0^* \cup \{y\}$ generates Π^*.

This definition enables us to classify the Fano subplanes of Π according to their being or not maximal. On the other hand, when a subplane of order two is contained in (extends to, or completes to) a Baer subplane, the latter needs not be unique as occurs in the Desarguesian plane. More precisely, there are Fano subplanes in Π which complete to one, two or three distinct Baer subplanes. Only the subplanes of type (i) seem to extend to more than one Baer subplane, provided that they are not maximal. However, there are subplanes of type (ii) which complete to Baer subplanes, the latter being unique. Finally, all the found subplanes of type (iii) are maximal. These behaviours of Fano subplanes of types (ii) and (iii) is in agreement with the fact that no Baer subplane seems to exist in Π which is tangent to $a0$.

Therefore, there are four inequivalent subtypes of Fano subplanes of type (i) when we look at the possible numbers of Baer subplanes they may extend to: zero, one, two or three, whereas there are just two subtypes of subplanes of type (ii) and one type only of those of type (iii).

Next, we analyze from this point of view the already written down Fano subplanes.

Among the listed subplanes of type (i), (2), (3), (4), (7), (8), and (10) are maximal and the same is true for (ii)(4) and (ii)(5). The remaining subplanes of type (i) and (ii) all extend to Baer subplanes. More precisely, (ii)(1) and (ii)(2) uniquely complete to the Baer subplane α, whereas (ii)(3) extends to β and (ii)(6) to γ. The incidences of these Baer subplanes as well as those of the following ones are given at the end of this section.

As to the subplanes of type (i), (1) is contained in three Baer subplanes, namely α, α', and α''; (5) uniquely completes to δ and (6) completes to α only. Notice the difference between (i)(1) and (i)(6) which both are subplanes of α. (9) uniquely completes to ε; (11) is contained in three Baer subplanes: η, η', and η''.

Finally, (12) extends to two Baer subplanes ω and ω'.

α:	$a0$	$: A0\ A1\quad A2\quad A3\quad A4$	α':	$a0$	$: A0\quad A1\quad A2\quad A5\quad A12$
	$a1$	$: A0\ B1\quad B2\quad B4\quad B10$		$a1$	$: A0\quad B1\quad B2\quad B9\quad B11$
	$a2$	$: A0\ C1\quad C2\quad C4\quad C10$		$a2$	$: A0\quad C1\quad C2\quad C9\quad C11$
	$a4$	$: A0\ F1\quad F2\quad F4\quad F10$		$a9$	$: A0\quad N1\quad N2\quad N9\quad N11$
	$a10$	$: A0\ P1\quad P2\quad P4\quad P10$		$a11$	$: A0\quad R1\quad R2\quad R9\quad R11$
	$b1$	$: A1\ B1\quad C1\quad F1\quad P1$		$b1$	$: A1\quad B1\quad C1\quad N1\quad R1$
	$b2$	$: A1\ B2\quad C2\quad F2\quad P2$		$b2$	$: A1\quad B2\quad C2\quad N2\quad R2$
	$b4$	$: A1\ B4\quad C4\quad F4\quad P4$		$b9$	$: A1\quad B9\quad C9\quad N9\quad R9$
	$b10$	$: A1\ B10\ C10\ F10\ P10$		$b11$	$: A1\quad B11\ C11\ N11\ R11$
	$c1$	$: A2\ B1\quad C2\quad F4\quad P10$		$c1$	$: A2\quad B1\quad C2\quad N9\quad R11$
	$c2$	$: A2\ B2\quad C1\quad F10\ P4$		$c2$	$: A2\quad B2\quad C1\quad N11\ R9$
	$c4$	$: A2\ B4\quad C10\ F1\quad P2$		$c9$	$: A2\quad B9\quad C11\ N1\quad R2$

$c10$: $A2$ $B10$ $C4$ $F2$ $P1$ $c11$: $A2$ $B11$ $C9$ $N2$ $R1$

$d1$: $A3$ $B1$ $C10$ $F2$ $P4$ $h1$: $A5$ $B1$ $C11$ $N2$ $R9$

$d2$: $A3$ $B2$ $C4$ $F1$ $P10$ $h2$: $A5$ $B2$ $C9$ $N1$ $R11$

$d4$: $A3$ $B4$ $C2$ $F10$ $P1$ $h9$: $A5$ $B9$ $C2$ $N11$ $R1$

$d10$: $A3$ $B10$ $C1$ $F4$ $P2$ $h11$: $A5$ $B11$ $C1$ $N9$ $R2$

$f1$: $A4$ $B1$ $C4$ $F10$ $P2$ $s1$: $A12$ $B1$ $C9$ $N11$ $R2$

$f2$: $A4$ $B2$ $C10$ $F4$ $P1$ $s2$: $A12$ $B2$ $C11$ $N9$ $R1$

$f4$: $A4$ $B4$ $C1$ $F2$ $P10$ $s9$: $A12$ $B9$ $C1$ $N2$ $R11$

$f10$: $A4$ $B10$ $C2$ $F1$ $P4$ $s11$: $A12$ $B11$ $C2$ $N1$ $R9$

α'' : $a0$: $A0$ $A1$ $A2$ $A6$ $A11$ β : $a0$: $A0$ $A1$ $A8$ $A13$ $A14$

$a1$: $A0$ $B1$ $B2$ $B13$ $B14$ $a2$: $A0$ $C2$ $C3$ $C5$ $C9$

$a2$: $A0$ $C1$ $C2$ $C13$ $C14$ $a6$: $A0$ $K2$ $K3$ $K5$ $K9$

$a13$: $A0$ $T1$ $T2$ $T13$ $T14$ $a8$: $A0$ $M2$ $M3$ $M5$ $M9$

$a14$: $A0$ $W1$ $W2$ $W13$ $W14$ $a13$: $A0$ $T2$ $T3$ $T5$ $T9$

$b1$: $A1$ $B1$ $C1$ $T1$ $W1$ $b2$: $A1$ $C2$ $K2$ $M2$ $T2$

$b2$: $A1$ $B2$ $C2$ $T2$ $W2$ $b3$: $A1$ $C3$ $K3$ $M3$ $T3$

$b13$: $A1$ $B13$ $C13$ $T13$ $W13$ $b5$: $A1$ $C5$ $K5$ $M5$ $T5$

$b14$: $A1$ $B14$ $C14$ $T14$ $W14$ $b9$: $A1$ $C9$ $K9$ $M9$ $T9$

$c1$: $A2$ $B1$ $C2$ $T13$ $W14$ $m1$: $A8$ $C5$ $K3$ $M2$ $T9$

$c2$: $A2$ $B2$ $C1$ $T14$ $W13$ $m6$: $A8$ $C9$ $K2$ $M3$ $T5$

$c13$: $A2$ $B13$ $C14$ $T1$ $W2$ $m11$: $A8$ $C3$ $K5$ $M9$ $T2$

$c14$: $A2$ $B14$ $C13$ $T2$ $W1$ $m16$: $A8$ $C2$ $K9$ $M5$ $T3$

$k1$: $A6$ $B1$ $C14$ $T2$ $W13$ $t4$: $A13$ $C5$ $K2$ $M9$ $T3$

$k2$: $A6$ $B2$ $C13$ $T1$ $W14$ $t12$: $A13$ $C9$ $K3$ $M5$ $T2$

$k13$: $A6$ $B13$ $C2$ $T14$ $W1$ $t13$: $A13$ $C3$ $K9$ $M2$ $T5$

$k14$: $A6$ $B14$ $C1$ $T13$ $W2$ $t15$: $A13$ $C2$ $K5$ $M3$ $T9$

$r1$: $A11$ $B1$ $C13$ $T14$ $W2$ $w7$: $A14$ $C9$ $K5$ $M2$ $T3$

$r2$: $A11$ $B2$ $C14$ $T13$ $W1$ $w8$: $A14$ $C2$ $K3$ $M9$ $T5$

$r13$: $A11$ $B13$ $C1$ $T2$ $W14$ $w10$: $A14$ $C5$ $K9$ $M3$ $T2$

$r14$: $A11$ $B14$ $C2$ $T1$ $W13$ $w14$: $A14$ $C3$ $K2$ $M5$ $T9$

γ : $a0$: $A0$ $A1$ $A2$ $A11$ $A12$ δ : $a0$: $A0$ $A9$ $A15$ $A10$ $A16$

$a1$: $A0$ $B1$ $B3$ $B4$ $B7$ $a1$: $A0$ $B1$ $B2$ $B4$ $B10$

$a3$: $A0$ $D1$ $D3$ $D4$ $D7$ $a2$: $A0$ $C3$ $C6$ $C7$ $C12$

$a4$: $A0$ $F1$ $F3$ $F4$ $F7$ $a4$: $A0$ $F5$ $F16$ $F8$ $F15$

$a7$: $A0$ $L1$ $L3$ $L4$ $L7$ $a10$: $A0$ $P9$ $P11$ $P13$ $P14$

$b1$: $A1$ $B1$ $D1$ $F1$ $L1$ $n1$: $A9$ $B1$ $C6$ $F16$ $P11$

$b3$: $A1$ $B3$ $D3$ $F3$ $L3$ $n2$: $A9$ $B2$ $C3$ $F5$ $P9$

$b4$: $A1$ $B4$ $D4$ $F4$ $L4$ $n4$: $A9$ $B4$ $C12$ $F15$ $P13$

$b7$: $A1$ $B7$ $D7$ $F7$ $L7$ $n10$: $A9$ $B10$ $C7$ $F8$ $P14$

$c1$: $A2$ $B1$ $D3$ $F4$ $L7$ $x1$: $A15$ $B1$ $C3$ $F8$ $P13$

$c3$: $A2$ $B3$ $D1$ $F7$ $L4$ $x2$: $A15$ $B2$ $C6$ $F15$ $P14$

$c4$: $A2$ $B4$ $D7$ $F1$ $L3$ $x4$: $A15$ $B4$ $C7$ $F5$ $P11$

$c7$: $A2$ $B7$ $D4$ $F3$ $L1$ $x10$: $A15$ $B10$ $C12$ $F16$ $P9$

$r1 : A11\ B1\ D7\ F3\ L4$
$r3 : A11\ B3\ D4\ F1\ L7$
$r4 : A11\ B4\ D3\ F7\ L1$
$r7 : A11\ B7\ D1\ F4\ L3$
$s1 : A12\ B1\ D4\ F7\ L3$
$s3 : A12\ B3\ D7\ F4\ L1$
$s4 : A12\ B4\ D1\ F3\ L7$
$s7 : A12\ B7\ D3\ F1\ L4$

$p1\ : A10\ B1\ C12\ F5\ P14$
$p2\ : A10\ B2\ C7\ F16\ P13$
$p4\ : A10\ B4\ C6\ F8\ P9$
$p10 : A10\ B10\ C3\ F15\ P11$
$z1\ : A16\ B1\ C7\ F15\ P9$
$z2\ : A16\ B2\ C12\ F8\ P11$
$z4\ : A16\ B4\ C3\ F16\ P14$
$z10 : A16\ B10\ C6\ F5\ P13$

$\varepsilon :$

$a0\ : A0\ A15\ A16\ A1\ A4$
$a1\ : A0\ B1\ B3\ B4\ B7$
$a8\ : A0\ M1\ M3\ M4\ M7$
$a2\ : A0\ C1\ C3\ C4\ C7$
$a15 : A0\ X1\ X3\ X4\ X7$
$b1\ : A1\ B1\ C1\ M1\ X1$
$b3\ : A1\ B3\ C3\ M3\ X3$
$b4\ : A1\ B4\ C4\ M4\ X4$
$b7\ : A1\ B7\ C7\ M7\ X7$
$x1\ : A15\ B1\ M7\ C3\ X4$
$x3\ : A15\ B3\ M4\ C1\ X7$
$x4\ : A15\ B4\ C7\ M3\ X1$
$x7\ : A15\ B7\ C4\ M1\ X3$
$z1\ : A16\ B1\ M4\ C7\ X3$
$z3\ : A16\ B3\ M7\ C4\ X1$
$z4\ : A16\ B4\ M1\ C3\ X7$
$z7\ : A16\ B7\ M3\ C1\ X4$
$f1\ : A4\ B1\ M3\ C4\ X7$
$f3\ : A4\ B3\ M1\ C7\ X4$
$f4\ : A4\ B4\ M7\ C1\ X3$
$f7\ : A4\ B7\ M4\ C3\ X1$

$\eta :$

$a0\ : A0\ A1\ A2\ A3\ A4$
$a13 : A0\ T8\ T15\ T5\ T16$
$a14 : A0\ W8\ W15\ W5\ W16$
$a10 : A0\ N5\ N16\ N8\ N15$
$a11 : A0\ R5\ R16\ R8\ R15$
$b8\ : A1\ T8\ W8\ N8\ R8$
$b15 : A1\ T15\ W15\ N15\ R15$
$b5\ : A1\ N5\ R5\ T5\ W5$
$b16 : A1\ N16\ R16\ T16\ W16$
$c3\ : A2\ T8\ W15\ N16\ R5$
$c6\ : A2\ T15\ W8\ N5\ R16$
$c7\ : A2\ N15\ R8\ T5\ W16$
$c12 : A2\ N8\ R15\ T16\ W5$
$d9\ : A3\ N15\ R5\ T16\ W8$
$d11 : A3\ N8\ R16\ T5\ W15$
$d13 : A3\ N5\ R15\ T8\ W16$
$d14 : A3\ N16\ R8\ T15\ W5$
$f1\ : A4\ N15\ R16\ T8\ W5$
$f2\ : A4\ N8\ R5\ T15\ W16$
$f4\ : A4\ N16\ R15\ T5\ W8$
$f10 : A4\ N5\ R8\ T16\ W15$

$\eta' :$

$a0\ : A0\ A1\ A2\ A5\ A12$
$a13 : A0\ T8\ T15\ T7\ T12$
$a14 : A0\ W8\ W15\ W7\ W12$
$a4\ : A0\ F7\ F12\ F8\ F15$
$a10 : A0\ P7\ P12\ P8\ P15$
$b8\ : A1\ T8\ W8\ F8\ P8$
$b15 : A1\ T15\ W15\ F15\ P15$
$b7\ : A1\ F7\ P7\ T7\ W7$
$b12 : A1\ F12\ P12\ T12\ W12$
$c3\ : A2\ T8\ W15\ F7\ P12$
$c6\ : A2\ T15\ W8\ F12\ P7$
$c5\ : A2\ F8\ P15\ T7\ W12$
$c16 : A2\ F15\ P8\ T12\ W7$
$h4\ : A5\ F12\ P15\ T8\ W7$

$\eta'' :$

$a0\ : A0\ A1\ A2\ A6\ A11$
$a13 : A0\ T8\ T15\ T3\ T6$
$a14 : A0\ W8\ W15\ W3\ W6$
$a1\ : A0\ B3\ B6\ B8\ B15$
$a2\ : A0\ C3\ C6\ C8\ C15$
$b8\ : A1\ T8\ W8\ B8\ C8$
$b15 : A1\ T15\ W15\ B15\ C15$
$b3\ : A1\ B3\ C3\ T3\ W3$
$b6\ : A1\ B6\ C6\ T6\ W6$
$c3\ : A2\ B3\ C6\ T8\ W15$
$c6\ : A2\ B1\ C3\ T15\ W8$
$c8\ : A2\ B3\ C15\ T3\ W6$
$c15 : A2\ F15\ C8\ T6\ W3$
$k3\ : A6\ b3\ C15\ T6\ W8$

$h10$:	$A5$	$F7$	$P8$	$T15$	$W12$	

$h10 : A5\ F7\ P8\ T15\ W12$ $\qquad$ $k6\ \ : A6\ B6\ C8\ T3\ W15$

$h13 : A5\ F15\ P12\ T7\ W8$ $\qquad$ $k8\ \ : A6\ B8\ C6\ T15\ W3$

$h14 : A5\ F8\ P7\ T12\ W15$ $\qquad$ $k15 : A6\ B15\ C3\ T8\ W6$

$s1\ \ : A12\ F7\ P15\ T12\ W8$ $\qquad$ $r3\ \ : A11\ B3\ C8\ T15\ W6$

$s2\ \ : A12\ F12\ P8\ T7\ W15$ $\qquad$ $r6\ \ : A11\ B6\ C15\ T8\ W3$

$s9\ \ : A12\ F15\ P7\ T8\ W12$ $\qquad$ $r8\ \ : A11\ B8\ C3\ T6\ W15$

$s11 : A12\ F8\ P12\ T15\ W7$ $\qquad$ $r15 : A11\ B15\ C6\ T3\ W8$

$\omega :$

$a0\ \ : A0\ A1\ A7\ A3\ A14$ $\qquad$ $\omega' : a0\ \ : A0\ A1\ A7\ A5\ A9$

$a1\ \ : A0\ B3\ B6\ B5\ B16$ $\qquad$ $a1\ \ : A0\ B3\ B6\ B8\ B15$

$a15 : A0\ X3\ X6\ X5\ X16$ $\qquad$ $a15 : A0\ X3\ X6\ X8\ X15$

$a4\ \ : A0\ F3\ F6\ F5\ F16$ $\qquad$ $a7\ \ : A0\ L3\ L6\ L8\ L15$

$a16 : A0\ Z3\ Z6\ Z5\ Z16$ $\qquad$ $a9\ \ : A0\ N3\ N6\ N8\ N15$

$b3\ \ : A1\ B3\ X3\ F3\ Z3$ $\qquad$ $b3\ \ : A1\ B3\ X3\ L3\ N3$

$b6\ \ : A1\ B6\ X6\ F6\ Z6$ $\qquad$ $b6\ \ : A1\ B6\ X6\ L6\ N6$

$b5\ \ : A1\ B5\ X5\ F5\ Z5$ $\qquad$ $b8\ \ : A1\ B8\ X8\ L8\ N8$

$b16 : A1\ B16\ X16\ F16\ Z16$ $\qquad$ $b15 : A1\ B15\ X15\ L15\ N15$

$\ell3\ \ : A7\ B3\ X6\ F5\ Z16$ $\qquad$ $\ell3\ \ : A7\ B3\ X6\ L15\ N8$

$\ell6\ \ : A7\ B6\ X3\ F16\ Z5$ $\qquad$ $\ell6\ \ : A7\ B6\ X3\ L8\ N15$

$\ell5\ \ : A7\ B5\ F3\ X16\ Z6$ $\qquad$ $\ell8\ \ : A7\ B8\ X15\ L6\ N3$

$\ell16 : A7\ B16\ F6\ X5\ Z3$ $\qquad$ $\ell15 : A7\ B15\ X8\ L3\ N6$

$d3\ \ : A3\ B3\ F6\ X16\ Z5$ $\qquad$ $h3\ \ : A5\ B3\ X15\ L8\ N6$

$d6\ \ : A3\ B6\ F3\ X5\ Z16$ $\qquad$ $h6\ \ : A5\ B6\ X8\ L15\ N3$

$d5\ \ : A3\ B5\ F16\ X6\ Z3$ $\qquad$ $h8\ \ : A5\ B8\ X6\ L3\ N15$

$d16 : A3\ B16\ F5\ X3\ Z6$ $\qquad$ $h15 : A5\ B15\ X3\ L6\ N8$

$w3\ \ : A14\ B3\ F16\ X5\ Z6$ $\qquad$ $n3\ \ : A9\ B3\ X8\ L6\ N15$

$w6\ \ : A14\ B6\ F5\ X16\ Z3$ $\qquad$ $n6\ \ : A9\ B6\ X15\ L3\ N8$

$w5\ \ : A14\ B5\ F6\ X3\ Z16$ $\qquad$ $n8\ \ : A9\ B8\ X3\ L15\ N6$

$w16 : A14\ B16\ F3\ X6\ Z5$ $\qquad$ $n15 : A9\ B15\ X6\ L8\ N3$

6. Small complete arcs. Recall that a k-arc K in a projective plane is a set of k points no three of which are collinear. Moreover, K is complete if every point of the plane lies on at least one secant to the arc. If q is the order of the plane, then $k \leq q + 1$ when q is odd and $k \leq q + 2$ when q is even. Such arcs of maximum possible sizes are obviously complete and a $(q + 2)$-arc will be referred to as a hyperoval. On the other hand, not much is known about the minimum possible size for a complete arc when $q \geq 11$ [7, 12]. In [9] it was conjectured that such a size is $3\sqrt{q} - 2$ for $q > 9$. Furthermore, when q is even and the plane is Desarguesian, every k-arc with $k > q - \sqrt{q} + 1$ completes to a hyperoval [12]. This is not the case for non-Desarguesian planes which may contain complete k-arcs for any $k \leq q + 1$ or $q + 2$ with just one exception. Namely, when q is even, every $(q + 1)$-arc completes to a hyperoval.

In the special case $q = 16$, the Desarguesian plane contains complete k-arcs other than hyperovals for $k = 10, 11, 12, 13$ [11, 17], the Hall plane contains complete 16-arcs [19], the Lorimer plane and its dual plane contain complete 14-arcs [10]. The semifield plane with kern GF(2) contains complete 13-arcs [8].

We call small the complete k-arcs in Π for $k = 10, 11, 12, 13$, i.e. those which occur also in PG(2,16). Complete 14-arcs will be dealt with in Section 7 and hyperovals in Section 8.

We observe that the lower bound 8 for the size of a complete arc in a plane of order 16 [10] seems not to be attained in any of the planes which so far have been investigated [8, 9, 10, 17].

It is very easy to find small complete arcs in Π, especially 12-arcs, but it seems hopeless to classify them beyond their behaviour w.r.t. the line at infinity. Therefore, we just list a few examples.

Complete 10-arcs:

$C3\ C6\ D3\ D6\ F8\ F15\ S8\ S15\ Ai\ Aj,\ i,\ j \neq 0,\ 1,\ 5,\ 7,\ 10,\ 15,\ 16;$

$B1\ B2\ C9\ C11\ F1\ F2\ P9\ P11\ Ai\ Aj,\ i,\ j \neq 0,\ 1,\ 3,\ 5,\ 9,\ 12,\ 16;$

$B1\ B2\ C4\ C10\ T1\ T2\ W4\ W10\ Ai\ Aj,\ i,\ j \neq 0,\ 1,\ 3,\ 4,\ 6,\ 7,\ 16;$

$A3\ A7\ B3\ B6\ C1\ C2\ D4\ D10\ Fi\ Fj,\ (i,\ j) = (7,\ 12)$ or $(8,\ 15);$

$B2\ B6\ D2\ D6\ C4\ C7\ H4\ H7\ P14\ P15;\quad B12\ C12\ B16\ C16\ D3$

$F3\ D8\ F8\ H6\ H15;$

$C5\ C7\ D5\ D7\ L3\ L9\ Z3\ Z9\ F10\ N8;\quad A2\ B1\ B2\ D1\ C3\ H8$

$K16\ F15\ K4\ F3.$

Complete 11-arcs:

$B1$	$B2$	$C1$	$C2$	$D4$	$D10$	$L4$	$L10$	$F8$	$F15$	$A4;$
$B1$	$B2$	$C1$	$C2$	$D4$	$D10$	$N4$	$N10$	$M7$	$M12$	$A5;$
$B1$	$B2$	$C1$	$C2$	$D4$	$D10$	$N4$	$N10$	$W3$	$W6$	$A9$
$B3$	$B6$	$Z3$	$Z6$	$L5$	$L16$	$W8$	$W15$	$C9$	$C11$	$A3;$
$B3$	$B6$	$Z3$	$Z6$	$L5$	$L16$	$W8$	$W15$	$H8$	$H15$	$A12;$
$B3$	$B6$	$C3$	$C6$	$D4$	$D10$	$F4$	$F10$	$L9$	$L11$	$A7;$
$B3$	$B6$	$D3$	$D6$	$F5$	$F16$	$W5$	$W16$	$K9$	$K11$	$A2;$
$B1$	$B2$	$C3$	$C6$	$D4$	$D10$	$H5$	$H16$	$K4$	$K10$	$A14;$
$B1$	$B2$	$H1$	$H2$	$C3$	$C6$	$F7$	$F12$	$L13$	$L14$	$A10;$
$D3$	$D6$	$K3$	$K6$	$H4$	$H10$	$T4$	$T10$	$C5$	$C16$	$A16;$
$D3$	$D6$	$K3$	$K6$	$H4$	$H10$	$T4$	$T10$	$S5$	$S16$	$A8;$
$B1$	$B3$	$C2$	$C6$	$D4$	$D7$	$F8$	$F13$	$M5$	$M11$	$A1;$
$A0$	$B1$	$C2$	$F3$	$H4$	$K16$	$L13$	$M4$	$P2$	$R7$	$D1;$
$B1$	$C2$	$Z15$	$X14$	$M3$	$N5$	$D8$	$W9$	$F7$	$K8$	$A0;$
$A3$	$A6$	$B1$	$B3$	$C4$	$F7$	$D5$	$K8$	$C7$	$P3$	$F9;$
$B1$	$C2$	$Z15$	$X14$	$M3$	$N5$	$D8$	$W9$	$F7$	$A0$	$A1;$
$B1$	$C2$	$F3$	$D6$	$H4$	$K10$	$M9$	$N11$	$P16$	$A0$	$A1;$
$B1$	$C2$	$Z15$	$X14$	$M3$	$N5$	$D8$	$W9$	$B3$	$C15$	$P8;$
$B1$	$C2$	$F3$	$D6$	$H4$	$K10$	$M9$	$N11$	$B10$	$M4$	$D11;$
$B1$	$C2$	$F3$	$D6$	$H4$	$K10$	$M9$	$N11$	$C7$	$R7$	$N2.$

Complete 12-arcs:

$B1$	$B2$	$C1$	$C2$	$D4$	$D10$	$L4$	$L10$	$F8$	$F15$	$P5$	$P16$;
$B1$	$B2$	$C1$	$C2$	$D4$	$D10$	$L4$	$L10$	$S3$	$S6$	$K7$	$K12$;
$B1$	$B2$	$C1$	$C2$	$D4$	$D10$	$N4$	$N10$	$M7$	$M12$	$P9$	$P11$;
$B1$	$B2$	$C1$	$C2$	$D4$	$D10$	$N4$	$N10$	$P5$	$P16$	$A3$	$A11$;
$B1$	$B2$	$C1$	$C2$	$D4$	$D10$	$S4$	$S10$	$H13$	$H14$	$A9$	$A10$;
$B1$	$B2$	$C1$	$C2$	$D4$	$D10$	$S4$	$S10$	$M9$	$M11$	$A15$	$A16$;
$B3$	$B6$	$Z3$	$Z6$	$L5$	$L16$	$W8$	$W15$	$C9$	$C11$	$F9$	$F11$;
$B3$	$B6$	$Z3$	$Z6$	$L5$	$L16$	$W8$	$W15$	$D13$	$D14$	$K13$	$K14$;
$B3$	$B6$	$Z3$	$Z6$	$L5$	$L16$	$W8$	$W15$	$H8$	$H15$	$S5$	$S16$;
$B3$	$B6$	$Z3$	$Z6$	$L5$	$L16$	$M5$	$M16$	$H7$	$H12$	$T9$	$T11$;
$B3$	$B6$	$Z3$	$Z6$	$L5$	$L16$	$M5$	$M16$	$H7$	$H12$	$A5$	$A7$;
$B3$	$B6$	$C3$	$C6$	$D4$	$D10$	$F4$	$F10$	$H8$	$H15$	$A3$	$A15$;
$C7$	$C12$	$D7$	$D12$	$K8$	$K15$	$L8$	$L15$	$W3$	$W6$	$A2$	$A10$;
$C3$	$C6$	$D3$	$D6$	$F4$	$F10$	$Z4$	$Z10$	$K9$	$K11$	$X9$	$X11$;
$B4$	$B10$	$C4$	$C10$	$Z13$	$Z14$	$W13$	$W14$	$F7$	$F12$	$A4$	$A15$;
$B4$	$B10$	$C4$	$C10$	$Z13$	$Z14$	$W13$	$W14$	$H5$	$H16$	$S5$	$S16$;
$L1$	$L2$	$M1$	$M2$	$D3$	$D6$	$F3$	$F6$	$B8$	$B15$	$X8$	$X15$;
$L1$	$L2$	$M1$	$M2$	$D3$	$D6$	$F3$	$F6$	$C13$	$C14$	$Z7$	$Z12$;
$L1$	$L2$	$M1$	$M2$	$D3$	$D6$	$F3$	$F6$	$K9$	$K11$	$A4$	$A6$;

Complete 13-arcs:

$B1$	$B2$	$H1$	$H2$	$C3$	$C6$	$F7$	$F12$	$D3$	$D6$	$A5$	$A7$;
$C3$	$C6$	$D3$	$D6$	$L4$	$L10$	$R4$	$R10$	$F9$	$F11$	$A9$	$A13$;
$C3$	$C6$	$D3$	$D6$	$L4$	$L10$	$R4$	$R10$	$H1$	$H2$	$M8$	$M15$;
$C3$	$C6$	$D3$	$D6$	$L4$	$L10$	$R4$	$R10$	$K9$	$K11$	$A11$	$A14$;
$L3$	$L6$	$M3$	$M6$	$N4$	$N10$	$P4$	$P10$	$F5$	$F16$	$A3$	$A16$;
$B2$	$B6$	$D2$	$D6$	$C4$	$C7$	$L4$	$L7$	$P9$	$P16$	$A7$	$A14$;
$B1$	$C2$	$Z15$	$X14$	$M3$	$N5$	$D8$	$W9$	$D2$	$X1$	$W13$	$A9$;
$A1$	$A3$	$B1$	$B3$	$C4$	$D5$	$F7$	$K8$	$M2$	$W11$	$X12$	$R10$;
$A1$	$A3$	$B1$	$B3$	$C4$	$F7$	$K8$	$D5$	$L6$	$N10$	$P9$	$C13$;
$B1$	$B13$	$C1$	$C13$	$D2$	$D14$	$F2$	$F14$	$H12$	$H16$	$Z3$	$Z8$.

$B1$	$B2$	$C1$	$C2$	$D4$	$D10$	$L4$	$L10$	$R13$	$R14$	$W9$	$W11$	$A4$;
$C7$	$C12$	$D7$	$D12$	$K8$	$K15$	$L8$	$L15$	$N4$	$N10$	$R5$	$R16$	$A15$;
$B7$	$B12$	$C7$	$C12$	$D3$	$D6$	$L3$	$L6$	$R5$	$R16$	$W8$	$W15$	$A4$;
$B7$	$B12$	$C7$	$C12$	$D3$	$D6$	$N3$	$N6$	$H5$	$H16$	$Z4$	$Z10$	$A11$;
$B7$	$B12$	$C7$	$C12$	$D3$	$D6$	$N3$	$N6$	$H5$	$H16$	$P8$	$P15$	$A11$;
$D7$	$D12$	$K7$	$K12$	$C8$	$C15$	$S8$	$S15$	$F5$	$F16$	$T5$	$T16$	$A15$;
$B1$	$B3$	$D1$	$D3$	$F2$	$F6$	$X2$	$X6$	$K4$	$K7$	$M4$	$M7$	$A7$;
$L1$	$L2$	$N1$	$N2$	$M3$	$M6$	$P3$	$P6$	$C4$	$C10$	$X5$	$X16$	$A13$.

We observe that, as in the semifield plane with kern GF(2) [8], the only complete 13-arcs we were able to find are tangent to the line at infinity. Notice that each of the listed 13-arcs is mapped onto itself by an involution, more precisely by a translation with centre $A0$.

7. Complete 14-arcs. We already observed that, for planes of order 16, complete 14-arcs can occur in non-Desarguesian planes only. The Dempwolff plane contains such arcs. Though no exhaustive search was carried out, just a thorough one, it seems that there is just one type of complete 14-arcs in Π. (Notice that the dual Lorimer plane contains at least three non-equivalent types of complete 14-arcs [10].)

An example of a complete 14-arc in Π is the following:

$$\Gamma : B3 \; B6 \; C3 \; C6 \; D4 \; D10 \; F4 \; F10 \; H1 \; H2 \; N13 \; N14 \; T7 \; T12.$$

The interior points (or 0-points, i.e. points on no tangent) of Γ are $A0$, the unique interior point on $l\infty$, $D5\,D16\,P5\,P16\,R3\,R6\,W8\,W15\,X3\,X6$. Moreover, there are just two 10-points, i.e. points on ten tangents; they are $R5$ and $R16$. The remaining points are 2-,4-, or 6-points. In particular, there is no 4-point on the line at infinity.

We observe that Γ is mapped onto itself by a translation with centre $A0$. This translation maps onto themselves also the interior points of Γ.

The ten finite 0-points lie on two lines; namely, on

$$x7 : A15 \; B7 \; C4 \; \underline{D16} \; F11 \; H8 \; K9 \; L10 \; M1 \; N2 \; \underline{P5} \; \underline{R6} \; S12 \; T14 \; \underline{W15} \; \underline{X3} \; Z13$$

and $x12$. $A15, B7, H8, M1, S12$, and $Z13$ are 6-points; $C4, K9, N2$, and $T14$ are 4-points; $F11$ and $L10$ are 2-points. The ten 0-points off $a0$ come into five pairs on five lines through $A0$ and one of them only, $a3$, is a secant to the arc. The remaining ones are exterior lines. There are four lines on $A1$ each of them containing two of the 0-points other than $A0$. Two of these lines contain also the two 10-points, one on each of them. They are $b5$ and $b16$. $A1$ is a 6-point and two of the mentioned lines, namely $b3$ and $b6$, are secant to Γ.

8. Hyperovals. Hyperovals in translation planes are thoroughly investigated and classified in [2]. Therefore, we just list a few examples among those we found. To begin with, consider the following four hyperovals:

$$B1 \; B2 \; C1 \; C2 \; D4 \; D10 \; K4 \; K10 \; H7 \; H12 \; Z7 \; Z12 \; N3 \; N6 \; R3 \; R6 \; A15 \; A16$$
$$B1 \; B2 \; C1 \; C2 \; D4 \; D10 \; K4 \; K10 \; M7 \; M12 \; X7 \; X12 \; T3 \; T6 \; W3 \; W6 \; A9 \; A10$$

$$H7 \; H12 \; Z7 \; Z12 \; N3 \; N6 \; R3 \; R6 \; F1 \; F2 \; L4 \; L10 \; P1 \; P2 \; S4 \; S10 \; A9 \; A10$$
$$M7 \; M12 \; X7 \; X12 \; T3 \; T6 \; W3 \; W6 \; F1 \; F2 \; L4 \; L10 \; P1 \; P2 \; S4 \; S10 \; A15 \; A16.$$

This quadruple of hyperovals shows that Π too contains a configuration of four hyperovals any two of which share either eight or two points. Such a configuration seems to be typical of (translation) planes of order 16. It occurs in the Desarguesian plane [17], in the Lorimer plane and in its dual plane [10], in the semifield plane with kern GF(2) [8].

Also the following pairs of hyperovals yield quadruples similar to the previous one:

*B*3 *B*6 *Z*3 *Z*6 *L*5 *L*16 *W*5 *W*16 *K*1 *K*2 *R*1 *R*2 *M*9 *M*11 *P*9 *P*11 *A*9 *A*12
*B*3 *B*6 *Z*3 *Z*6 *L*5 *L*16 *W*5 *W*16 *K*4 *K*10 *R*4 *R*10 *M*13 *M*14 *P*13 *P*14 *A*3 *A*8

*C*3 *C*6 *D*3 *D*6 *F*4 *F*10 *S*4 *S*10 *B*7 *B*12 *K*7 *K*12 *L*1 *L*2 *P*1 *P*2 *A*10 *A*15
*C*3 *C*6 *D*3 *D*6 *F*4 *F*10 *S*4 *S*10 *B*8 *B*15 *K*8 *K*15 *L*9 *L*11 *P*9 *P*11 *A*5 *A*7

*B*7 *B*12 *L*7 *L*12 *D*5 *D*16 *F*5 *F*16 *C*1 *C*2 *S*1 *S*2 *K*13 *K*14 *P*13 *P*14 *A*12 *A*14
*B*7 *B*12 *L*7 *L*12 *D*5 *D*16 *F*5 *F*16 *N*8 *N*15 *X*8 *X*15 *W*3 *W*6 *Z*3 *Z*6 *A*5 *A*6

*B*7 *B*12 *C*7 *C*12 *D*3 *D*6 *K*3 *K*6 *H*1 *H*2 *N*4 *N*10 *R*4 *R*10 *Z*1 *Z*2 *A*15 *A*16
*B*7 *B*12 *C*7 *C*12 *D*3 *D*6 *K*3 *K*6 *M*1 *M*2 *T*4 *T*10 *W*4 *W*10 *X*1 *X*2 *A*9 *A*10

Notice that all the listed hyperovals are translation hyperovals [2]. As a matter of fact, they are constructed with the help of a translation. By using different translations (involutions) we get hyperovals which look different. For instance, compare the following ones:

*B*3 *B*6 *D*3 *D*6 *F*5 *F*16 *L*5 *L*16 *C*1 *C*2 *K*1 *K*2 *P*9 *P*11 *S*9 *S*11 *A*11 *A*13

*B*1 *B*3 *T*1 *T*3 *C*2 *C*6 *W*2 *W*6 *D*10 *D*12 *M*10 *M*12 *K*4 *K*7 *X*4 *X*7 *A*12 *A*13

*D*2 *D*8 *N*7 *N*9 *F*11 *F*12 *X*1 *X*15 *K*7 *K*9 *M*11 *M*12 *P*1 *P*15 *R*2 *R*8 *A*7 *A*14

*C*5 *C*7 *D*5 *D*7 *L*3 *L*8 *P*3 *P*8 *H*1 *H*13 *N*1 *N*13 *W*4 *W*11 *M*4 *M*11 *A*9 *A*12
*C*5 *C*7 *D*5 *D*7 *L*3 *L*8 *P*3 *P*8 *R*2 *R*14 *Z*2 *Z*14 *T*9 *T*10 *X*9 *X*10 *A*3 *A*13.

The intersection properties of the hyperovals in Π are similar to those of the hyperovals in the semifield plane with kern GF(2) from which Π may be derived [8].

Finally, we observe that some hyperovals in Baer subplanes of Π do extend to hyperovals in Π, but not all of them. For instance, *A*0 *A*1 *B*1 *C*2 *F*10 *P*4 is a hyperoval in the Baer subplane α. It extends to the following hyperoval in Π:

*A*0 *A*1 *B*1 *C*2 *F*10 *P*4 *H*7 *Z*12 *D*16 *K*5 *L*8 *S*15 *N*13 *R*14 *T*11 *W*9.

On the other hand, the hyperoval *A*3 *A*4 *B*4 *B*10 *C*4 *C*10, again in the Baer subplane α, completes to a 10-arc in Π with any of the following quadruples of points: *H*5 *H*16 *Z*5 *Z*16, *F*3 *F*6 *P*3 *P*6, *D*3 *D*6 *K*3 *K*6. Such a behaviour resembles the behaviour of hyperovals in Fano subplanes of PG(2,8) which can be extended to either complete 6-arcs or hyperovals [12].

REFERENCES

1. R. C. Bose, J. W. Freeman and D. G. Glynn, *On the intersection of two Baer subplanes in a finite projective plane*, Utilitas Math. **17** (1980), 65–77.
2. W. Cherowitzo, *Hyperovals in translation planes*, to appear.
3. P. Dembowski, *Finite Geometries*, Springer Verlag, Berlin, Heidelberg, New York, 1968.

4. U. Dempwolff, *Einige Translationsebenen der Ordnung* 16 *und ihre Kollineationen*, unpublished.

5. U. Dempwolff and A. Reifart, *The classification of the translation planes of order* 16 , *I*, Geometriae Dedicata **15** (1983), 137–153.

6. _____, *Translation planes of order* 16 *admitting a Baer* 4-*group*, J. Comb. Th. 4 **A 32** (1982), 119–124.

7. R. H. F. Denniston, *On arcs in projective planes of order* 9 , Manuscr. Math. **4** (1971), 61-90.

8. M. J. de Resmini, *On the semifield plane of order* 16 *with kern GF* (2), Ars Combin. **25b** (Proc. 11th British Comb. Conf. July 1987). (1988), 75–92.

9. _____, *Some combinatorial properties of a semi translation plane*, Congressus numerantium **59** (1987), 5–12.

10. M. J. de Resmini and L. Puccio, *Some combinatorial properties of the dual Lorimer plane*, Ars Combinatoria **24A** (1987), 131–148.

11. J. C. Fisher, J. W. P. Hirschfeld and J. A. Thas, *Complete arcs in planes of square order*, Ann. Discr. Math. **30** (1986), 243-250.

12. J. W. P. Hirschfeld, *Projective Geometries over Finite Fields*, Clarendon Press, Oxford, 1979.

13. D. R. Hughes and F. C. Piper, *Projective Planes*, Springer Verlag, Berlin, Heidelberg, New York, 1973.

14. N. L. Johnson, *The translation planes of order* 16 *that admit* $SL(2, 4)$, Ann. Discr. Math. **14** (1982), 225–236.

15. N. L. Johnson, *Representation of* $SL(2, 4)$ *on translation planes of even order*, J. Geometry **21** (1983), 184–200.

16. _____, *The translation planes of order* 16 *that admit non-solvable collineation groups*, Math. Z. **185** (1984), 355–372.

17. L. Lunelli and M. Sce, *k-archi completi nei piani proiettivi Desarguesiani di rango* 8 *e* 16 , Centro di Calcoli Numerici, Politecnico di Milano, 1958.

18. H. Lüneburg, *Translation Planes*, Springer Verlag, Berlin, Heidelberg, New York, 1981.

19. G. Menichetti, *q-archi completi nei piani di Hall di ordine* $q = 2^k$, Rend. Acc. Naz. Lincei, VIII **56** (1974), 1–8.

20. T. G. Ostrom, *Finite Translation Planes*, Lecture Notes in Math. 158, Springer Verlag, Berlin, Heidelberg, New York, 1970.

DIPARTIMENTO DI MATEMATICA, UNIVERSITA' DI ROMA "LA SAPIENZA", I-00185 ROME, ITALY

Contemporary Mathematics
Volume **111**, 1990

Difference Sets in 2-Groups

J. F. DILLON

ABSTRACT. We give a simple construction for difference sets in the rank 2 abelian groups $Z_{2^{s+1}} \times Z_{2^{s+1}}$ and $Z_{2^s} \times Z_{2^{s+2}}$. We also establish a number of composition theorems which provide difference sets in many other groups not necessarily abelian.

1. Introduction. The k-subset D of the group G of order v is a difference set with parameters (v, k, λ) if, for every nonidentity element g in G, the equation

$$g = xy^{-1}$$

has exactly λ solutions (x, y) with x and y in D; the related parameter n is defined to be equal to $k - \lambda$. If we identify the subset D of G with the element of the group ring $Z[G]$ which is its characteristic function, then we have that D is a difference set precisely when it satisfies in $Z[G]$ the equation

$$DD^{(-1)} = n + \lambda G,$$

where "n" stands for "n times the identity element of G" and, for any element

$$A = \sum_{g \in G} a_g g$$

of $Z[G]$, $A^{(-1)}$ denotes the element

$$\sum_{g \in G} a_g g^{-1}.$$

The associated (± 1)-valued function

$$D^* = G - 2D,$$

then satisfies

$$D^* D^{*(-1)} = 4n + (v - 4n)G,$$

1980 *Mathematics Subject Classification* (1985 *Revision*). Primary 05B05,05B10.
This paper is in final form and no version of it will be submitted for publication elsewhere.

so that if either $v = 1$ or $v = 4n$, we have

(1.1) $$D^* D^{*(-1)} = v.$$

In such an event the difference set D is called a *Hadamard* difference set since the $v \times v\,(\pm 1)$-incidence matrix (whose rows and columns are indexed by the elements of G and whose (g, h) th entry is the coefficient of gh^{-1} in D^*) is Hadamard. In case the group G is abelian the condition (1.1) is equivalent to

(1.2) $$\left| \chi(D^*) \right|^2 = v \text{ for all characters } \chi \text{ of } G.$$

An outstanding problem in combinatorics is that of determining the class **H** of groups which contain a Hadamard difference set. It is well-known that **H** is closed under direct products. Indeed, **H** is closed under arbitrary products in the following sense.

THEOREM 1.1. *Let* G_1 *and* G_2 *belong to* **H** *and let G contain both* G_1 *and* G_2 *with* $|G| = |G_1||G_2|$ *and* $|G_1 \cap G_2| = 1$. *Then G also belongs to* **H**.

PROOF. Let D_1 and D_2 be Hadamard difference sets in G_1 and G_2 respectively, and let D be the subset of $G = G_1 G_2$ whose associate D^* is given by $D^* = D_1^* D_2^*$ where each D_i^* is an element of $Z[G_i]$ embedded in $Z[G]$ in the natural way. Then

$$
\begin{aligned}
D^* D^{*(-1)} &= (D_1^* D_2^*)(D_1^* D_2^*)^{(-1)} \\
&= D_1^* D_2^* D_2^{*(-1)} D_1^{*(-1)} \\
&= D_1^* |G_2| D_1^{*(-1)} \\
&= D_1^* D_1^{*(-1)} |G_2| \\
&= |G_1||G_2| \\
&= |G|
\end{aligned}
$$

so that D is a Hadamard difference set in G.   □

In this paper we are concerned only with 2-groups and, for the most part, only with the question of which 2-groups can belong to **H**. H. B. Mann [5] has shown that every nontrivial difference set in a 2-group (indeed, every nontrivial *SBIBD* having v a power of 2) must, up to complementation, have parameters of the form

(1.3) $$(v, k, \lambda) = (2^{2s+2}, 2^{2s+1} - 2^s, 2^{2s} - 2^s).$$

Since these parameters satisfy the relation $v = 4n$ any s ich difference set must be Hadamard. Now the group of order 1 and both groups of order 4 contain Hadamard difference sets trivially (in the groups c f order 4 these are singletons and their complements), but the product construction produces from these trivial difference sets distinctly nontrivial difference sets with parameters (1.3) in any group which is a direct product of $s + 1$ groups of order 4. This observation is due to R. J. Turyn and P. Kesava Menon who worked independently and from different perspectives. In [4] we obtained the following generalization.

THEOREM 1.2. **H** *contains every group of order* 2^{2s+2} *which is a direct product of* $s + 1$ *nontrivial subgroups.*

A further generalization [3] is:

THEOREM 1.3. *Let* G *belong to* **H** *and let* $E = Z_2^{s+1}$. *Let* M *be any group of order* $2^{2s+2}|G|$ *which contains the direct product* $G \times E$ *and whose center contains* E. *Then* M *also belongs to* **H**.

In the next section we shall give a new proof of this result which makes it clear how certain highly structured difference sets can give rise to difference sets in many other groups. In the case of abelian groups our Theorem 1.2 shows that **H** contains every group of order 2^{2s+2} and rank at least $s + 1$. We also have the following result of R. J. Turyn [7].

THEOREM 1.4. *If the abelian group* G *of order* 2^{2s+2} *belongs to* **H** *then* G *has exponent at most* 2^{s+2}.

In particular, if $s > 0$ then **H** cannot contain the cyclic group of order 2^{2s+2}. Application of our Theorem 1.2 to the group

$$G = Z_2^s \times Z_{2^{s+2}}$$

shows that Turyn's bound is sharp for all s.

It is our main purpose in this paper to consider the rank 2 abelian groups

$$Z_{2^{s+1}} \times Z_{2^{s+1}} \quad \text{and} \quad Z_{2^s} \times Z_{2^{s+2}}.$$

Until this year no abelian Hadamard group of order 2^{2s+2} and rank less than $s + 1$ was known. R. J. Turyn (private communication) did a computer search for difference sets in $Z_8 \times Z_8$ and discovered the following beautiful example:

	0	1	2	3	4	5	6	7
0	1	1	1	1	1	1	1	1
1	1	1	1	1	–	–	–	–
2	1	1	–	–	1	1	–	–
$D^* = 3$	1	1	–	1	–	–	1	–
4	1	–	1	–	1	–	1	–
5	1	–	1	–	–	1	–	1
6	1	–	–	1	1	–	–	1
7	1	–	–	–	–	1	1	1

Recently James A. Davis [1] has established the existence of difference sets in *all* these rank 2 groups. His method, however, involves a rather complicated induction and it is not at all clear what his difference sets look like. We shall give in Section 3 a simple direct construction for difference sets in the groups $Z_{2^{s+1}} \times Z_{2^{s+1}}$. Turyn's set turns out to be a special case. Moreover, our sets will have a structure which will allow us to transfer them to the groups $Z_{2^s} \times Z_{2^{s+2}}$. The various composition theorems may then be invoked to produce myriad Hadamard groups which need not be abelian.

2. Some composition theorems. We begin with a proof of Theorem 1.3. Let G be any Hadamard group and let H denote the direct product $E \times G$ where $E = Z_2^{s+1}$. Let M be any group which contains H as a subgroup of index 2^{s+1} and whose center contains E. Let

$$g_0, g_1, g_2, \ldots, g_{2^{s+1}-1}$$

be representatives of the cosets of H in M so that

$$M = g_0 H + g_1 H + g_2 H + \cdots + g_{2^{s+1}-1} H$$
$$= g_0 EG + g_1 EG + g_2 EG + \cdots + g_{2^{s+1}-1} EG.$$

If $\Delta_0, \Delta_1, \Delta_2, \ldots, \Delta_{2^{s+1}-1}$ are any Hadamard difference sets in G, we may define a set D in M whose associate D^* is given by

$$D^* = \sum_{i=0}^{2^{s+1}-1} g_i \chi_i \Delta_i^*,$$

where $\chi_0, \chi_1, \chi_2, \ldots, \chi_{2^{s+1}-1}$ are the distinct characters of E. Note that since E is an elementary abelian 2-group its characters are (± 1)-valued.

Then,

$$D^*D^{*(-1)} = \sum_{i,j}(g_i\chi_i\Delta_i^*)(g_j\chi_j\Delta_j^*)^{(-1)}$$

$$= \sum_{i,j}(g_i\chi_i\Delta_i^*)(\Delta_j^{*(-1)}\chi_jg_j^{-1})$$

$$= \sum_{i,j}g_i(\chi_i\chi_j)(\Delta_i^*\Delta_j^{*(-1)})g_j^{-1}$$

$$= \sum_i g_i(|E|\chi_i)(|G|)g_i^{-1}$$

$$= |E||G|\sum_i \chi_i$$

$$= |E|^2|G|$$

$$= |M|,$$

and it follows that D is a Hadamard difference set in M. □

Notice that the function D^* is comprised of pieces $g_i\chi_i\Delta_i^*$ which are pairwise "orthogonal" and this suggests that a similar construction may be possible in the case that the elementary abelian group E is not a direct factor of the containing subgroup H. Now, in general, if G is any group containing a subgroup H of index 2 and g is any element of G which is not in H, so that

$$G = H + gH,$$

then any element D^* in $Z[G]$ may be written as

$$D^* = A^* + gB^*$$

with A* and B* in Z[H] and we have

$$D^*D^{*(-1)} = \{A^*A^{*(-1)} + g(B^*B^{*(-1)})g^{-1}\} + g\{B^*A^{*(-1)} + g^{-2}\cdot g(A^*B^{*(-1)})g^{-1}\}.$$

The following is then clear.

THEOREM 2. *Suppose that A^* and B^* are ± 1-valued functions on a group H satisfying*

$$A^*A^{*(-1)} + B^*B^{*(-1)} = 2|H| \quad and \quad A^*B^{*(-1)} = 0.$$

Let $G = \langle H, g\rangle$ be any group containing H as a subgroup of index 2 and whose center contains g. Let D be the subset of G whose associate D^* is given by

$$D^* = A^* + gB^*.$$

Then D is a difference set in G.

COROLLARY. *If the Hadamard group $G_0 = \langle H, g_0\rangle$ is a central extension of a subgroup H of index 2 and if G_0 contains a difference set D with associate $D^* = A^* + g_0B^*$ satisfying $A^*B^{*(-1)} = 0$, then every central extension $G = \langle H, g\rangle$ of H of order $2|H|$ is a Hadamard group.*

It is this last result that we shall exploit in the next section to transfer to $Z_{2^s} \times Z_{2^{s+2}}$ our difference sets in $Z_{2^{s+1}} \times Z_{2^{s+1}}$.

3. Rank-2 groups. According to Turyn's exponent bound the only rank 2 abelian groups of order 2^{2s+2} which could possibly have nontrivial difference sets are $Z_{2^{s+1}} \times Z_{2^{s+1}}$ and $Z_{2^s} \times Z_{2^{s+2}}$. We shall now give a very simple construction of difference sets in the square case and then transfer these sets to the other class of rank 2 groups. Let $G = Z_{2^{s+1}} \times Z_{2^{s+1}}$ where here we want to regard $Z_{2^{s+1}}$ as the additive group of integers mod 2^{s+1}.
Let

$$f : Z_{2^{s+1}} \to \{\pm 1\}$$

be any function with the property that

$$f(x + 2^s) = -f(x),$$

for all x in $Z_{2^{s+1}}$. There are of course 2^{2^s} such functions one of which is the "high order bit" function given by

$$f(\sum_{i=0}^{s} x_i 2^i) = (-1)^{x_s},$$

or, equivalently,

$$f(x) = (-1)^{\tau(x)} \quad \text{where } \tau(x) = \lfloor \frac{x}{2^s} \rfloor, \qquad 0 \le x < 2^{s+1}.$$

Those functions with $f(0) = 1$ will produce difference sets with parameters (1.3) while the others will produce difference sets with the complementary parameters. Notice that f takes both values 1 and -1 on every coset of the subgroup of order 2 so, in particular, f sums to 0 on every subgroup of $Z_{2^{s+1}}$ of order greater than 1. Let

$$\pi : Z_{2^{s+1}} \longrightarrow Z_{2^{s+1}}$$

be the permutation which maps each residue $2^r t$, t odd, onto the residue $2^r \bar{t}$ where $t\bar{t} \equiv 1 \pmod{2^{s+1}}$. Let

$$F : Z_{2^{s+1}} \times Z_{2^{s+1}} \longrightarrow \{\pm 1\}$$

be defined by

$$F(x, y) = f(\pi(x)y),$$

where $\pi(x)y$ is the *product* of $\pi(x)$ and y taken mod 2^{s+1}. To show that F is the (± 1)-valued characteristic function of a difference set it is convenient to invoke the criterion (1.2); i.e. that the character sums be of constant magnitude. Indeed, if for each $(u, v) \in G$ we denote by $\chi_{u,v}$ the character on G given by

$$\chi_{u,v} : (x, y) \longrightarrow \zeta^{ux+vy},$$

where $\zeta = \exp(2\pi i/2^{s+1})$, and if we denote by $F^\wedge(u, v)$ the value of the character $\chi_{u,v}$ on F; i.e.

$$F^\wedge(u, v) = \sum_{x,y} F(x, y)\zeta^{ux+vy},$$

then it is easy to show that

(3.1) $$F^\wedge(x, y) = 2^{s+1}F(y, -x),$$

for all $(x, y) \in G$. Thus, not only are the character sums of the right magnitude, but they are in fact all equal to $\pm 2^{s+1}$. Of course the fact that the character sums are real follows immediately from the observation that the function F is invariant under inversion; i.e. $F(-x, -y) = F(x, y)$. Indeed, for any odd t

$$F(tx, ty) = f(\pi(tx)ty) = f(t^{-1}\pi(x)ty) = f(\pi(x)y) = F(x, y),$$

so that the corresponding difference set D is fixed by the group of automorphisms $(x, y) \rightarrow (tx, ty)$, t odd. Thus, the difference set D has $U_{2^{s+1}} \simeq Z_2 \times Z_{2^{s-1}}$ as a group of multipliers. The (± 1)-valued function

$$\frac{1}{2^{s+1}}F^\wedge(x, y)$$

defines another difference set which is "dual" to D but (3.1) shows that it is equivalent to D under the automorphism

$$(x, y) \rightarrow (y, -x).$$

Note that when $s \leq 2$ the permutation π is the identity map and a difference set may be defined by

$$D = \{(x, y) : 0 \leq x, y < 2^{s+1} \text{ and } \lfloor\frac{xy}{2^s}\rfloor = 1\}$$

where xy is interpreted as a residue in the set $\{0, 1, 2, \ldots, 2^{s+1} - 1\}$; i.e. $(x, y) \in D$ if the residue xy is greater than or equal to 2^s. For $s = 2$ this is precisely Turyn's example.

Finally, reverting to multiplicative notation

$$G = \langle a \rangle \times \langle b \rangle, \qquad a^{2^{s+1}} = 1 = b^{2^{s+1}},$$

for the group $Z_{2^{s+1}} \times Z_{2^{s+1}}$ by way of the usual isomorphism

$$a^i b^j \leftrightarrow (i, j),$$

we may observe that our difference set D has the form

$$D^* = A^* + aB^*,$$

where A^* and B^* are (± 1)-valued functions on the subgroup $H = \langle a^2 \rangle \times \langle b \rangle$, and that A^* and B^* are in fact orthogonal since they may be factored as

$$A^* = (1 + b^{2^s})\Delta_0^* \quad \text{and} \quad B^* = (1 - b^{2^s})\Delta_1^*,$$

72   J. F. DILLON

where Δ_0^* and Δ_1^* are (± 1)-valued functions on $C = <a^2> \{1, b, b^2, \ldots,$ $b^{2^s-1}\}$ which is a set of representatives of the cosets in H of the subgroup $\langle b^{2^s} \rangle$. Thus, by the corollary to Theorem 2 we may transfer the difference set D in $\langle a \rangle \times \langle b \rangle$ to a difference set E in

$$\langle a^2 \rangle \times \langle c \rangle, \qquad c^2 = b,$$

which is isomorphic to $Z_{2^s} \times Z_{2^{s+2}}$. The new difference set E is given by

$$E^* = A^* + cB^*.$$

We reiterate the foregoing in the

THEOREM 3. *Let* $G = Z_{2^{s+1}} \times Z_{2^{s+1}}$ *where* $Z_{2^{s+1}} = \{0, 1, 2, \ldots, 2^{s+1} - 1\}$ *is the additive group of integers* mod 2^{s+1}. *Let* $\tau : Z_{2^{s+1}} \longrightarrow \{0, 1\}$ *be the characteristic function of the set of residues greater than or equal to* 2^s, *and let* $\pi : Z_{2^{s+1}} \longrightarrow Z_{2^{s+1}}$ *be the permutation which maps the residue* $2^r t$, t *odd, to the residue* $2^r \bar{t}$ *where* $t\bar{t} \equiv 1 \pmod{2^{s+1}}$. *Let D be the subset of G given by*

$$D = \{(x, y) : \tau\{\pi(x)y\} = 1\}.$$

Then D is a difference set in G. Moreover, this difference set may also be interpreted as a difference set in the group $Z_{2^s} \times Z_{2^{s+2}}$.

REFERENCES

1. James A. Davis, *Difference Sets in Abelian 2-Groups*, Thesis, U. of Virginia, 1987.
2. J. F. Dillon, *Elementary Hadamard Difference Sets*, Congressus Numerantium **XIV** (1975).
3. _____, *On Hadamard Difference Sets*, Ars Combinatoria **1** (1976).
4. _____, *Variations on a Scheme of McFarland for Noncyclic Difference Sets*, J. Combinatorial Theory (A) **40** (1985).
5. H. B. Mann, *Addition Theorems*, Wiley Interscience, N.Y., 1965.
6. R. L. McFarland, *A family of Difference Sets in Noncyclic Groups*, J. Combinatorial Theory (A) **15** (1973).
7. R. J. Turyn, *Character Sums and Difference Sets*, Pacific J. Math. **15** (1965).

NATIONAL SECURITY AGENCY, FORT GEORGE G. MEADE, MARYLAND 20755

Contemporary Mathematics
Volume **111**, 1990

On the Existence of Room Squares with Subsquares

J. H. DINITZ AND D. R. STINSON

ABSTRACT. We prove several results on the existence of Room squares containing Room subsquares. For example, we prove that if v and w are odd integers, $w \geq 23085$, and $v \geq 3w + 40$, then there is a Room square of side v containing a Room subsquare of side w. This result is very close to best possible, since existence of a Room square of side v containing a Room subsquare of side w requires that v and w be odd, and $v \geq 3w + 2$. We also prove that if v and w are odd integers, $w \geq 127$, and $v \geq 3w + 240$, then there is a Room square of side v containing a Room subsquare of side w. Results are also presented for values of $w < 127$.

1. Introduction. Let S be a set of $v + 1$ elements called *symbols*. A *Room square* (on symbol set S) is a v by v array, F, which satisfies the following properties:

(1) every cell of F either is empty or contains an unordered pair of symbols from S,

(2) each symbol of S occurs once in each row and column of F, and

(3) every unordered pair of symbols occurs in precisely one cell of F.

The spectrum for Room squares was determined in 1975 by Mullin and Wallis, who proved the following in [**9**].

THEOREM 1.1. *There exists a Room square of side v if and only if v is an odd positive integer, $v \neq 3$ or 5.*

Now, suppose F is a Room square of side v on symbol set S. A w by w subarray G of F is said to be a *Room subsquare* of side w if it is itself a Room square of side w on a subset of $w + 1$ symbols. In view of Theorem 1.1, no Room square can contain a Room subsquare of side 3 or 5. However, we can construct Room squares *missing* subsquares of these sides. We have the following formal definition.

1980 *Mathematics Subject Classification* (1985 *Revision*). Primary 05B15.
The second author was supported by NSERC Grant A9287.
This paper is in final form and no version of its will be submitted for publication elsewhere.

Let S be a set of $v+1$ symbols and let T be a subset of S of cardinality $w+1$. A (v, w)-IRS (*incomplete Room square*) is a v by v array F which satisfies the following:

(1) every cell of F either is empty or contains an unordered pair of symbols of S,

(2) there is an empty w by w subarray G of F,

(3) each symbol of $S \setminus T$ occurs once in each row and column of F,

(4) each symbol of T occurs once in each row and column not meeting G, but not in any row or column meeting G, and

(5) the pairs occurring in F are precisely those $\{x, y\}$ where $(x, y) \in (S \times S) \setminus (T \times T)$.

We refer to the subarray T as the *hole*. Observe that the hole can be filled in with any Room square of side w (provided $w \neq 3$ or 5), thereby constructing a Room square of side v containing a subsquare of side w.

In this paper we seek to answer the question: for which ordered pairs (v, w) does there exist a (v, w)-IRS? Of course, v and w must be odd, and it is not difficult to see that $v \geq 3w+2$ (see [2]). It has been conjectured that these necessary conditions are sufficient for the existence of a (v, w)-IRS, with the single exception $(v, w) \neq (5, 1)$.

The following theorem summarizes known existence results for (v, w)-IRS.

THEOREM 1.2.

(1) *For all odd* $w \geq 3$, *there is a* $(3w+2, w)$-IRS. ([12])

(2) *For all odd* $v \geq 41$, *there is a* $(v, 3)$-IRS. ([7])

(3) *For all odd* $v \geq 69$, *there is a* $(v, 5)$-IRS. ([7])

(4) *For all odd* $v \geq 55$, *there is a* $(v, 7)$-IRS. ([7])

(5) *For all odd* $w \geq 7$ *and all odd* $v \geq 6w+41$, *there is a* (v, w)-IRS. ([11])

(6) *For all odd* $w \geq 129$ *and all odd* $v \geq 4w+29$, *there is a* (v, w)-IRS. ([11])

In this paper, we obtain significant improvements to (2)–(6), as indicated in the following Theorem.

THEOREM 1.3.

(1) *For* $w = 3, 5, 7$, *and for all odd* $v \geq 3w+2$, *there is a* (v, w)-IRS.

(2) *For all odd* $w \geq 37$ *and all odd* $v \geq (7w-5)/2$, *there is a* (v, w)-IRS.

(3) *For all odd* $w \geq 127$ *and all odd* $v \geq 3w+240$, *there is a* (v, w)-IRS.

(4) *For all odd* $w \geq 23085$ *and all odd* $v \geq 3w+40$, *there is a* (v, w)-IRS.

2. Room frames. Our main constructions require the use of a generalization of a Room square called a Room frame. Let S be a set, and let $\{S_1, \ldots, S_n\}$ be a partition of S. An $\{S_1, \ldots, S_n\}$-*Room frame* is an $|S|$ by $|S|$ array, F, indexed by S, which satisfies the following properties:

(1) every cell of F either is empty or contains an unordered pair of symbols of S,
(2) the subarrays $S_i \times S_i$ are empty, for $1 \le i \le n$ (these subarrays are referred to as *holes*),
(3) each symbol of $S \setminus S_i$ occurs once in row (or column) s, for any $s \in S_i$, and
(4) the pairs occurring in F are precisely those $\{s, t\}$, where

$$(s, t) \in (S \times S) \setminus \bigcup_{1 \le i \le n} (S_i \times S_i).$$

The *type* of F is defined to be the multiset $\{|S_i| : 1 \le i \le n\}$. We usually use an "exponential" notation to describe types: a type $t_1^{u_1} \ldots t_k^{u_k}$ denotes u_i occurrences of t_i, $1 \le i \le k$. In Figure 1, we present a Room frame of type $2^5 4^1$.

A Room frame can be thought of as a Room square from which a spanning set of subsquares has been removed. A Room frame of type 1^v gives rise to a Room square of side v by filling in each diagonal cell (s, s) with the pair $\{\infty, s\}$, where ∞ is a new symbol. More generally, a (v, w)-IRS is equivalent to a Room frame of type $1^{v-w} w^1$.

We remark that Room frames have been used extensively in the literature as a tool in the construction of Room squares (see, for example, [10] and [11]). In the past, they have usually been referred to simply as "frames". However, frames for other types of combinatorial designs have recently been studied by several researchers. Hence, we will use the term "Room frame" in this paper.

We shall make use of the following Room frames.

THEOREM 2.1 [3]. *There exists a Room frame of type 4^u if and only if $u \ge 4$, and there exists a Room frame of type 2^u if and only if $u \ge 5$.*

For our main constructions, we require Room frames of types $2^i 4^{6-i}$, for $0 \le i \le 6$. Theorem 2.1 gives us those of types 4^6 and 2^6. A Room frame of type $2^1 4^5$ was constructed by the intransitive starter-adder method, as follows.

Let G be an abelian group and let H be a subgroup of G, where $g = |G|$, $h = |H|$, and suppose that $g - h$ is even. An *intransitive starter* in $G \setminus H$ is

1 – 2		10, 12		3, 13			8, 11	7, 14		4, 6	5, 9		
		7, 11		9, 13		6, 14	4, 12		8, 10	3, 5			
9, 12		3 – 4		8, 14		5, 11		7, 13		1, 6			2, 10
	9, 14				10, 11			6, 12	1, 13	5, 7		2, 8	
		8, 12	1, 14	5 – 6		10, 13	9, 11			2, 3		4, 7	
4, 13	3, 12	10, 14						2, 11		8, 9			1, 7
3, 11		5, 13		2, 12	4, 14	7 – 8				1, 10			6, 9
		2, 13	1, 11					3, 14	5, 12	4, 9			6, 10
5, 14	4, 11			1, 12	6, 13			9 – 10		2, 7			3, 8
	8, 13	6, 11		7, 12		2, 14						1, 3	4, 5
6, 8	5, 10			3, 7	2, 9			1, 4					
	6, 7	1, 9	5, 8			2, 4	3, 10				11 – 14		
			7, 9	4, 10	1, 8			2, 5	3, 6				
7, 10		2, 6				3, 9	1, 5		4, 8				

FIGURE 1. A Room frame of type $2^5 4^1$

defined to be a triple (S, R, C), where

$$S = \bigcup_{1 \le i \le (g-h-2)/2} (\{\{s_i, t_i\} : 1 \le i \le (g-h-2)/2\}) \cup \{\{u\}, \{v\}\},$$

$$C = \{\{p, q\}\}, \quad \text{and}$$

$$R = \{\{p', q'\}\},$$

satisfying

(1) $\bigcup_{1 \le i \le (g-h-2)/2}(\{s_i\} \cup \{t_i\}) \cup \{u, v, p, q\} = G \backslash H$, and

(2) $\bigcup_{1 \le i \le (g-h-2)/2}(\{\pm(s_i - t_i)\} \cup \{\pm(p - q)\} \cup \{\pm(p' - q')\}) = G \backslash H$, and

(3) both $p - q$ and $p' - q'$ have even order in G.

An *adder* for (S, C, R) is an injection $A : S \to G \backslash H$, such that

$$\bigcup_{1 \le i \le (g-h-2)/2} (\{s_i + a_i\} \cup \{t_i + a_i\}) \cup \{u + A(u), v + A(v), p', q'\} = G \backslash H,$$

where $a_i = A(s_i, t_i)$, $1 \le i \le (g - h)/2$.

THEOREM 2.2 [3, 10]. *If there is an intransitive starter and adder in* $G \backslash H$, *where* $g = |G|$ *and* $h = |H|$, *then there is a Room frame of type* $h^{g/h} 2^1$.

In Example 2.1, we present an intransitive starter and adder in $G \backslash H$, where $G = Z_{20}$ and $H = \{0, 10\}$. Hence, there is a Room frame of type $2^1 4^5$.

Example 2.1. A Room frame of type $2^1 4^5$.

	S(starter)	A(adder)		S+A	
	∞_1 1	1		∞_1	2
	∞_2 6	3		∞_2	9
	7 9	9		16	18
	14 18	19		13	17
	11 17	17		8	14
	12 19	12		4	11
	8 16	11		19	7
	4 13	8		12	1
C=	2 3				
R=				3	6

The constructions of the remaining Room frames were accomplished by means of the computer. In order to describe the algorithms used, we require some more definitions. Let G be a graph. A *one-factor* of G is a set of disjoint edges which spans the vertex set (i.e. a perfect matching). A *one-factorization* of G is a set of one-factors which partitions the edge set. Two one-factors of G are *orthogonal* if they contain at most one common edge. A one-factor f is *orthogonal* to a one-factorization $\mathscr{F}$ if f is orthogonal to every one-factor in $\mathscr{F}$. Finally, two one-factorizations $\mathscr{F}_1$ and $\mathscr{F}_2$ are *orthogonal* if every one-factor in $\mathscr{F}_1$ is orthogonal to $\mathscr{F}_2$. It is well-known (see [8]) that a Room square of side n is equivalent to two orthogonal one-factorizations of the complete graph K_{n+1}.

We give an analogous characterization of Room frames. Let $K(t_1^{u_1} \cdots t_k^{u_k})$ denote the complete multipartite graph having u_i parts of size t_i, $1 \leq i \leq k$. We shall refer to the parts of this graph as *holes*. A *holey one-factor* of $K(t_1^{u_1} \cdots t_k^{u_k})$ is a set of disjoint edges, f, that partitions the vertices not in some hole, which we call the hole *corresponding* to f. A *holey one-factorization* of $K(t_1^{u_1} \cdots t_k^{u_k})$ is a set of holey one-factors that partition the edge set. It is not difficult to prove that if $\mathscr{F}$ is a holey one-factorization of $K(t_1^{u_1} \cdots t_k^{u_k})$, then for every hole H, there are precisely $|H|$ holey one-factors corresponding to H.

Next, we define the idea of orthogonality. Two holey one-factors of $K(t_1^{u_1} \cdots t_k^{u_k})$ are *orthogonal* if they correspond to different holes and contain at most one common edge, or if they correspond to the same hole and are disjoint. Then, as before, a holey one-factor f is *orthogonal* to a holey one-factorization $\mathscr{F}$ if f is orthogonal to every holey one-factor in $\mathscr{F}$;

and two holey one-factorizations $\mathscr{F}_1$ and $\mathscr{F}_2$ are *orthogonal* if every holey one-factor in $\mathscr{F}_1$ is orthogonal to $\mathscr{F}_2$. Then we have the following character-ization of frames in terms of orthogonal holey one-factorizations. The proof is left to the reader.

THEOREM 2.3. *A Room frame of type* $t_1^{u_1} \cdots t_k^{u_k}$ *is equivalent to the exis-tence of two orthogonal holey one-factorizations of* $K(t_1^{u_1} \cdots t_k^{u_k})$.

In [4], a hill-climbing algorithm is presented for the construction of one-factorizations and orthogonal one-factorizations of K_n. It is not difficult to modify this algorithm to search for (orthogonal) holey one-factorizations of $K(t_1^{u_1} \cdots t_k^{u_k})$. Using the modified algorithm, we found Room frames of types $2^2 4^4$ (presented in Appendix 1 at the end of this paper) and $2^5 4^1$ (presented in Figure 1). However, after more than 30,000 attempts, we were unable to find Room frames of types $2^4 4^2$ and $2^3 4^3$ by this method.

We were able to handle the final two cases by augmenting the hill-climbing algorithm with a backtracking algorithm, as follows.

Algorithm.

(1) Construct a holey one-factorization $\mathscr{F}$ of the graph $K(t_1^{u_1} \cdots t_k^{u_k})$ using the hill-climbing algorithm.

(2) Construct a partial holey one-factorization $\mathscr{F}_1$ of $K(t_1^{u_1} \cdots t_k^{u_k})$ which is orthogonal to $\mathscr{F}$ using the hill-climbing algorithm.

(3) Construct the set $\mathscr{S}$ of all holey one-factors which are orthogonal to $\mathscr{F}$ and disjoint from every holey one-factor in $\mathscr{F}_1$.

(4) Using a backtracking algorithm, attempt to extend $\mathscr{F}_1$ to a holey one-factorization $\mathscr{F}_2$ by using holey one-factors in $\mathscr{S}$. (Note that $\mathscr{F}_2$ will then be orthogonal to $\mathscr{F}$.)

For the graph $K(2^3 4^3)$, we constructed an $\mathscr{F}_1$ (in step 2) consisting of 5 holey one-factors. Then $\mathscr{S}$ contained about 4500 holey one-factors. The exhaustive search in step 4 could then be performed in a reasonable amount of time. On the first four attempts, the algorithm failed. That is, there did not exist any extension of $\mathscr{F}_1$ to a holey one-factorization $\mathscr{F}_2$ orthogonal to $\mathscr{F}$. On the fifth attempt, the algorithm succeeded in finding the desired Room frame.

The case of the graph $K(2^4 4^2)$ was similar. Here, we were able to take $\mathscr{F}_1$ to be empty, and the resulting set $\mathscr{S}$ was still small enough to perform the exhaustive backtrack. We found the desired $\mathscr{F}_2$ on the seventh attempt. This means that there exist holey one-factorizations of $K(2^4 4^2)$ for which it is impossible to find an orthogonal holey one-factorization. The authors found this somewhat surprising.

These last two Room frames are presented in Appendix 1. Summarizing, we have the following.

THEOREM 2.4. *There exist Room frames of types* $2^6, 2^5 4^1, 2^4 4^2, 2^3 4^3,$ $2^2 4^4, 2^1 4^5,$ *and* 4^6.

3. Room squares with holes of size 3, 5, or 7.

In the following three theorems, we show that the necessary conditions for the existence of $(v, 3)$-, $(v, 5)$-, and $(v, 7)$-IRS are sufficient. We use these particular IRS for our main constructions in Section 4.

THEOREM 3.1. *A* $(v, 3)$-IRS *exists if and only if* v *is odd and* $v \geq 11$.

PROOF. It was shown in [7] that there exists a $(v, 3)$-IRS if v is odd, $v \geq 11$, and $v \neq 15, 17, 19, 21, 23, 25, 27, 29, 31, 37$, or 39. A $(39, 3)$-IRS follows from the existence of a $(13, 3)$-IRS and [7, Lemma 2.11]. For these 10 remaining values of v, we constructed $(v, 3)$-IRS using the hill-climbing algorithms described in [4]. For $v = 15, 17, 19$, and 21, the squares are presented in Appendix 2. The remaining squares are displayed in the research report [6].

THEOREM 3.2. *A* $(v, 5)$-IRS *exists if and only if* v *is odd and* $v \geq 17$.

PROOF. It was shown in [7] that there exists a $(v, 5)$-IRS if v is odd, $v \geq 17$, and $v \neq 19, 21, 23, 25, 27, 29, 31, 33, 35, 37, 39, 41, 43, 45, 47,$ $49, 57, 59, 61, 63$, or 67. For 19 of these remaining values of v (excluding $v = 57$ and $v = 63$), we constructed $(v, 5)$-IRS using the hill-climbing algorithms described in [4]. These squares are presented in the research report [5]. The existence of $(57,5)$- and $(63,5)$-IRS follow from the existence of $(19,5)$- and $(21,5)$-IRS, using [7, Lemma 2.11].

THEOREM 3.3. *A* $(v, 7)$-IRS *exists if and only if* v *is odd and* $v \geq 23$.

PROOF. It was shown in [7] that there exists a $(v, 7)$-IRS if v is odd, $v \geq 23$, and $v \neq 25, 27, 33, 41, 45, 51$, or 53. For these 7 remaining values of v, we constructed $(v, 7)$-IRS using the hill-climbing algorithms described in [4]. These squares are presented in the research report [5].

4. Main results.

In this section, we prove our main results using recursive constructions for Room frames. We first need to define some design-theoretic terminology. A *pairwise balanced design* (or, PBD) of index 1 is a pair $(X, \mathscr{A})$, such that X is a set of elements (called *points*) and $\mathscr{A}$ is a set of subsets of X (called *blocks*), each of cardinality at least two, such that every unordered pair of points is contained in a unique block of $\mathscr{A}$. If v is a positive integer and K is a set of positive integers, each of which is greater than or equal to 2, then we say that $(X, \mathscr{A})$ is a (v, K)-PBD if $|X| = v$, and $|A| \in K$ for every $A \in \mathscr{A}$. The integer v is called the *order* of the PBD.

Using this notation, a $(v, k, 1)$-BIBD (*balanced incomplete block design*) can be defined to be a $(v, \{k\})$-PBD.

A *parallel class* in a PBD is a set of blocks that partition the point set. A PBD is *resolvable* if the block set can be partitioned into parallel classes.

A *group-divisible design* (or GDD) is a triple $(X, \mathscr{G}, \mathscr{A})$, which satisfies the following properties:

(1) $\mathscr{G}$ is a partition of X into subsets called *groups*,

(2) $\mathscr{A}$ is a set of subsets of X (called *blocks*) such that a group and a block contain at most one common point, and

(3) every pair of points from distinct groups occurs in a unique block.

The *group-type* of a GDD $(X, \mathscr{G}, \mathscr{A})$ is the multiset $\{|G| : G \in \mathscr{G}\}$. As with Room frames, we shall use an exponential notation for group-types, for convenience. We will say that a GDD is a K-GDD if $|A| \in K$ for every $A \in \mathscr{A}$

A *transversal design* TD (k, m) can be defined to be a $\{k\}$-GDD of type m^k. It is well-known that a TD (k, m) is equivalent to $k - 2$ mutually orthogonal Latin squares of order m. For results on the existence of transversal designs we refer to [1].

The following is our main recursive construction for Room frames.

CONSTRUCTION 4.1. [**10**, Construction 2.2] *Let* $(X, \mathscr{G}, \mathscr{A})$ *be a GDD, and let* $w : X \to Z^+ \cup \{0\}$ *(we say that* w *is a weighting). For every* $A \in \mathscr{A}$, *suppose there is a Room frame of type* $\{w(x) : x \in A\}$. *Then there is a Room frame of type* $\{\sum_{x \in G} w(x) : G \in \mathscr{G}\}$.

Once we have constructed Room frames, we fill in the holes (see [**10**, Corollary 4.9]).

LEMMA 4.2. (i) *Suppose there is a Room frame of type* $t_1^{u_1} \dots t_k^{u_k}$. *Let* $a \geq 0$. *For* $1 \leq i \leq k$, *suppose there is a* $(t_i + a, a)$-IRS. *Then, for* $1 \leq i \leq k$, *there is a* $(\sum_{1 \leq i \leq k} t_i u_i + a, t_i + a)$-IRS, *and a* $(\sum_{1 \leq i \leq k} t_i u_i + a, a)$-IRS.
(ii) *Suppose there is a Room frame of type* $t_1^{u_1} \dots t_k^{u_k}$, *where* $u_1 = 1$. *Let* $a \geq 0$. *For* $2 \leq i \leq k$, *suppose there is a* $(t_i + a, a)$-IRS. *Then there is a* $(\sum_{1 \leq i \leq k} t_i u_i + a, t_1 + a)$-IRS.

We apply these constructions as follows.

THEOREM 4.3. *Suppose there is a resolvable* $(20m + 5, 5, 1)$-BIBD. *Let* $a = 1$ *or* 3; *let* $5m \leq t \leq 10m$; *and let* $20m + 5 \leq s \leq 40m + 10$. *Then there exists a* $(2s + 2t + a, 2t + a)$-IRS.

PROOF. Adjoin infinite points to all but one parallel class of the BIBD, giving rise to a $\{6\}$-GDD of group-type $5^{4m+1}(5m)^1$. Give the points in the group G of size $5m$ weights 2 and 4 in such a way that

$$\sum_{x \in G} w(x) = 2t,$$

and give the other points in the GDD weights 2 and 4 in such a way that

$$\sum_{x \notin G} w(x) = 2s.$$

Apply Construction 4.1. Now adjoin a infinite points to the Room frame, and apply Lemma 4.2 (ii). We fill in the appropriate $(v, 1)$-IRS ($v = 11, 13$, $15, 17, 19, 21$), if $a = 1$; or the appropriate $(v, 3)$-IRS ($v = 13$, $15, 17, 19, 21, 23$), if $a = 3$.

THEOREM 4.4. *For any odd* $w \geq 127$, *and for any odd* $v \geq 3w + 240$, *there is a* (v, w)-IRS.

PROOF. Write $w = 20m_0 + 3 - 2\alpha$, where $0 \leq \alpha \leq 9$. Now, there is an m, $m_0 \leq m \leq m_0 + 5$, such that a resolvable $(20m + 5, 5, 1)$-BIBD exists (see Table 1 of [13]). Apply Theorem 4.3 with $a = 3$. In order to construct a (v, w)-IRS, we need to have

$$10m + 3 \leq w \leq 20m + 3.$$

The second inequality is clearly true, and the first inequality will be true provided

$$10(m_0 + 5) + 3 \leq 20m_0 + 3 - 18,$$

or

$$10m_0 \geq 68.$$

But $w \geq 127$, so $m_0 \geq 7$. By Theorem 4.3, we can handle all odd v where

$$40m + 10 + w \leq v \leq 80m + 20 + w.$$

Now, we have

$$40m + 10 \leq 40m_0 + 210 \leq 2w + 240;$$

and

$$80m + 20 \geq 80m_0 + 20 \geq 4w + 8.$$

So, we can handle all odd v such that

$$3w + 240 \leq v \leq 5w + 8.$$

For odd $w \geq 127$ and odd $v \geq 5w + 2$, there is a (v, w)-IRS by [11, Theorem 3.13]. This completes the proof.

By similar methods, we can prove

THEOREM 4.5. *For any odd* $w \geq 23085$, *and for any odd* $v \geq 3w + 40$, *there is a* (v, w)-IRS.

PROOF. As before, write $w = 20m_0 + 3 - 2\alpha$, where $0 \leq \alpha \leq 9$. Since $w \geq 23085$, we have $m_0 \geq 1155$, and hence a resolvable $(20m_0 + 5, 5, 1)$-BIBD exists, by [13, Theorem 1.1]. Proceed as in Theorem 4.4

For some values of w, the existence of (v, w)-IRS is completely determined. For example, we have the following.

THEOREM 4.6. *For any* $w \geq 23103$ $w \equiv 3$ *modulo* 20, *and for any odd* $v \geq 3w + 2$, *there is a* (v, w)-IRS.

PROOF. Here, we can write $w = 20m_0 + 3$, where $m_0 \geq 1155$. Proceed as in Theorem 4.5. This constructs (v, w)-IRS for all odd $v \geq 3w + 4$. However, there is a $(3w + 2, w)$-IRS from [12].

THEOREM 4.7. *Suppose there is a* $TD(6, m)$, *and suppose there exists a* $(2r + a, a)$-*IRS for all* r *such that* $m \le r \le 2m$. *Let* $m \le t \le 2m$ *and let* $5m \le s \le 10m$. *Then there exists a* $(2s + 2t + a, 2t + a)$-*IRS*.

PROOF. Give the points in one group G of the TD weights 2 and 4 in such a way that

$$\sum_{x \in G} w(x) = 2t,$$

and give the other points in the TD weights 2 and 4 in such a way that

$$\sum_{x \notin G} w(x) = 2s.$$

Apply Construction 4.1. Now adjoin a infinite points to the Room frame, filling in $(2r + a, a)$-IRS for the required values of r (Lemma 4.2 ii)). This constructs the desired Room square.

COROLLARY 4.8. *Suppose there is a* $TD(6, m)$, *where* $m \ge 8$. *Let* $a = 1, 3, 5$ *or* 7; *let* $m \le t \le 2m$; *and let* $5m \le s \le 10m$. *Then there is a* $(2s + 2t + a, 2t + a)$-*IRS*.

PROOF. Apply Theorem 4.7 and Theorems 3.1–3.3.

THEOREM 4.9. *For any odd* $w \ge 37$, *and for any odd* $v \ge (7w - 5)/2$, *there is a* (v, w)-*IRS*.

PROOF. First, write $w = 4m_0 + a$, where m is odd and $a = 1, 3, 5$, or 7. Second, write $w = 2m_1 + a$, where m is odd and $a = 1$ or 3. We apply Corollary 4.8 for any odd value of m such that $m_0 \le m \le m_1$. Note that a $TD(6, m)$ exists for all odd $m > 5$; see [1]. This enables us to construct a (v, w)-IRS for all odd v where

$$10m_0 + w \le v \le 20m_1 + w.$$

Now, we have $m_0 \le (w - 1)/4$ and $m_1 \ge (w - 3)/2$. So, we can handle all odd v such that

$$(7w - 5)/2 \le v \le 11w - 30.$$

By [11,Corollary 3.7], there exists a (v, w)-IRS for all odd $v \ge 6w + 41$, and $11w - 30 \ge 6w + 41$ when $w \ge 15$. But $w \ge 37$, so we are done. This completes the proof.

5. Room squares with holes of size ≤ 35. We have already shown in Section 3 that for $w = 3, 5$, or 7, there is a (v, w)-IRS if and only if v is odd and $v \ge 3w + 2$. For $9 \le w \le 31$, we now updat bounds given in [11]. In Table 1, we list for each such w values v_0 such hat a (v, w)-IRS is known to exist for all $v \ge v_0$. The old bound is taken from [11] and the new bound is proved in this section.

TABLE 1

Bounds on the existence of Room squares with subsquares

w	old bound v_0	new bound v_0	w	old bound v_0	new bound v_0
1	7	7	3	41	11
5	69	17	7	55	23
9	85	33	11	87	43
13	89	63	15	77	65
17	71	67	19	97	69
21	137	71	23	117	73
25	127	95	27	137	97
29	195	99	31	197	101
33	135	103	35	177	115

THEOREM 5.1. *A* $(v, 9)$-IRS *exists if* $v = 29$, *or if* v *is odd and* $v \geq 33$.

PROOF. It was shown in [11] that there exists a $(v, 9)$-IRS if v is odd, $v \geq 85$. For $v = 37, 39, 45, 49, 57, 59, 63, 69, 73, 79$, and 81, the existence of $(v, 9)$-IRS follow easily from [7, Lemma 2.4 and 2.7–2.10]. For $v = 33, 35, 41, 43, 47, 51, 53, 55, 61, 65, 67, 71, 75, 77$, and 83, we constructed $(v, 9)$-IRS using the hill-climbing algorithms described in [4]. These squares are presented in the research report [5].

THEOREM 5.2. *A* $(v, 11)$-IRS *exists if* $v \geq 43$.

PROOF. It was shown in [11] that there exists a $(v, 11)$-IRS if v is odd, $v \geq 87$. For $61 \leq v \leq 85$, the existence of $(v, 11)$-IRS follows from Theorem 4.7 with $m = 5, t = 5, a = 1$. For $v = 43, 45, 51$, and 55, the existence of $(v, 11)$-IRS follow easily from [7, Lemmata 2.4, 2.5, 2.7 and 2.8]. For $v = 47, 49$, and 53, we constructed $(v, 11)$-IRS using the hill-climbing algorithms described in [4]. These squares are presented in the research report [5]. It remains to construct $(57, 11)$- and $(59, 11)$-IRS. These are done by means of a frame construction. To construct a $(57, 11)$-IRS, start with a TD$(6, 5)$ and give every point weight 2, except for 3 points in one group, which get weight 0. Apply Construction 4.1, using Room frames of types 2^5 and 2^6, which exist from Theorem 2.1. A Room frame of type $10^5 6^1$ results. Now Apply Lemma 4.2 i) with $a = 1$. To construct a $(59, 11)$-IRS, proceed similarly, but give only two points weight 0. We get a Room frame of type $10^5 8^1$, which we fill in as before.

We prove the remaining new bounds given in Table 1 using Theorem 4.7. We present several applications in Table 2.

TABLE 2

Applications of Theorem 4.7

w	m	t	a	interval covered
13	5	6	1	$63 \leq v \leq 113$
15	5	7	1	$65 \leq v \leq 115$
17	5	8	1	$67 \leq v \leq 117$
19	5	9	1	$69 \leq v \leq 119$
21	5	10	1	$71 \leq v \leq 121$
21	7	10	1	$91 \leq v \leq 161$
23	5	10	3	$73 \leq v \leq 123$
25	7	12	1	$95 \leq v \leq 165$
27	7	13	1	$97 \leq v \leq 167$
29	7	14	1	$99 \leq v \leq 169$
29	9	14	1	$119 \leq v \leq 209$
31	7	14	3	$101 \leq v \leq 171$
31	9	15	1	$121 \leq v \leq 211$
33	7	14	5	$103 \leq v \leq 173$
35	8	16	3	$115 \leq v \leq 195$

REFERENCES

1. Th. Beth, D. Jungnickel, and H. Lenz, *Design Theory*, Bibliographisches Institut, Zurich, 1985.
2. R. J. Collens and R. C. Mullin, *Some properties of Room squares—a computer search*, Proc. First Louisiana Conf. on Combin., Graph Theory and Comput., Baton Rouge (1970), 87–111.
3. J. H. Dinitz and D. R. Stinson, *Further results on frames*, Ars Combinatoria **11** (1981), 275–288.
4. _____, *A hill-climbing algorithm for the construction of one-factorizations and Room squares*, SIAM J. Algebraic and Discrete Methods **8** (1987), 430–438.
5. _____, *Some small Room squares with holes of sizes 5, 7, 9, and 11*, Research report 87–19, Dept. of Mathematics, University of Vermont.
6. _____, *On the existence of Room squares with subsquares*, Research report 87–22, Dept. of Mathematics, University of Vermont.
7. J. H. Dinitz, D. R. Stinson and W. D. Wallis, *Room squares with holes of size 3, 5, and 7*, Discrete Math. **47** (1983), 221–228.
8. J. D. Horton, *Room squares and one-factorizations*, Aequationes Math. **22** (1981), 56-63.
9. R. C. Mullin and W. D. Wallis, *The existence of Room squares*, Aequationes Math. **13** (1975), 1–7.
10. D. R. Stinson, *Some constructions for frames, Room squares, and subsquares*, Ars Combinatoria **12** (1981), 229–267.
11. _____, *Room squres and subsquares*, Lecture Notes in Math **1036** (1983), 86–95.
12. W. D. Wallis, *All Room squares have minimal supersquares*, Cong. Numer. **36** (1982), 3–14.
13. Zhu Lie, Chen Demeng, and Du Beiliang, *On the existence of (v, 5, 1) resolvable BIBD*, J. Suzlon Univ. **3** (1987), 115–129.

Appendix 1

								12, 13	7, 11	8, 9	14, 15	5, 16			6, 10
	1 – 4			10, 16				5, 15		8, 13			6, 12	7, 14	9, 11
				11, 14				8, 16	10, 15		6, 9			5, 13	7, 12
			9, 12		10, 13			7, 15		5, 14	6, 16			8, 11	
12, 16			10, 14							4, 9	3, 15	1, 11			2, 13
	12, 15				5 – 8			2, 14	3, 13		11, 16	4, 10	1, 9		
11, 15	9, 14							1, 13	2, 16	3, 10				4, 12	
		13, 16						4, 11	3, 12			9, 15	2, 10		1, 14
7, 13		5, 11		3, 16		12, 14	4, 15		9 – 10	1, 6		2, 8			
	7, 16	6, 13	8, 15	4, 14	1, 12	3, 11				2, 5					
5, 9		7, 10		13, 15		4, 16	3, 14	2, 6		11 – 12		1, 8			
	5, 10			1, 15	9, 13	14, 16						2, 7		3, 6	4, 8
	8, 12	9, 16		2, 15		10, 11		4, 6		1, 7		13 – 14			3, 5
8, 10	6, 15		5, 12	2, 11	3, 9		1, 16			4, 7					
6, 14	11, 13		7, 9	1, 10		2, 12	3, 8					4, 5		15 – 16	
	8, 14	6, 11		4, 13	2, 9		1, 5					10, 12	3, 7		

A Room frame of type $2^4 4^2$

				11, 17	9, 15	13, 16	8, 18		5, 14			6, 12		7, 10			
	1 – 4						6, 13	16, 18		7, 15	9, 17	5, 12	11, 14				8, 10
				9, 18			12, 17		14, 16		7, 13	8, 11	5, 15	6, 10			
				12, 16	10, 13		7, 14		15, 18						6, 17	8, 9	5, 11
		11, 16						15, 17				3, 10		1, 9	13, 18	2, 14	4, 12
	14, 17				5 – 8			3, 16			2, 12	1, 10	11, 18	4, 9			13, 15
14, 18	9, 13	10, 15							4, 17		2, 11				3, 12		1, 16
	11, 15	12, 13						2, 17		1, 14		9, 16	3, 18				4, 10
5, 13		8, 14		1, 17		2, 16						4, 18				6, 15	3, 7
6, 18	16, 17	8, 15		3, 14	4, 13				9 – 12						2, 5	1, 7	
6, 14		3, 13		1, 18	4, 15							5, 17	8, 16	2, 7			
7, 16	5, 18	13, 17		1, 15								4, 14	3, 8				2, 6
8, 17		7, 9	10, 18		3, 11		4, 6		2, 15			13 – 14		1, 12	5, 16		
12, 15	5, 10			2, 9		1, 11	7, 18	4, 8	6, 16						3, 17		
6, 9			4, 11	12, 18		10, 14	2, 13	1, 8				7, 17	3, 5	15 – 16			
	8, 12				10, 17	2, 18	1, 5		3, 6			4, 7				11, 13	9, 14
7, 11		5, 9	2, 10	4, 16	12, 14			3, 15				1, 6		8, 13		17 – 18	
10, 16		6, 11	7, 12	14, 15			3, 9		4, 5	1, 13	2, 8						

A Room frame of type $2^3 4^3$

1	2	3	4	5	6	7	8	9	10	11	12	13	14	15	16	17	18	19	20
1 – 2		12, 13		10, 16						8, 17	9, 18	5, 19	11, 20		6, 14	4, 7			3, 15
					13, 20	10, 17		14, 19			4, 18	6, 9	3, 8		7, 16	5, 12	11, 15		
		3 – 4		11, 17		9, 19		6, 13	7, 18	16, 20		8, 12			2, 10	1, 15			5, 14
				11, 18	10, 15	1, 13	8, 14		16, 19	2, 5	6, 20			12, 17			7, 9		
13, 18		2, 16		5 – 8				15, 17	3, 20		4, 12	1, 9		11, 19			10, 14		
12, 18		11, 14						3, 19		2, 15		10, 20	1, 17			9, 16			4, 13
14, 17		3, 16								4, 20			12, 19	2, 18	9, 15	11, 13			1, 10
12, 20		9, 13	10, 18						2, 17		15, 19	1, 11				3, 14	4, 16		
8, 15	13, 19	1, 6		2, 14		16, 18		9 – 12				3, 17		4, 5	7, 20				
6, 19		8, 20	1, 7	15, 18		16, 17	4, 14						2, 3		5, 13				
	14, 20		5, 16	1, 3	4, 15							7, 19	6, 17	8, 18			2, 13		
	14, 18				13, 17	1, 19	15, 20						2, 7		4, 8		3, 5	6, 16	
4, 10	6, 11		9, 17		2, 19	3, 18	5, 20					13 – 16						1, 8	7, 12
	7, 10		8, 19		1, 20	4, 11	3, 12	6, 18		5, 17									2, 9
3, 7	4, 17	10, 19		9, 20				5, 18								1, 12	2, 6		8, 11
5, 9		7, 17	2, 20	4, 19							1, 18					3, 11	8, 10	6, 12	
11, 16			6, 15		3, 9	2, 12			1, 4	8, 13	7, 14	5, 10				17 – 20			
	5, 15			10, 13	12, 14			1, 16		3, 6		2, 8	7, 11		4, 9				
	8, 9	5, 11		12, 16				2, 4	7, 15	1, 14	3, 13				6, 10				
		12, 15				9, 14	2, 11	7, 13	8, 16	4, 6				3, 10	1, 5				

A Room frame of type $2^2 4^4$

J. H. DINITZ AND D. R. STINSON

Appendix 2

							14, 12	5, 9			6, 10	16, 15	8, 11		7, 13
				9, 10	12, 7	14, 16						5, 11	6, 13	8, 15	
				11, 7			8, 13		5, 6			14, 10	12, 15	9, 16	
15, 9	10, 11	5, 14		16, 12			7, 1				8, 2	6, 4		3, 13	
6, 16							12, 9	10, 15	13, 2	14, 11	5, 4	1, 8	7, 3		
	12, 5	9, 6		13, 14	15, 3		16, 4			10, 8		2, 7			11, 1
	7, 15		6, 12		2, 10	5, 8	9, 11	1, 16	13, 4	14, 3					
		16, 3	15, 2				10, 4	9, 7				13, 12	5, 1	6, 11	8, 14
			8, 4	14, 6	1, 13						15, 11	9, 3	16, 10	7, 5	12, 2
11, 13		12, 10	15, 5	6, 1				8, 3		16, 7				14, 2	9, 4
5, 10		7, 8	14, 1		15, 4	6, 3	11, 12	16, 2	9, 13						
14, 7		13, 15	8, 9		4, 11	6, 2		12, 3						10, 1	16, 5
	9, 14		13, 10		16, 8	11, 3	5, 2	6, 7	1, 15					12, 4	
8, 12	16, 13		11, 2	5, 3	1, 9	7, 10							14, 4		6, 15
	6, 8	16, 11	7, 4		13, 5				15, 14		12, 1		9, 2		10, 3

A (15, 3)-incomplete Room square

			5, 15				18, 17	7, 11	8, 10	12, 9	14, 13			6, 16		
		10, 14		11, 15		12, 6		13, 16				9, 17	18, 5	7, 8		
		17, 7			11, 13		12, 16	5, 14	8, 18	6, 9	10, 15					
	6, 14	15, 7		16, 4	8, 1	9, 5	12, 10			13, 17					11, 2	3, 18
12, 14	8, 17		13, 15	18, 1	6, 10		5, 2		11, 4			3, 7		16, 9		
		16, 2			13, 18	10, 3	15, 9	8, 5				11, 1	4, 12	7, 6		17, 14
6, 11	13, 9		12, 3	16, 18	10, 7	17, 2						1, 5			15, 14	8, 4
	15, 18	13, 12	6, 8		9, 3	5, 4		1, 14					11, 16	10, 17		2, 7
5, 17		8, 9		1, 6		15, 12	14, 18	16, 7				10, 2			3, 11	13, 4
	10, 16		9, 4	14, 11	6, 17			13, 3	18, 2			5, 7	8, 15	12, 1		
	16, 14		7, 9		1, 17	8, 11			18, 12	6, 2				4, 15	3, 5	13, 10
9, 10		18, 11	4, 17	12, 8		2, 15		13, 6	1, 7			3, 14				5, 16
	17, 11		8, 3	2, 14			4, 10			16, 1	12, 5	7, 13			9, 18	15, 6
18, 7			5, 10		9, 11	16, 8		15, 1	6, 3	14, 4			2, 13			17, 12
	7, 12	5, 6	13, 1		3, 16			9, 14	17, 15	10, 11	18, 4	2, 8				
16, 15				5, 13		7, 4		17, 3	2, 9				18, 6	8, 14	1, 10	11, 12
8, 13	11, 5	10, 18		2, 12		7, 14		6, 4			3, 15	17, 16				9, 1

A (17, 3)-incomplete Room square

			20, 15	13, 5	8, 12	18, 16			7, 6			11, 10	17, 9			14, 19		
					14, 20	19, 10			8, 13		12, 7				18, 11	17, 5	9, 15	6, 16
		5, 16	9, 7			14, 8	12, 6		11, 15				20,13				10, 17	18, 19
			15, 8	17, 16		18, 5		20, 3		13, 19	9, 1			2, 14	4, 6	11, 12	7, 10	
	17, 7		14, 10		3, 6	2, 8	1, 20		15, 19		11, 5	4, 16			16, 12			13, 9
13, 15		3, 9	12, 16	11, 6				17, 2	14, 7	10, 18		4, 19					8, 20	5, 1
	5, 15	20, 4		1, 14	2, 19		9, 11	18, 17		10, 3		8, 6		12, 13		16, 7		
6, 20	12, 14	16, 8		19, 11	2, 10		1, 17				4, 9	13, 18		5, 3	15, 7			
		12, 10	3, 8	18, 4			16, 9		17, 13	6, 15	19, 5		1, 7	11, 14				20, 2
5, 14	9, 10	13, 11		3, 16	15, 18		7, 19	8, 4	12, 1		20, 17					2, 6		
12, 9		19, 8	18, 6		11, 1	7, 3	10, 4	14, 13		20, 5	16, 2							15, 17
8, 10		7, 11	13, 1		5, 9		17, 6	15, 2		19, 16		12, 20	3, 18					4, 14
16, 11	20, 18		17, 14		19, 9		6, 5	12, 2					7, 8	15, 4	10, 1	13, 3		
7, 18	8, 5		13, 6		17, 12			20, 10	4, 11	2, 9		15, 14	16, 1				19, 3	
	6, 19		15, 1	17, 4		13, 10	14, 9		7, 5		8, 18			2, 11		16, 20		12, 3
		19, 12		2, 5		7, 4	11, 20	13, 16		9, 18	6, 1	3, 14		10, 15	8, 17			
	7, 13	9, 20	18, 2			12, 15		1, 19			17, 3	14, 6	10, 16		5, 4	11, 8		
19, 17	16, 15	14, 18		20, 7			11, 3	10, 5	4, 12	8, 1			6, 9	13, 2				
	17, 11	10, 6			13, 4	15, 3		14, 16			7, 2	12, 5	20, 19	8, 9	1, 18			

A (19, 3)-incomplete Room square

			18,15	21,5		9,12	14,6			17,10				22,7		16,13	8,20	11,19	
				18,14						19,16	21,10	22,11	15,20	5,6		13,7	12,8	17,9	
					6,21	7,15	19,10		17,22		5,14	12,13	20,16		8,9	11,18			
	6,19						1,16	4,9	20,13		11,15	17,18			22,8	10,2	21,12	7,5	3,14
	11,14	13,18	8,1			7,12	17,19	22,2		21,3					16,10		5,9	6,15	4,20
10,22		9,19			5,4	12,20	13,17	1,15			16,13	7,6	21,14	11,2					8,18
	17,21	22,12		20,18	2,13		11,1	14,4						10,6	15,5	16,8	3,9	7,19	
	13,6	14,15	7,21		3,22		19,5	18,16		12,2	10,9		4,8	17,1				11,20	
	5,20		10,3	15,16		11,8	18,12			1,9	2,7			19,4		14,17	22,6	21,13	
17,7	12,19		4,6		5,1			14,16	9,2	8,13	22,20	21,11					18,3		10,15
						20,1	18,9	21,15	3,7		6,17	19,22	8,2	5,10	13,14	12,4			11,16
		5,16	17,11	4,22	21,8				10,12	9,6	13,3	14,20		7,1	19,15			2,18	
		11,10	5,3	2,19		15,13			14,1	7,8		18,4			17,12	20,9	16,6		22,21
12,11	8,10		9,13		3,6	5,22			20,7		15,4	19,1		21,18	14,2			16,17	
12,16	7,18	20,2	8,14	9,11								17,5	12,6	15,22		19,3		10,4	13,1
14,9	22,16		12,1		13,11			7,10	6,8	5,2	19,18	15,3				4,21		17,20	
	15,9	8,17	13,22		16,2		4,7	20,6	21,19			12,14	3,11			18,5		10,1	
19,20		10,14	17,15				21,2	3,8		18,22		1,6		16,9		7,11		4,13	15,12
5,13			11,6	19,8	10,18				12,15	16,4		9,21	17,2	20,3			7,14	22,1	
8,15		21,20		16,7	17,4			5,11			13,10	1,18			12,3	9,22		19,14	6,2
18,6		9,7	12,16		20,10	22,14	17,3			11,4		8,5			19,13		21,1	15,2	

A (21, 3)-incomplete Room square

Department of Mathematics, University of Vermont, Burlington, Vermont 05405

Department of Computer Science, University of Manitoba, Winnipeg, Manitoba, Canada

Current address: Department of Computer Science and Engineering, Ferguson Hall, University of Nebraska, Lincoln, Nebraska 68588

Contemporary Mathematics
Volume **111**, 1990

A Bound for Blocking Sets in Finite Projective Planes

DAVID A. DRAKE

Introduction. A point set B of a projective plane Π is called a *blocking set* of Π if $B \cap L$ is a proper subset of L for every line L of Π. Suppose that Π has finite order $n = m^2 - t$ where $0 \leq t \leq 2m - 2$. A. Bruen proved [3], [4] that every blocking set B of Π has cardinality $n + m + 1$ or more; in case $t = 0$, he also proved that the blocking sets of cardinality $n + m + 1$ are just the Baer subplanes of Π. Under the assumptions $t \neq 0$, $n \neq 3, 5$, J. Bierbrauer [1] raised the Bruen bound to $n + m + 2$.

Let B be a blocking set of cardinality v in a projective plane Π of order n, B^C be the set of points of Π not in B. Call a line L of Π an *i-line* if $|B \cap L| = i$, and write b_i to denote the number of i-lines. If P is a point of B^C, then no one of the $n + 1$ lines through P can contain more than $v - n$ points of B; and no two can contain more than $v - n + 1$. Thus

(1) $b_i = 0$ for $i > v - n$; and

(2) if $i + j \geq v - n + 2$, every i-line meets every j-line in a point of B.

Following Bruen and Silverman [5], we say that B is of *Rédei type* if $b_{v-n} \neq 0$. In this note we prove

THEOREM 1. *Let B be a blocking set of cardinality $n + m + 2$ in a projective plane of order $n = m^2 - t$. Then B is of Rédei type if any one of the following conditions holds*:

    (i) $t = 1$, $m \geq 4$;

    (ii) $t = 2$, $m = 4$ *or* $m \geq 6$;

    (iii) $t = 3$, $m = 6$ *or* 8 *or* $m \geq 10$;

    (iv) $t = 4$, $m = 14$ *or* $m \geq 16$.

C. Kitto [7] has proved that there are no blocking sets of Rédei type of cardinality $n + m + 2$ in planes of order $n = m^2 - t$ if $1 \leq t \leq 2m - 2$ and $m \geq 7$. Thus one obtains:

1980 *Mathematics Subject Classification* (1985 *Revision*). Primary 05B25.

This paper is in final form and no version of it will be submitted for publication elsewhere.

COROLLARY 2. *In every projective plane of order* $n = m^2 - t$, *every blocking set has cardinality at least* $n + m + 3$ *if any of the following conditions holds*: $t = 1$ *or* 2 *and* $m \geq 7$; $t = 3$ *and* $m = 8$ *or* $m \geq 10$; $t = 4$ *and* $m = 14$ *or* $m \geq 16$.

We obtain Theorem 1 as a corollary to the following result.

PROPOSITION 3. *Let* Π *be a projective plane of order* $n = m^2 - t$, $m \geq 4$, $t \geq 1$. *Let* B *be a blocking set of* Π *which is not of Rédei type and whose cardinality is* $n + m + 2$. *If* $t < m(1 - \frac{1}{\sqrt{2}})$ *or if* m *is even and* $t \leq m/2$, *then* $b_i = 0$ *for all* i *with*

$$3 \leq i < m - \frac{(t-3)}{2} + \frac{t(t+3)}{2m(m-1)} - \frac{(2t-1)}{(m-1)} .$$

Corollary 2 is a small improvement on the known bound for blocking sets in arbitrary projective planes. Blokhuis and Brouwer [2] have obtained the better bound $n + \sqrt{2n} + 1$ for Desarguesian planes of odd non-square order $n \geq 7$, $n \neq 27$. Raising the bound for arbitrary planes appears to be more difficult. In [6], for example, Drake and Ho consider certain blocking sets of size 22 in a putative projective plane of order 15.

Main section. Throughout this section, Π denotes a projective plane of order $n = m^2 - t$, $m \geq 4$, $t \geq 1$; and B denotes a blocking set of Π which is not of Rédei type. It is assumed that $|B| = v = m^2 + m + 2 - t$. Counting lines of Π, flags of Π whose points lie in B, and pairs of points of B, one obtains the following three equations.

$$(3) \qquad \sum_{i=1}^{m+1} b_i = n^2 + n + 1.$$

$$(4) \qquad \sum_{i=1}^{m+1} i b_i = v(n+1).$$

$$(5) \qquad \sum_{i=1}^{m+1} i(i-1) b_i = v(v-1).$$

One multiplies equation (3) by 2, adds equation (5) and subtracts twice equation (4) to obtain

$$(6) \qquad \sum_{i=1}^{m+1} (i-1)(i-2) b_i = m^4 - 2tm^2 + m + t^2 + t.$$

The combination $3(3) + (5) - 3(4)$ yields

$$(7) \quad \sum_{i=1}^{m+1} (i-1)(i-3) b_i = m^4 - m^3 - 2m^2 - 2tm^2 + tm + t^2 + 3t - 1.$$

By (2), none of the $n^2 + n + 1 - v$ points of B^C lies in more than one line which intersects B in $(m + 4)/2$ or more points. Thus

$$(8) \qquad \sum_{(m+4)/2}^{m+1} (m^2 + 1 - t - i) b_i \leq m^4 - 2tm^2 - m + t^2 - 1,$$

where the sum begins with the smallest integer value of i which is at least $(m + 4)/2$.

Next, we derive an upper bound for b_{m+1}. Thus, assume $b_{m+1} > 0$. Let P be any point of B^C which lies on an $(m + 1)$-line. The remaining $m^2 - t$ lines through P join P to the remaining $m^2 - t + 1$ lines of B, so P lies on a 2-line. Thus

$$(m^2 - m - t)b_{m+1} \leq (m^2 - 1 - t)b_2.$$

Into this inequality, substitute the value of b_2 given by equation (7) to obtain

$$(9) \qquad (-m^2 + m + t)\, b_{m+1} + (m^2 - 1 - t) \sum_{i=4}^{m+1}(i - 1)(i - 3)b_i \geq$$
$$m^6 - m^5 - (3t + 3)m^4 + (2t + 1)m^3 + (3t^2 + 7t + 1)m^2$$
$$-(t^2 + t)m - (t^3 + 4t^2 + 2t - 1).$$

(10) We claim that $b_{m+1} \leq jm$ if $b_j \neq 0$ for some $j \geq 3$.

To prove (10), let L be a j-line. By (2), every $(m + 1)$-line meets L in one of the j points of $L \cap B$. As $|B \setminus L| < m^2 + m$, at most m of the $(m + 1)$-lines contain any given point of $L \cap B$. Next, we claim that

$$(11) \qquad \sum_{(m+4)/2}^{m+1} b_i \leq m^2 + m - t,$$

if $m \geq 2t$ and if there is a j-line for some j with $3 \leq j \leq m$.

To prove (11), apply (8) and (10). One obtains

$$(m^2 + 1 - t - m) \sum_{(m+4)/2}^{m+1} b_i \leq jm + (m^4 - 2tm^2 - m + t^2 - 1).$$

Since the b_i are integers, the preceding inequality yields (11).

A number of long and tedious computations are now required to complete the proofs of the main results. These computations come in pairs: one computation in each pair treats all values of m; the other, only even values. The untrusting reader may wish to repeat the computations. He or she can reduce the work of such a repetition by omitting the treatment of the even-value case. The general argument suffices to prove the conclusion of Theorem 1 for all but the pairs $(t, m) = (2,4), (2,6), (3,6), (3,8), (3,10)$ and the conclusion of Corollary 2 for all but the pairs $(3,8)$ and $(3,10)$.

(12) LEMMA. *Let* $m \geq 2t$, $b_j \neq 0$ *for some* j *with* $3 \leq j \leq m$. *Then the left hand side of inequality* (9) *is at most* $U + U_o$ *if* m *is odd and at most* $U + U_e$ *if* m *is even, where*

$$U = jm[(-m^2 + m + t) + (m^2 - 1 - t)(m^2 - 2m)]$$
$$+ (m^2 + m - t - jm)(m^2 - 1 - t)(m - 1)(m - 3),$$
$$V = (m^2 - 1 - t)(2m^3 - 2jm^2 - tm^2 + m^2 + 2jm - 3tm - m + t^2 + 3t),$$
$$U_o = V(m - 3)/(m - 1), \qquad U_e = V(m - 4)/(m - 2).$$

PROOF. Let $b_1, b_2, \ldots, b_{m+1}$ be any non-negative real numbers which satisfy equation (6); and suppose that $b_\ell \geq d > 0$ for some ℓ with $2 \leq \ell \leq m$. If b_ℓ is reduced by d and $b_{\ell+1}$ is increased by $d(\ell - 2)/\ell$, then a new sequence of non-negative b_i's is produced which satisfies equation (6). Computations reveal that no such change decreases the left hand side of inequality (9). Application of (10) and (11) now shows that the left hand side of (9) attains its maximum value when $b_{m+1} = jm$, $b_m = m^2 + m - t - jm$ and all other b_i are zero except for b_k where $k = (m + 3)/2$ if m is odd and $k = (m + 2)/2$ if m is even. The values of b_k are now determined by equation (6). Insertion of the values of the b_i's into the left hand side of (9) completes the proof.

An onerous calculation using (9) and (12) yields

(13) LEMMA. *Let* $m \geq 2t$, $m \geq 4$. *If* $b_j \neq 0$ *for some* j *with* $3 \leq j \leq m$, *then*

(13a) $j(2m^4 - m^3 - 4m^2 + 3m - 2tm^2 + 2tm) \geq$
$$2m^5 - 2tm^4 + 2m^4 - 5tm^3 - 3m^3 + 4t^2m^2 + 3tm^2$$
$$- m^2 + 3t^2m + 4tm + m - 2t^3 - 5t^2 - 4t - 1.$$

(13b) If m is even, then also

$$j(2m^4 + m^3 - 5m^2 + 2m - 2tm^2) \geq$$
$$2m^5 - 2tm^4 + 4m^4 - 5tm^3 - m^3 + 4t^2m^2 - 2tm^2$$
$$- 2m^2 + 3t^2m - 18tm - m - 2t^3 - 2t^2 - 2t - 2.$$

(14) LEMMA. *Let* $m \geq 4$. *Suppose either that* $t < m(1 - 1/\sqrt{2})$ *or that* m *is even and* $t \leq m/2$. *Then* $b_i = 0$ *for* $3 \leq i \leq m - t + 1$.

PROOF. Assume first that $t < m(1 - 1/\sqrt{2})$. Since $4 \leq m$, the left hand side of (13a) is positive; and thus, for each pair (m, t), it is an increasing function of j. Hence, it suffices to verify that (13a) does not hold with $j = m - t + 1$; i.e., to verify that

$$0 < f_m(t) := -2t^3 + (2m^2 + 5m - 5)t^2 - (4m^3 + m^2 - 5m + 4)t$$

(15) $$+ (m^4 + 2m^3 - 2m - 1).$$

If $m \geq 4$, computation with the derivative $f'_m(t)$ shows that f_m is a decreasing function for $t \leq m - 1$ and that $f_m(t) > 0$ for $t = m(1 - 1/\sqrt{2})$. Thus, (15) holds for all specified pairs (m, t). The similar proof for even values of m utilizes (13b) in place of (13a).

PROOF OF PROPOSITION 3. Let j be the smallest integer i greater than 2 with $b_i \neq 0$. Since $|B| > n + 2$, B is not an arc; so j is defined. Conditions (6) and (10) yield

$$(16) \qquad (m^2 - 3m + 2) \sum_{i=j}^{m+1} b_i + (2m - 2)mj \geq m^4 - 2tm^2 + m + t^2 + t;$$

By (14), $j \geq m - t + 2 \geq (m + 4)/2$. Either $j = m + 1$, and we are done; or (11) implies

$$(17) \qquad\qquad\qquad m^2 + m - t \geq \sum_{i=j}^{m+1} b_i.$$

Application of (16) and (17) finishes the proof of Proposition 3.

PROOF OF THEOREM 1. Assume, by way of contradiction, that B is not of Rédei type. If $t = 1$ or 2, Lemma (14) implies that $b_i = 0$ for all i with $3 \leq i \leq m - 1$. The same conclusion follows from Proposition 3 if $t = 3$ or 4. Thus, equation (6) implies that $m - 1$ divides $m^4 - 2tm^2 + m + t^2 + t$ and, hence, that $m - 1$ divides $t^2 - t + 2$. The latter divisibility requirement fails for all pairs (m, t) listed in Theorem 1.

REFERENCES

1. J. Bierbrauer, *On minimal blocking sets*, Archiv der Math. **35** (1980), 394–400.
2. A. Blokhuis and A. E. Brouwer, *Blocking sets in Desarguesian projective planes*, Bull. London Math. Soc. **18** (1986), 132–134.
3. A. A. Bruen, *Baer subplanes and blocking sets*, Bull. Amer. Math. Soc. **76** (1970), 342–344.
4. _____, *Blocking sets in finite projective planes*, SIAM J. Appl. Math. **21** (1971), 380–392.
5. A. A. Bruen and R. Silverman, *Arcs and blocking sets*, Finite Geometries and Designs, London Math. Soc. Lecture Note Series, eds. P. J. Cameron, J. W. P. Hirschfeld, and D. R. Hughes, **49**, 1981, pp. 52–60.
6. D. A. Drake and C. Y. Ho, *Projective extensions of Kirkman systems as substructures of projective planes*, J. Combinatorial Theory, series A **48** (1988), 197–208.
7. C. L. Kitto, III, *A bound for blocking sets of maximal type in finite projective planes*, Arch. Math. **52** (1989), 203–208.

DEPARTMENT OF MATHEMATICS, UNIVERSITY OF FLORIDA, GAINESVILLE, FLORIDA 32611

Contemporary Mathematics
Volume **111**, 1990

Sets With More Than One Representation as an Algebraic Curve of Degree Three

J. W. P. HIRSCHFELD AND J. A. THAS

1. The problem. A general problem in the geometry of finite spaces is to describe or characterize subsets with given properties. In particular, in a finite plane, it is sometimes desirable to characterize a set as the zeros of an algebraic curve. This leads to the question of deciding whether, given a set of points that is the set of zeros of an algebraic curve, the curve is unique. Let us put this in a precise way.

In PG(2, q), *the projective plane over the field* GF(q), *find all sets* $\mathscr{K}$ *for which there exist two forms* F *and* G *of the same degree* d *over* GF(q), *where* F *and* G *are absolutely irreducible* ([**3**], p.50) *and* $F \neq \lambda G$, *such that* $\mathscr{K}$ *is precisely the set of zeros of both* F *and* G.

Let $|\mathscr{K}| = k$. Then to avoid trivial cases, assume that k is big enough to define a curve of degree d; so, by [**2**], §10.2,

$$(1.1) \qquad\qquad k \geq \frac{1}{2}d(d+3).$$

However, if F and G have no non-trivial common factor, then Bézout's theorem ([**2**], §10.3) gives that

$$(1.2) \qquad\qquad k \leq d^2.$$

As a start to the problem, we consider the smallest d for which both (1.1) and (1.2) hold, namely $d = 3$, in which case $k = 9$ and there is equality in both (1.1) and (1.2).

To proceed further, it is necessary to be precise about the nature of a curve. Given a homogeneous non-constant ternary form F over GF(q), without multiple factors, define the *curve* $\mathscr{F}$ to be the pair (V(F), (F)); here V(F) is the set of points in PG(2, q) which are zeros of F and (F) is the ideal of all multiples of F in the polynomial ring GF(q) [X, Y, Z]. The *order* of

---

1980 *Mathematics Subject Classification* (1985 *Revision*). Primary 11G20.
This paper is in final form and no version of it will be submitted for publication elsewhere.

the curve is the degree of the polynomial F; a curve of order three is a *cubic* curve. If $\mathcal{F} = (\mathcal{K}, (F))$ and $\mathcal{G} = (\mathcal{K}, (G))$ are distinct cubic curves, with F and G absolutely irreducible, and each with $k = |\mathcal{K}| = 9$, then neither can be singular, as any singularity belongs to $\mathcal{K}$ ([3], p.254) and at a singularity the intersection multiplicity would be at least two; thus Bézout's theorem would be contradicted. The problem now becomes the following:

In PG$(2, q)$, *find all sets $\mathcal{K}$ of 9 points such that $\mathcal{K}$ is precisely the set of points of two distinct elliptic cubics.*

2. Properties of elliptic cubics. Consider $\mathcal{F} = (\mathcal{K}, (F))$, where $\mathcal{K} = V(F)$, deg $F = 3$ and F is elliptic. Some properties of elliptic cubics are listed.

2.1. If $P \in \mathcal{K}$ the tangent at P to $\mathcal{F}$ meets $\mathcal{K}$ again at P', the *tangential* of P. If $P' = P$ then P is an *inflexion*.

2.2. $\mathcal{K}$ is an abelian group under the following operation ([2],§10.6). Choose a point O in $\mathcal{K}$. For P, Q in $\mathcal{K}$, let R be the third point in $\mathcal{K}$ of the line PQ; then $P + Q$ is the third point in $\mathcal{K}$ of OR. The special cases are natural:

(a) if $PQ(P \neq Q)$ is the tangent to $\mathcal{F}$ at P, then $R = P$;
(b) if $P = Q$, then R is the tangential of P;
(c) if $OR(O \neq R)$ is the tangent at R, then $P + Q = R$;
(d) if $OR(O \neq R)$ is the tangent at O, then $P + Q = O$.

2.3. Three points P_1, P_2, P_3 are collinear if and only if $P_1 + P_2 + P_3 = O'$, where O' is the tangential of O. So, if O is an inflexion, then P_1, P_2, P_3 are collinear if and only if $P_1 + P_2 + P_3 = O$.

2.4. Let $\mathcal{K}'$ be the set of tangentials of $\mathcal{F}$ and suppose O is an inflexion. Then $\mathcal{K}'$ is a subgroup of $\mathcal{K}$; for, if $2P + P' = O$ and $2Q + Q' = O$, then $2(P - Q) + (P' - Q') = O$.

2.5. With O an inflexion, the map $\phi : \mathcal{K} \to \mathcal{K}'$ given by $\phi(P) = P'$ is a homomorphism, whose kernel is the set of points with tangential O. Hence the inverse image of every point P' in $\mathcal{K}'$ has the same size $t = |\ker \phi|$; that is, the number of points with a given tangential P' is t. Since ϕ is surjective,

$$|\mathcal{K}| = t|\mathcal{K}'|.$$

2.6. Still with O an inflexion, $2P \in \mathcal{K}'$ for every P in $\mathcal{K}$, since $2(-P) + 2P = O$. Hence every non-identity element of $\mathcal{K}/\mathcal{K}'$ has order 2. So $\mathcal{K}/\mathcal{K}'$ is elementary abelian and $t = 2^s$ for some $s \geq 0$.

2.7. The conclusion $|\mathcal{K}| = t|\mathcal{K}'|$ with $t = 2^s$ holds also when $\mathcal{F}$ has no inflexion in PG$(2, q)$, [9], [10]. However, to prove this, a different structure than the group structure is required on $\mathcal{K}$.

2.8. When $|\mathcal{K}| = 9$, then it follows that $t = 1$ and $\mathcal{K}' = \mathcal{K}$.

2.9. Let O be an inflexion. The other inflexions are the elements of order 3 of the group $\mathcal{K}$. So the inflexions form an elementary abelian group of order 3^s. So, if n is the number of inflexions, then $n = 0, 1, 3$ or 9. Let $n \neq 0$. Since $|\mathcal{H}| = 9$, the group $\mathcal{H}$ has at least one element of order 3; hence $n \neq 1$. We conclude that $n = 0, 3$ or 9.

3. The configuration $\mathcal{H}$. With $\mathcal{F} = (\mathcal{H}, (F))$ an elliptic cubic, we consider the structure $\mathcal{H}$ when $|\mathcal{H}| = 9$. Let τ_i be the number of i-secants of $\mathcal{H}$, that is the number of lines ℓ in $PG(2, q)$ such that $|\ell \cap \mathcal{H}| = i$. Let $\rho_i = \rho_i(P)$ be the number of i-secants through a point P of $\mathcal{H}$. Then

(3.1)
$$\tau_0 + \tau_1 + \tau_2 + \tau_3 = q^2 + q + 1,$$
$$\tau_1 + 2\tau_2 + 3\tau_3 = 9(q + 1),$$
$$\tau_2 + 3\tau_3 = 36,$$

[3], p. 320. As any 2-secant is a tangent, so $\tau_2 \leq 9$. Since any tangent is a 2-secant if and only if the point of contact is not an inflexion, so

(3.2)
$$\tau_2 = 9 - n.$$

Thus 2.9, (3.1) and (3.2) give the possibilities

n	0	3	9
τ_2	9	6	0
τ_3	9	10	12.

By 2.8,

$$\rho_2 = 0 \quad \text{if } P \text{ is an inflexion},$$
(3.3)
$$\rho_2 = 2 \quad \text{if } P \text{ is not an inflexion}.$$

A consequence is that the non-inflexions are the vertices of a number of disjoint polygons whose sides are the bisecants of $\mathcal{H}$. For each P,

(3.4)
$$\rho_2 + 2\rho_3 = 8.$$

Hence there are the possibilities

ρ_2	0	2
ρ_3	4	3.

(3.5)

Also,

$$\sum_{P \in K} \rho_2 = 2\tau_2,$$

(3.6)
$$\sum_{P \in K} \rho_3 = 3\tau_3,$$

[3], p. 320.

A set of 9 points on a cubic curve is an *associated set* if there is another cubic curve containing the 9 points.

The three possibilities for n are now examined separately.

I. $n = 0$. By (3.3) we have $\rho_2 = 2$ for each point P of $\mathcal{H}$.

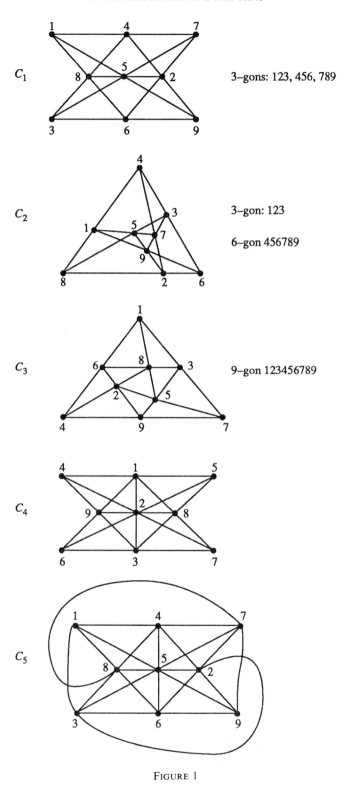

C_1 3–gons: 123, 456, 789

C_2 3–gon: 123

6–gon 456789

C_3 9–gon 123456789

C_4

C_5

FIGURE 1

LEMMA 1. *When $n = 0$, the bisecants form* (1) *three 3-gons,* (2) *a 3-gon and a 6-gon, or* (3) *a 9-gon. The configurations are respectively C_1, C_2, C_3* (*See Figure* 1). *Only C_1 gives an associated set.*

PROOF. Suppose the bisecants of $\mathcal{H}$ are sides of a 4-gon $\mathcal{N} = P_1 P_2 P_3 P_4$ and a 5-gon $\mathcal{N}' = Q_1 Q_2 Q_3 Q_4 Q_5$. Then $\mathcal{N}'$ is a 5-arc and $\mathcal{N}$ is a 4-arc. Through Q_1 there are the bisecants $Q_1 Q_2$ and $Q_1 Q_5$ of $\mathcal{H}$; hence $Q_1 Q_3$ and $Q_1 Q_4$ are 3-secants and each contain some P_i. The remaining 3-secant of $\mathcal{H}$ through Q_1 lies on a diagonal of $\mathcal{N}$. Hence each vertex of $\mathcal{N}'$ lies on a diagonal of $\mathcal{N}$; this is a contradiction. This leaves the possibilities (1), (2), (3). It is then straightforward to obtain the corresponding unique configurations C_1, C_2, C_3. In case (1), each pair of 3-gons is in triple perspective from the vertices of the remaining one. In case (2), each of the 9 diagonals of the 6-gon contains a vertex of the 3-gon. In case (3), the 3-secants of $\mathcal{H}$ are i, $i+2$, $i+6$ (mod 9). That C_1, C_2, C_3 are the only configurations obtained was first proved by Kantor [4] in 1881. See Steinitz [8] for a historical account.

The configuration C_1 is an associated set since there are three cubic curves, each consisting of three lines, through it; namely {147, 258, 369}, {168, 249, 357}, {159, 267, 348}. If C_2 were an associated set, the three lines 148, 259, 37 contain eight of the points and would contain the ninth; that is, 6 would lie on the line 37, a contradiction. Similarly, if C_3 were an associated set, the three lines 146, 359, 28 contain eight of the points and would contain the ninth; that is, 7 would lie on the line 28, a contradiction.

LEMMA 2. *For $n = 0$, the configuration C_2 does not occur.*

PROOF. Using the notation in Lemma 1, consider the group $\mathcal{H}$ with identity element 1. Then {1, 2, 3} is a subgroup of $\mathcal{H}$. Let 12 be the tangent to the cubic curve at 1; therefore 13 is the tangent at 3 and 23 the tangent at 2. Then $-4 = 7$, $5 + 7 = 2$, $9 + 7 = 3$; so $4 + 1 = 4$, $4 + 2 = 5$, $4 + 3 = 9$. Hence {4, 5, 9} is a coset of {1, 2, 3}. Further, $-5 = 9$, $6 + 9 = 2$, $7 + 9 = 3$; so $5 + 1 = 5$, $5 + 2 = 6$, $5 + 3 = 7$. Hence {5, 6, 7} is also a coset of {1, 2, 3}. As the cosets {4, 5, 9} and {5, 6, 7} have an element in common, we have a contradiction.

II. $n = 3$. From (3.3) at each of the 3 inflexions $\rho_2 = 0$ and $\rho_3 = 4$; at each of the 6 non-inflexions, $\rho_2 = 2$ and $\rho_3 = 3$. Thus through each inflexion there is the 3-secant of $\mathcal{H}$ of the inflexions plus three other 3-secants, giving all 10. Hence there is no 3-secant of $\mathcal{H}$ containing 3 non-inflexions. Therefore, the 6 non-inflexions form a 6-arc.

There are three ways of partitioning the 9 points by three 3-secants. By the theorem of the nine associated points, the 6 non-inflexions lie on a conic. $\mathcal{H}$ is an associated set and the configuration of points and 3-secants is denoted C_4.

LEMMA 3. *For* n $= 3$ *the non-inflexions are the vertices of two disjoint* *triangles whose sides are the bisecants of $\mathcal{K}$.*

PROOF. Let 1, 2, 3 be the inflexions and consider the group $\mathcal{K}$ with identity element 1. Then $\{1, 2, 3\}$ is a subgroup of $\mathcal{K}$. Suppose that the non-inflexions are the vertices of a 6-gon whose sides are the bisecants of $\mathcal{K}$. Let the vertices of the 6-gon be 4, 5, 6, 7, 8, 9 and let the line joining i and $i + 1$ be the tangent at i, where $10 = 4$. Without loss of generality suppose 461, 472, 483 are lines. Then $-4 = 6$, $4 + 7 + 2 = 1$, $4 + 8 + 3 = 1$. As $2 + 2 = 3$ and $3 + 3 = 2$, we have $6 + 1 = 6$, $6 + 3 = 7$, $6 + 2 = 8$. So $\{6, 7, 8\}$ is a coset of $\{1, 2, 3\}$. Similarly considering the diagonals of the 6-gon through 5, we find that $\{7, 8, 9\}$ is also a coset of $\{1, 2, 3\}$. This contradiction gives the result.

In the diagram of C_4, the inflexions are 1, 2, 3 and the non-inflexions 4, 5, 6, 7, 8, 9.

III. $n = 9$. The join of any two inflexions contains a third. Hence $\rho_2 = 0$, for all P in $\mathcal{K}$, as in (3.3). So $\rho_3 = 4$ and the configuration formed by the points of $\mathcal{K}$ and its twelve 3-secants is AG(2,3). Also $\mathcal{K}$ is an associated set and the configuration of points and 3-secants is denoted C_3.

These configurations have also been obtained by Keedwell [6].

THEOREM 1. *If $\mathcal{K}$ with $|\mathcal{K}| = 9$ is the set of zeros of more than one cubic with n inflexions, then one of the following holds:*

  (i) $n = 0$ and $\mathcal{K} = C_1$;
  (ii) $n = 3$ and $\mathcal{K} = C_4$;
  (iii) $n = 9$ and $\mathcal{K} = C_5$.

4. The curves with nine points. Since an elliptic curve over GF(q) with N points has the property that

$$(\sqrt{q} - 1)^2 \le N \le (\sqrt{q} + 1)^2,$$

so, for $N = 9$,

$$4 \le q \le 16.$$

The number $\mathcal{P}(q)$ of projectively distinct elliptic curves with 9 points for these values of q is given in Table 1; they are taken from [1] and [7].

TABLE 1

q	4	5	7	8	9	11	13	16
$\mathcal{P}(q)$	2	2	5	2	4	4	5	2

So it is only necessary to look at the configurations of the points on the 26 curves.   Each curve is denoted $\mathcal{G}_n^r$ or $\mathcal{E}_n^r$, where n is the number of inflexions over the ground field and r is the number of inflexional triangles over the ground field. The latter notion is not applicable for characteristic 3 and r is omitted. The notation $\overline{\mathcal{E}}_3^2$ indicates that the three inflexional tangents are concurrent, whereas $\mathcal{E}_0^4$ and $\overline{\mathcal{E}}_0^4$ are distinguished as having, for q odd, a Hessian which in the first case is reducible over $GF(q^3)$ but not over GF(q) and in the second case is reducible over GF(q) for q even, $\mathcal{E}_0^4$ and $\overline{\mathcal{E}}_0^4$ are the same. See [3], chapter 11, for further details and for the forms of the curves as in Table 2.

For the fields needed, the following notation is used.

$$GF(4) = \{0,\, 1,\, \omega,\, \omega^2\},$$
$$GF(5) = \{0,\, \pm 1,\, \pm 2\},$$
$$GF(7) = \{0,\, \pm 1,\, \pm 2,\, \pm 3\},$$
$$GF(8) = \{0,\, 1,\, \varepsilon^i \mid 1 < i \leq 6,\, \varepsilon^3 + \varepsilon^2 + 1 = 0\},$$
$$GF(9) = \{0,\, \pm 1,\, \pm\sigma,\, \pm\sigma^2,\, \pm\sigma^3 \mid \sigma^2 - \sigma - 1 = 0\},$$
$$GF(11) = \{0,\, \pm 1,\, \pm 2,\, \pm 3,\, \pm 4,\, \pm 5\},$$
$$GF(13) = \{0,\, \pm 1,\, \pm 2,\, \pm 3,\, \pm 4,\, \pm 5,\, \pm 6\},$$
$$GF(16) = \{0,\, 1,\, \eta^i \mid 1 \leq i \leq 14,\, \eta^4 + \eta + 1 = 0\}.$$

If $\mathcal{K}$ is one of the configurations C_1, C_4, C_5, it is an associated set and therefore is the base of a pencil of cubic curves.

LEMMA 4. *If $\mathcal{K}$ with $|\mathcal{K}| = 9$ has a bisecant, then there are at most two irreducible curves in the pencil with base $\mathcal{K}$ having $\mathcal{K}$ as their set of points.*

PROOF. If $\ell \cap \mathcal{K} = \{P, P'\}$ with $\mathcal{F} = (\mathcal{K}, (\mathrm{F}))$ and $\mathcal{F}' = (\mathcal{K}, (F'))$, then ℓ is tangent to $\mathcal{F}$ at P, say. However, ℓ is also tangent to $\mathcal{F}'$ and it cannot be at P, as Bézout's theorem would be contradicted. So ℓ is tangent to $\mathcal{F}'$ at P'. Now, a third curve $\mathcal{F}'' = (\mathcal{K}, (F''))$ would have ℓ as a tangent at either P or P', again contradicting Bézout's theorem.

LEMMA 5. *When $q \equiv 1 \pmod 6$, an equianharmonic cubic with 9 inflexions has three other curves equivalent to it in its syzygetic pencil.*

PROOF. If $\mathcal{F}(\mu) = (\mathcal{K}, (F_\mu))$ is a curve with

$$F_\mu = X^3 + Y^3 + Z^3 + \mu XYZ,$$

then $\mathcal{F}(\mu)$ is projectively equivalent to $\mathcal{F}(0)$ when $q \equiv 1 \pmod 3$ for $\mu = 6,\, 6c,\, 6c^2$, where $c^2 + c + 1 = 0$, [3], p. 278.

Lemma 5 explains the two occurrences of four curves with the same $\mathcal{K}$ in Table 3.

Now, we examine each of the configurations $\mathcal{K}$ determined by the 26 curves. If $\mathcal{K}$ is an associated set, the number of points on each curve in the pencil with base $\mathcal{K}$ is determined. Table 2 lists the configuration $\mathcal{K}$ and the form F for each projectively distinct $\mathcal{F} = (\mathcal{K}, (\mathrm{F}))$ with $|\mathcal{K}| = 9$.

In the examination of the 26 cases, for each $\mathcal{K}$ that is an associated set, let $\mathcal{F}(\lambda) = (V(F + \lambda G)), (F + \lambda G))$, where G is some cubic form such that $\mathcal{K} \subset V(G)$. Then, as λ varies in $GF(q) \cup \{\infty\}$, all members of the pencil are obtained: here $\mathscr{F}(\infty) = (V(G), (G))$. Also, $|\mathscr{F}(\lambda)|$ is written instead of $|V(F + \lambda G)|$. A check is provided by the equation $9 + \sum(|\mathscr{F}(\lambda)| - 9) = q^2 + q + 1$.

THEOREM 2. *If* $\mathscr{K}$ *with* $|\mathscr{K}| = 9$ *is precisely the set of zeros of more than one cubic, then* $\mathscr{K}$ *is one of the* 11 *cases listed in Table* 3, *up to projective equivalence.*

COROLLARY. *If* $\mathscr{K}$ *with* $|\mathscr{K}| = 9$ *is precisely the set of zeros of more than two cubics, then* $q = 7$ *or* 13 *and for each of these* q *there is a unique* $\mathscr{K}$, *up to projective equivalence.*

TABLE 2

No.	q	Curve $\mathscr{K}$		Form F
1	4	$\mathscr{E}_9^4$	C_5	$X^3 + Y^3 + Z^3$
2	4	$\overline{\mathscr{E}}_0^4$	C_1	$X^3 + \omega Y^3 + \omega^2 Z^3$
3	5	$\mathscr{G}_3^2$	C_4	$XYZ - (X + Y + Z)^3$
4	5	$\mathscr{G}_0^2$	C_3	$Z^3 - (X^2 - XY + Y^2)Z - (X^3 + 2XY^2 + Y^3)$
5	7	$\mathscr{E}_9^4$	C_5	$X^3 + Y^3 + Z^3$
6	7	$\mathscr{G}_3^1$	C_4	$XYZ - (X + Y + Z)^3$
7	7	$\mathscr{E}_0^4$	C_1	$X^3 + 2Y^3 - 3Z^3 + 3XYZ$
8	7	$\overline{\mathscr{E}}_0^4$	C_1	$X^3 + 2Y^3 - 3Z^3$
9	7	$\mathscr{G}_0^1$	C_3	$XY^2 + X^2Z - 2YZ^2 + X^3 - 2Y^3 - 3Z^3 - XYZ$
10	8	$\overline{\mathscr{E}}_3^2$	C_4	$XY(X + Y) + Z^3$
11	8	$\mathscr{E}_0^2$	C_3	$X^3 + XY^2 + \varepsilon^3 Y^3 + Z^3$
12	9	$\mathscr{G}_3$	C_4	$XYZ + \sigma^2(X + Y + Z)^3$
13	9	$\mathscr{G}_3$	C_4	$XYZ - \sigma^2(X + Y + Z)^3$
14	9	$\mathscr{G}_0$	C_3	$X^3 + Y^3 - \sigma Z^3 - X^2Z - XY^2 + XZ^2 - YZ^2$
15	9	$\mathscr{G}_0$	C_3	$X^3 + Y^3 - \sigma^3 Z^3 - X^2Z - XY^2 + XZ^2 - YZ^2$
16	11	$\mathscr{G}_3^2$	C_4	$XYZ + 4(X + Y + Z)^3$
17	11	$\mathscr{G}_3^2$	C_4	$XYZ - 5(X + Y + Z)^3$
18	11	$\mathscr{G}_0^2$	C_3	$Z^3 - (X^2 - XY + Y^2)Z - (X^3 - 3XY^2 + Y^3)$
19	11	$\mathscr{G}_0^2$	C_3	$Z^3 + 4(X^2 - XY + Y^2)Z - (X^3 - 3XY^2 + Y^3)$

20   13 $\mathscr{E}_9^4$ C_5 $X^3 + Y^3 + Z^3$

21   13 $\mathscr{G}_3^1$ C_4 $XYZ + 5(X + Y + Z)^3$

22   13 $\mathscr{E}_0^4$ C_1 $X^3 + 2Y^3 + 4Z^3 - XYZ$

23   13 $\overline{\mathscr{E}}_0^4$ C_1 $X^3 + 3Y^3 - 4Z^3$

24   13 $\mathscr{G}_0^1$ C_3 $XY^2 + X^2Z + 4YZ^2 - 2(X^3 + 4Y^3 + 3Z^3 + XYZ)$

---

25   16 $\mathscr{E}_9^4$ C_5 $X^3 + Y^3 + Z^3$

26   16 $\overline{\mathscr{E}}_0^4$ C_1 $X^3 + \eta^5 Y^3 + \eta^{10} Z^3$

---

1. $q = 4$, $F + \lambda G = X^3 + Y^3 + Z^3 + \lambda XYZ$, $|\mathscr{F}(0)| = 9$, $|\mathscr{F}(\infty)| = |\mathscr{F}(1)| = |\mathscr{F}(\omega)| = \left|\mathscr{F}(\omega^2)\right| = 12$. Each $\mathscr{F}(\lambda)$, $\lambda \neq 0$, is a triangle.

2. $q = 4$, $F + \lambda G = X^3 + \omega Y^3 + \omega^2 Z^3 + \lambda(X^3 + \omega^2 Y^3 + \omega Z^3)$. $|\mathscr{F}(0)| = |\mathscr{F}(\infty)| = 9$, $|\mathscr{F}(1)| = |\mathscr{F}(\omega)| = |\mathscr{F}(\omega^2)| = 13$. Each $\mathscr{F}(\lambda)$, $\lambda \neq 0$, ∞, is a triad (three concurrent lines).

3. $q = 5$, $F + \lambda G = XYZ - (X + Y + Z)^3 + \lambda(X^3 + Y^3 + Z^3 - XYZ)$. $|\mathscr{F}(0)| = |\mathscr{F}(\infty)| = 9$, $|\mathscr{F}(1)| = |\mathscr{F}(-1)| = 15$, $|\mathscr{F}(-2)| = 16$, $|\mathscr{F}(2)| = 12$. $\mathscr{F}(1)$ and $\mathscr{F}(-1)$ are triangles; $\mathscr{F}(-2)$ is a triad; $\mathscr{F}(2)$ is a conic plus an external line.

4. $q = 5$, $\mathscr{K} = C_3$.

5. $q = 7$, $F + \lambda G = X^3 + Y^3 + Z^3 + \lambda XYZ$. $|\mathscr{F}(0)| = |\mathscr{F}(3)| = |\mathscr{F}(-1)| = |\mathscr{F}(-2)| = 9$, $|\mathscr{F}(\infty)| = |\mathscr{F}(1)| = |\mathscr{F}(2)| = |\mathscr{F}(-3)| = 21$. The cubics with 21 points are triangles.

6. $q = 7$, $F + \lambda G = XYZ - (X + Y + Z)^3 + \lambda(X + Y + Z)(X^2 + Y^2 + Z^2)$. $|\mathscr{F}(0)| = |\mathscr{F}(-2)| = 9$, $|\mathscr{F}(1)| = |\mathscr{F}(-3)| = 12$, $|\mathscr{F}(\infty)| = 14$, $|\mathscr{F}(2)| = |\mathscr{F}(3)| = 21$, $|\mathscr{F}(-1)| = 22$. $\mathscr{F}(\infty)$ is a conic plus a bisecant; $\mathscr{F}(2)$ and $\mathscr{F}(3)$ are triangles; $\mathscr{F}(-1)$ is a triad.

7. $q = 7$, $F + \lambda G = X^3 + 2Y^3 - 3Z^3 + 3XYZ + \lambda(X^2Z + XY^2 + 2YZ^2)$. $|\mathscr{F}(0)| = 9$, $|\mathscr{F}(\infty)| = |\mathscr{F}(3)| = |\mathscr{F}(-1)| = |\mathscr{F}(-2)| = 12$, $|\mathscr{F}(1)| = |\mathscr{F}(2)| = |\mathscr{F}(-3)| = 21$. $\mathscr{F}(1)$, $\mathscr{F}(2)$ and $\mathscr{F}(-3)$ are triangles.

8. $q = 7$, $F + \lambda G = X^3 + 2Y^3 - 3Z^3 + \lambda(X^3 - 3Y^3 + 2Z^3)$. $|\mathscr{F}(0)| = |\mathscr{F}(\infty)| = 9$, $|\mathscr{F}(-1)| = |\mathscr{F}(3)| = |\mathscr{F}(-2)| = 22$, $|\mathscr{F}(1)| = |\mathscr{F}(2)| = |\mathscr{F}(-3)| = 12$. $\mathscr{F}(-1)$, $\mathscr{F}(3)$ and $\mathscr{F}(-2)$ are triads.

9. $q = 7$, $\mathscr{K} = C_3$.

10. $q = 8$, $F + \lambda G = XY(X + Y) + Z^3 + \lambda Z(X^2 + Y^2 + XY + Z^2)$. $|\mathscr{F}(0)| = 9$, $|\mathscr{F}(1)| = 10$, $|\mathscr{F}(\varepsilon)| = \left|\mathscr{F}(\varepsilon^2)\right| = \left|\mathscr{F}(\varepsilon^4)\right| = 12$, $|\mathscr{F}(\infty)| = 18$, $\left|\mathscr{F}(\varepsilon^3)\right| = \left|\mathscr{F}(\varepsilon^5)\right| = \left|\mathscr{F}(\varepsilon^6)\right| = 24$. $\mathscr{F}(1)$ is singular with an isolated double point; $\mathscr{F}(\infty)$ is conic plus an external line;

$\mathcal{F}(\varepsilon^3)$, $\mathcal{F}(\varepsilon^5)$ and $\mathcal{F}(\varepsilon^6)$ are triangles.

11. $q = 8$, $\mathcal{K} = C_3$.

12. $q = 9$, $F + \lambda G = XYZ + \sigma^2 (X + Y + Z)^3 + \lambda (Y + Z)(X + Z)(X + Y)$. $|\mathcal{F}(0)| = \left|\mathcal{F}(\sigma^3)\right| = 9$, $|\mathcal{F}(\sigma)| = |\mathcal{F}(-\sigma)| = 15$, $|\mathcal{F}(1)| = 19$, $|\mathcal{F}(-1)| = \left|\mathcal{F}(-\sigma^2)\right| = 12$, $|\mathcal{F}(\infty)| = \left|\mathcal{F}(\sigma^2)\right| = \left|\mathcal{F}(-\sigma^3)\right| = 27$. $\mathcal{F}(1)$ is a conic plus a tangent; $\mathcal{F}(\infty)$, $\mathcal{F}(\sigma^2)$ and $\mathcal{F}(-\sigma^3)$ are triangles.

13. $q = 9$, $F + \lambda G = XYZ - \sigma^2(X + Y + Z)^3 + \lambda(Y + Z)(X + Z)(X + Y)$. $|\mathcal{F}(0)| = |\mathcal{F}(\sigma)| = 9$, $\left|\mathcal{F}(\sigma^3)\right| = \left|\mathcal{F}(-\sigma^3)\right| = 15$, $|\mathcal{F}(1)| = 19$, $|\mathcal{F}(-1)| = \left|\mathcal{F}(\sigma^2)\right| = 12$, $|\mathcal{F}(\infty)| = \left|\mathcal{F}(-\sigma^2)\right| = |\mathcal{F}(-\sigma)| = 27$. $\mathcal{F}(1)$ is a conic plus a tangent; $\mathcal{F}(\infty)$, $\mathcal{F}(-\sigma^2)$ and $\mathcal{F}(-\sigma)$ are triangles. This case is equivalent to the previous one under the collineation $(X, Y, Z) \to (X^3, Y^3, Z^3)$.

14. $q = 9$, $\mathcal{K} = C_3$.

15. $q = 9$, $\mathcal{K} = C_3$. This case is equivalent to the previous one under the collineation $(X, Y, Z) \to (X^3, Y^3, Z^3)$.

16. $q = 11$, $F + \lambda G = XYZ + 4(X + Y + Z)^3 + \lambda(Y + Z)(X + Z)(X + Y)$. $|\mathcal{F}(0)| = 9$, $|\mathcal{F}(-4)| = 13$, $|\mathcal{F}(3)| = |\mathcal{F}(-1)| = |\mathcal{F}(-2)| = 15$, $|\mathcal{F}(5)| = |\mathcal{F}(-5)| = 12$, $|\mathcal{F}(1)| = 24$, $|\mathcal{F}(\infty)| = |\mathcal{F}(2)| = |\mathcal{F}(-3)| = 33$, $|\mathcal{F}(4)| = 18$. $\mathcal{F}(-4)$ is singular with an isolated double point; $\mathcal{F}(1)$ is a conic plus an external line; $\mathcal{F}(\infty)$, $\mathcal{F}(2)$ and $\mathcal{F}(-3)$ are triangles.

17. $q = 11$, $F + \lambda G = XYZ - 5(X + Y + Z)^3 + \lambda(X + Y + Z)(X^2 + Y^2 + Z^2 - 2XY - 2XZ - 2YZ)$. $|\mathcal{F}(0)| = |\mathcal{F}(2)| = 9$, $|\mathcal{F}(5)| = |\mathcal{F}(-3)| = 12$, $|\mathcal{F}(-4)| = |\mathcal{F}(-5)| = 15$, $|\mathcal{F}(3)| = |\mathcal{F}(-1)| = 18$, $|\mathcal{F}(\infty)| = 24$, $|\mathcal{F}(-2)| = |\mathcal{F}(4)| = 33$, $|\mathcal{F}(1)| = 34$. $\mathcal{F}(\infty)$ is a conic plus an external line; $\mathcal{F}(-2)$ and $\mathcal{F}(4)$ are triangles; $\mathcal{F}(1)$ is a triad.

18. $q = 11$, $\mathcal{K} = C_3$.

19. $q = 11$, $\mathcal{K} = C_3$.

20. $q = 13$, $F + \lambda G = X^3 + Y^3 + Z^3 + \lambda XYZ$. $|\mathcal{F}(0)| = |\mathcal{F}(2)| = |\mathcal{F}(5)| = |\mathcal{F}(6)| = 9$, $|\mathcal{F}(1)| = |\mathcal{F}(3)| = |\mathcal{F}(-2)| = |\mathcal{F}(-4)| = |\mathcal{F}(-5)| = |\mathcal{F}(-6)| = 18$, $|\mathcal{F}(\infty)| = |\mathcal{F}(4)| = |\mathcal{F}(-1)|$ and $|\mathcal{F}(-3)| = 39$. $\mathcal{F}(\infty)$, $\mathcal{F}(4)$, $\mathcal{F}(-1)$ and $\mathcal{F}(-3)$ are triangles.

21. $q = 13$, $F + \lambda G = XYZ + 5(X + Y + Z)^3 + \lambda(X + Y + Z)(X^2 + Y^2 + Z^2 + 6XY + 6XZ + 6YZ)$. $|\mathcal{F}(0)| = 9$, $|\mathcal{F}(2)| = 12$, $|\mathcal{F}(3)| = 13$, $|\mathcal{F}(6)| = |\mathcal{F}(-2)| = 15$, $|\mathcal{F}(4)| = |\mathcal{F}(5)| = |\mathcal{F}(-4)| = |\mathcal{F}(-5)| = 18$, $|\mathcal{F}(1)| = 21$, $|\mathcal{F}(\infty)| = 26$, $|\mathcal{F}(-1)| = |\mathcal{F}(-3)| = |\mathcal{F}(-6)| = 39$. $\mathcal{F}(3)$ is singular with a node; $\mathcal{F}(\infty)$ is a conic plus a bisecant; $\mathcal{F}(-1)$, $\mathcal{F}(-3)$ and $\mathcal{F}(-6)$ are triangles.

22. $q = 13$, $F + \lambda G = X^3 + 2Y^3 + 4Z^3 - XYZ + \lambda(2X + Y + 3Z)(3X - Y + 5Z)(X + Y - Z)$. $|\mathcal{F}(0)| = 9$, $|\mathcal{F}(4)| = |\mathcal{F}(5)| = |\mathcal{F}(-4)| =$

15, $|\mathscr{F}(1)| = |\mathscr{F}(6)| = |\mathscr{F}(-2)| = |\mathscr{F}(-3)| = |\mathscr{F}(-6)| = |\mathscr{F}(-5)| = $
18, $|\mathscr{F}(2)| = 21$, $|\mathscr{F}(\infty)| = |\mathscr{F}(3)| = |\mathscr{F}(-1)| = 39$. $\mathscr{F}(\infty), \mathscr{F}(3)$
and $\mathscr{F}(-1)$ are triangles.

23. $q = 13$, $F + \lambda G = X^3 + 3Y^3 - 4Z^3 + \lambda(X^3 - 4Y^3 + 3Z^3)$. $|\mathscr{F}(0)| = $
$|\mathscr{F}(\infty)| = 9$, $|\mathscr{F}(1)| = |\mathscr{F}(3)| = |\mathscr{F}(-4)| = 12$, $|\mathscr{F}(2)| = |\mathscr{F}(5)| = $
$|\mathscr{F}(6)| = |\mathscr{F}(-2)| = |\mathscr{F}(-5)| = |\mathscr{F}(-6)| = 21$, $|\mathscr{F}(4)| = |\mathscr{F}(-1)| = $
$|\mathscr{F}(-3)| = 40$. $\mathscr{F}(4), \mathscr{F}(-1)$, and $\mathscr{F}(-3)$ are triads.

24. $q = 13$, $\mathscr{K} = C_3$.

25. $q = 16$, $F + \lambda G = X^3 + Y^3 + Z^3 + \lambda XYZ$. $|\mathscr{F}(0)| = 9$, $|\mathscr{F}(\infty)| = $
$|\mathscr{F}(1)| = \left|\mathscr{F}(\eta^5)\right| = \left|\mathscr{F}(\eta^{10})\right| = 48$, $\left|\mathscr{F}(\eta^i)\right| = 18$, $1 \leq i \leq 14$, $i \neq$
5, 10. $\mathscr{F}(\infty), \mathscr{F}(1), \mathscr{F}(\eta^5), \mathscr{F}(\eta^{10})$ are triangles.

26. $q = 16$, $F + \lambda G = X^3 + \eta^5 Y^3 + \eta^{10} Z^3 + \lambda(X^3 + \eta^{10}Y^3 + \eta^5 Z^3)$.
$|\mathscr{F}(0)| = |\mathscr{F}(\infty)| = 9$, $|\mathscr{F}(1)| = \left|\mathscr{F}(\eta^5)\right| = \left|\mathscr{F}(\eta^{10})\right| = 49$,
$\left|\mathscr{F}(\eta^i)\right| = 21$, $1 \leq i \leq 14$, $i \neq 5, 10$. $\mathscr{F}(1), \mathscr{F}(\eta^5)$ and $\mathscr{F}(\eta^{10})$
are triads.

TABLE 3

No.	q	$\mathscr{K}$	n	Curves with point set $\mathscr{K}$
2	4	C_1	0	$\mathscr{F}(0), \mathscr{F}(\infty)$
3	5	C_4	3	$\mathscr{F}(0), \mathscr{F}(\infty)$
5	7	C_5	9	$\mathscr{F}(0), \mathscr{F}(3), \mathscr{F}(-1), \mathscr{F}(-2)$
6	7	C_4	3	$\mathscr{F}(0), \mathscr{F}(-2)$
8	7	C_1	0	$\mathscr{F}(0), \mathscr{F}(\infty)$
12	9	C_4	3	$\mathscr{F}(0), \mathscr{F}(\sigma^3)$
13	9	C_4	3	$\mathscr{F}(0), \mathscr{F}(\sigma)$
17	11	C_4	3	$\mathscr{F}(0), \mathscr{F}(2)$
20	13	C_5	9	$\mathscr{F}(0), \mathscr{F}(2), \mathscr{F}(5), \mathscr{F}(6)$
23	13	C_1	0	$\mathscr{F}(0), \mathscr{F}(\infty)$
26	16	C_1	0	$\mathscr{F}(0), \mathscr{F}(\infty)$

Acknowledgments. We are grateful to R. De Groote and H. M. Tabrizi for checking some calculations.

REFERENCES

1. R. De Groote and J. W. P. Hirschfeld, *The number of points on an elliptic curve over a finite field*, European J. Combin. **1** (1980), 327–333.
2. W. Fulton, *Algebraic Curves*, Benjamin, Reading, Mass., 1969.
3. J. W. P. Hirschfeld, *Projective Geometries Over Finite Fields*, Oxford University Press, Oxford, 1979.
4. S. Kantor, *Über die Configurationen* (3, 3) *mit den Indices* 8, 9 *under ihren Zusammenhang mit den Curven dritter Ordnung*, Sitzungsberichte der math.natur. Classe der Kaiserl. Akad. der Wiss. zu Wien **84** (1881), 915–932.
5. ———, *Die Configurationen* (3, 3)$_{10}$, Sitzungsberichte der math.-natur. Classe der Kaiserl. Akad. der Wiss. zu Wien **84** (1881), 1291–1314.

6. A. D. Keedwell, *Simple constructions for elliptic cubic curves with specified small numbers of points*, European J. Combin. **9** (1988), 463–481.

7. R. Schoof, *Nonsingular plane cubic curves over finite fields*, J. Combin. Theory Ser. A **46** (1987), 183–211.

8. E. Steinitz, *Konfigurationen der projektiven Geometrie*, Enzyk. der Math. Wiss. III AB5a, 1910, pp. 481–516.

9. F. Zirilli, *C-struttura associata ad una cubica piana e C-struttura astratta*, Ricerche Mat. **16** (1967), 202–232.

10. _____, *Su una classe di k-archi di un piano di Galois*, Atti Accad. Naz. Lincei Rend. **54** (1973), 393–397.

MATHEMATICS DIVISION, UNIVERSITY OF SUSSEX, BRIGHTON BN1 9QH, UNITED KINGDOM

SEMINARIE VOOR MEETKUNDE EN KOMBINATORIEK, RIJKSUNIVERSITEIT-GENT, B-9000 GENT, BELGIUM

Contemporary Mathematics
Volume 111, 1990

Flocks and Partial Flocks of Quadric Sets

N. L. JOHNSON

1. Flocks, generalized quadrangles and translation planes. Recently, connections have been established between flocks of quadric sets and certain generalized quadrangles and translation planes (viz. [3], [12], [14]). In this article, we observe some connections with *partial* flocks and certain translation planes.

(1.1) DEFINITION. Let $\Sigma = PG(3, q)$ when q is a prime power. Let $\mathcal{C}, \mathcal{H}, \mathcal{E}$ denote a quadratic cone, hyperbolic quadric or elliptic quadric, respectively. A *flock* of $\mathcal{C}, \mathcal{H}, \mathcal{E}$ is a maximal set of mutually disjoint irreducible conics on $\mathcal{C}, \mathcal{H}, \mathcal{E}$ respectively. If $Q \in \{\mathcal{C}, \mathcal{H}, \mathcal{E}\}$ then a flock of Q contains $q, q + 1, q - 1$ conics respectively as Q is $\mathcal{C}$, $\mathcal{H}$, or $\mathcal{E}$.

A *partial flock of deficiency* t of Q is a non-empty set of $q - t, (q + 1) - t, (q - 1) - t$ mutually disjoint irreducible conics of $\mathcal{C}, \mathcal{H}, \mathcal{E}$ respectively.

In this article, we shall only be concerned with quadratic cones and hyperbolic quadrics.

(1.2) Flocks and translation planes—the construction of Thas-Walker. Embed $\Sigma = PG(3, q)$ in $PG(5, q)$ so that the quadric set Q in question lies in the Klein quadric $\mathcal{K}$. If the partial flock $\mathcal{F}_Q^t$ of Q of deficiency t is $\{\mathcal{C}_1, \ldots, \mathcal{C}_s\}$ where $s = q - t$ or $(q + 1) - t$ respectively as Q is $\mathcal{C}$ or $\mathcal{H}$, let $\pi_i \geq \mathcal{C}_i$ denote the plane containing $\mathcal{C}_i$ for $i = 1, \ldots, s$, and let π_i^* denote the polar plane of π_i with respect to $\mathcal{K}$. Then $\{\pi_i^* \cap \mathcal{K} = \mathcal{C}_i^* | i = 1, \ldots, s\}$ is a set of $(q^2 + 1) - tq, (q^2 + 1) - t(q - 1)$ points of $\mathcal{K}$, respectively as Q is $\mathcal{C}$ or $\mathcal{H}$, no two of which are joined by a line in $\mathcal{K}$. With these points correspond lines of $PG(3, q)$ using Plücker coordinates, and in this way there arises a partial spread of degree $(q^2 + 1) - tq$ resp. $(q^2 + 1) - t(q - 1)$.

(1.3) THEOREM (see Thas [14], Gevaert and Johnson [3], [4], Johnson [12]).

(1) *A partial flock of deficiency* t *of a quadratic cone produces a partial*

1980 *Mathematics Subject Classification* (1985 *Revision*). Primary 51B10, 51B15, 51A35.
This paper is in final form and no version of it will be submitted for publication elsewhere.

spread S^t of $PG(3, q)$ where S^t consists of $q - t$ reguli that mutually share a line.

(2) A partial flock of deficiency t of a hyperbolic quadric produces a partial spread P^t of $PG(3, q)$ where P^t consists of $(q + 1) - t$ reguli that mutually share two lines.

Using Gevaert, Johnson, Thas [5] and Gevaert, Johnson [4] the converse of (1.3) may be established.

Hence,

(1.4) THEOREM. (1) There is a $1 - 1$ correspondence between partial flocks of quadratic cones in $PG(3, q)$ of deficiency t and partial spreads in $PG(3, q)$ consisting of $q - t$ reguli mutually sharing a line.

(2) There is a $1 - 1$ correspondence between partial flocks of hyperbolic quadrics in $PG(3, q)$ of deficiency t and partial spreads in $PG(3, q)$ consisting of $(q + 1) - t$ reguli mutually sharing two lines.

The partial spreads of (1.4) may be characterized by their central collineation groups.

For the following, we identify partial spreads in $PG(3, q)$ and the corresponding translation nets in the 4-dimensional vector space V_4 over $GF(q)$.

(1.5) THEOREM (see [3], [4], [12]). (1) Let N denote a finite translation net in $V_4/GF(q)$. If N admits an elation group $\mathcal{E}$ fixing a component $\mathcal{L}$ of N pointwise and one of the component orbits union $\mathcal{L}$ is a regulus then each of the component orbits of $\mathcal{E}$ union $\mathcal{L}$ is a regulus. Hence, N is a translation net of degree $1 + q(q - t)$ whose partial spread consists of $q - t$ reguli mutually sharing $\mathcal{L}$.

(2) If N admits a homology group $\mathcal{H}$ fixing a component $\mathcal{L}$ pointwise and also fixing a component M and one of the component orbits union $\{\mathcal{L}, M\}$ is a regulus then each of the component orbits of $\mathcal{H}$ union $\{\mathcal{L}, M\}$ is a regulus. Thus, N is a translation net of degree $2 + (q - 1)((q + 1) - t)$ whose partial spread consists of $(q + 1) - t$ reguli mutually sharing $\mathcal{L}$ and M.

(3) Conversely, partial spreads consisting of $q - t$ reguli mutually sharing a line or $(q + 1) - t$ reguli mutually sharing two lines correspond to translation nets admitting central collineation groups whose component orbits union the axis (axis and coaxis) define the reguli.

(1.6) DEFINITION. Regulus inducing groups. Let N be a translation net in $V_4/GF(q)$. let $\mathcal{E}, \mathcal{H}$ denote central collineation groups of N of orders $q, q - 1$ respectively. If the component orbits of $\mathcal{E}$ or $\mathcal{H}$ union the axis or axis-coaxis respectively are reguli, we shall say that $\mathcal{E}$ or $\mathcal{H}$ is a regulus inducing group.

Hence,

(1.7) THEOREM. (1) Partial flocks of quadratic cones of deficiency t in $PG(3, q)$ are equivalent to finite translation nets in $V_4/GF(q)$ admitting a regulus inducing elation group.

(2) *Partial flocks of hyperbolic quadrics of deficiency t in $PG(3, q)$ are equivalent to finite translation nets $V_4/GF(q)$ admitting a regulus inducing homology group.*

Although we have not dealt with this, there are similar connections with flocks (of cones) and certain generalized quadrangles. (See Thas [14] and/or Gevaert, Johnson [3].)

2. Baer groups. Let $\mathcal{F}_Q^t$ be a partial flock of deficiency t of Q a quadratic cone or a hyperbolic quadric in $PG(3, q)$. Let $\pi_{\mathcal{F}_Q^t}$ denote the corresponding translation net as in (1.7) admitting the central collineation group $\mathcal{G}$ ($\mathcal{G}$ is an elation group of order q or homology group of order $q - 1$ respectively as Q is a quadratic cone or hyperbolic quadric).

Let $\mathcal{L}$ be any component of $\pi_{\mathcal{F}_Q^t}$ which is not fixed by $\mathcal{G}$. Then

$$\{\text{Fix}\,\mathcal{G},\ \text{CoFix}\,\mathcal{G}(\text{if }\ \mathcal{G}\ \text{ is a homology group})\} \cup \mathcal{G}\mathcal{L}$$

is a regulus $\mathcal{R}_{\mathcal{L}}$.

Now derive $\mathcal{R}_{\mathcal{L}}$—replace the partial spread by its oposite regulus $\mathcal{R}_{\mathcal{L}}$. Then $\bar{\pi}_{\mathcal{F}_Q^t} = (\pi_{\mathcal{F}_Q^t} - \mathcal{R}_{\mathcal{L}}) \cup \mathcal{R}_{\mathcal{L}}$ defines a translation net admitting a Baer group in the sense that $\mathcal{L}$ is now a Baer subplane of the translation net determined by $\mathcal{R}_{\mathcal{L}}$. The degree of $\pi_{\mathcal{F}_Q^t}$ is $1+q(q-t)$ or $2+(q-1)(q+1-t)$ respectively as Q is a quadratic cone or a hyperbolic quadric. If $t \neq 0$, $\pi_{\mathcal{F}_Q^t}$-component of the net fixed pointwise by $\mathcal{E}$ may be extended to a translation net

$$\pi_{\mathcal{F}_Q^t}^+ \text{ of degree } q \text{ or } q - 1$$

larger than the degree of $\pi_{\mathcal{F}_Q^t}$ and which also admits $\mathcal{G}$ as a Baer group.

(2.1) THEOREM. *Let $\mathcal{F}_Q^t$ for t an integer > 0 be a partial flock of deficiency t of a quadratic cone in $PG(3, q)$. Let $\pi_{\mathcal{F}_Q^t}$ denote the corresponding translation net of degree $1+q(q-t)$ whose partial spread is defined by $q-t$ reguli mutually sharing a line (component of the net). Let $\mathcal{E}$ denote the (elation) group of order q such that if $\mathcal{L}$ is a component $\neq$ Fix $\mathcal{E}$ of $\pi_{\mathcal{F}_Q^t}$ then $\mathcal{E}\mathcal{L} \cup$ Fix $\mathcal{E}$ is a regulus. Let K denote the field isomorphic to $GF(q)$ over which the components of $\pi_{\mathcal{F}_Q^t}$ are 2-spaces.*

Let $\{T_i(i = 1, \ldots, q + 1)\}$ denote the set of 1-spaces over of K of Fix $\mathcal{E}$. Let $V_4 - \pi_{\mathcal{F}_Q^t}$ denote the set of vectors of V_4 over K which do not lie on a component of $\pi_{\mathcal{F}_Q^t}$. For each T_i of Fix E, there exist exactly q $\mathcal{E}$-orbits of 1-spaces of $V_4 -$ Fix $\mathcal{E}$ such that the subspace generated is a 2-K-space containing T_i. There are exactly t such orbits in $V_4 - \pi_{\mathcal{F}_Q^t}$. For each i, $i = 1, \ldots, q+1$, choose any such $\mathcal{E}$-orbit of 1-spaces and denote the set by D. Then $(\pi_{\mathcal{F}_Q^t} - Fix\ \mathcal{E}) \cup D = \pi_{\mathcal{F}_Q^t}^+$ is a translation net of degree $1+q(q-t)+q$

which contains $\pi_{\mathcal{F}_Q^i} - Fix\mathcal{E}$ *and admits* $\mathcal{E}$ *as a collineation group which fixes a Baer subplane of* $\pi_{\mathcal{F}_Q^i}^+$ *pointwise.*

The proof is a natural extension of ideas in Johnson [11] and Foulser [2]. The first assertion is in (2.1) [11].

There is also an analogous theorem for partial flocks of hyperbolic quadrics.

We now concentrate on partial flocks of deficiency "one" although most of what we will say is valid for arbitrary deficiency.

(2.2) THEOREM (Johnson [11]). *Let* π *be any translation plane of order* q^2 *and kernel* $\geq GF(q)$ *that admits a Baer group* B_q *(or* B_{q-1}*) of order* q *(or* $q-1$*) in the translation component.*

(1) *If* $\mathcal{L}$ *is any component of* π *which is not fixed by* B_q *then* $B_q\mathcal{L} \cup Fix\, B_q$ *is a regulus in* $PG(3,q)$.

(2) *If* M *is any component of* π *which is not fixed by* B_{q-1} *then the net of* π *containing Fix* B_{q-1} *contains another* B_{q-1} *invariant Baer subplane Co Fix* B_{q-1} *and* $B_{q-1}M \cup \{Fix\, B_{q-1}, Co\, Fix\, B_{q-1}\}$ *is a regulus in* $PG(3,q)$.

If π *is a plane satisfying* (2.2) *then there are* $q^2 - q$ *components of* π *which are not fixed by* B_q (B_{q-1}) *in* $\frac{q^2-q}{q(\text{or } q-1)} = q - 1$ (*or* q) *orbits.*

Therefore,

(2.3) THEOREM (Johnson [11]). (1) *A translation plane* π *of order* q^2 *and kernel* $\supseteq GF(q)$ *that admits a Baer group of order* q *gives rise to a partial flock of a quadratic cone of deficiency one.*

(2) *A translation plane* π *of order* q^2 *and kernel* $\supseteq GF(q)$ *that admits a Baer group of order* $q - 1$ *gives rise to a partial flock of a hyperbolic quadric of deficiency one.*

We have seen in (2.1) that the converse of (2.3)(1) is also valid.

(2.3) DEFINITION. A partial flock $\mathcal{F}$ of deficiency t of a quadric set Q (quadratic cone, hyperbolic quadric, elliptic quadric) in $PG(3,q)$ is *maximal* if and only if there is no partial flock of deficiency $t - 1$ containing $\mathcal{F}$.

It turns out for deficiency $t = 1$ that maximality may be interpreted in the corresponding translation plane using the net defined by the associated Baer group.

(2.5) THEOREM (Johnson [11]). (1) *A partial flock* $\mathcal{F}^1$ *of a quadratic cone in* $PG(3,q)$ *of deficiency one is equivalent to a translation plane* $\pi_{\mathcal{F}^1}^+$ *of order* q^2 *and kernel* $\supseteq GF(q)$ *that admits a Baer subgroup* B_q *of order* q. $\mathcal{F}^1$ *is maximal if and only if the net of* $\pi_{\mathcal{F}^1}^+$ *containing the compc\,ents of the Baer subplane Fix* B_q *is not derivable.*

(2) *A partial flock* $\mathcal{P}^1$ *of a hyperbolic quadric in* $PG(3,q)$ *of deficiency one is equivalent to a translation plane* $\pi_{\mathcal{P}^1}^+$ *of order* q^2 *and kernel* $\supseteq GF(q)$

that admits a Baer group $\mathcal{B}_{q-1}$ of order $q - 1$. $\mathcal{P}^1$ is maximal if and only if the net of $\pi_{\mathcal{P}^1}^+$ containing the components of Fix $\mathcal{B}_{q-1}$ is not a regulus.

3. Examples. (see [11], Section 4) (3.1) Let $V_4 = \{(x_1, x_2, y_1, y_2)\, x_i, y_i \in K \cong GF(4), i = 1, 2\}$. Let $x = 0, y = x \begin{bmatrix} \beta^2 & \omega\alpha^2 \\ \alpha & \beta \end{bmatrix}$ for $\alpha, \beta \in K$ define a translation plane of order 16 admitting a Baer group $\mathcal{B}_3$ of order 3. The net containing Fix $\mathcal{B}_3$ is defined by $x = 0, y = x \begin{bmatrix} \beta^2 & 0 \\ 0 & \beta \end{bmatrix}$ and is not a regulus over K (in $PG(3, K)$). Hence, the corresponding partial flock is a maximal partial flock of deficiency one of a hyperbolic quadric in $PG(3, 4)$. There are no known examples of maximal partial flocks of deficiency one of a quadratic cone in $PG(3, q)$. In fact, Payne and Thas have shown during this conference that for q even none are possible. We consider this in the next section.

4. Baer-Elation planes and partial flocks. A Baer-Elation plane of order q^2 and type $(2^b, 2^e)$ is a translation plane that admits a Baer 2-group $\mathcal{B}$ of order 2^b and an elation group $\mathcal{E}$ of order 2^e, $b, e \geq 1$ such that $\mathcal{B}$ and $\mathcal{E}$ normalize each other. By Jha-Johnson ([8], [9]) type $(q, 2^e)$ planes are type $(q, 2)$ and type $(2^b, q)$ planes are type $(2, q)$.

The only known Baer-Elation planes of order q^2, kernel $\geq GF(q)$ and type $(q, 2)$ are the Hall planes. Furthermore, it was shown in Johnson [10] that the only derivable $(2, q)$ or $(q, 2)$ planes of order q^2 and kernel $\geq GF(q)$ are the Desarguesian and Hall planes respectively. However, Huang and Johnson [6] found seven non-Desarguesian $(2, q)$ planes for $q = 8$.

The purpose of this section is to observe

THEOREM A. *If π is a Baer-Elation plane of order q^2, kernel $\geq GF(q)$, and of type $(q, 2)$ then π is Hall.*

Theorem A follows from the more general:

THEOREM B. *If π is a translation plane of even order q^2 and kernel $\supseteq K \cong GF(q)$ that admits a Baer group $\mathcal{B}$ of order q then the net defined by the subplane Fix $\mathcal{B}$ is a regulus (in $PG(3, K)$).*

Recently, Payne and Thas proved:

THEOREM C (Payne, Thas [13]). *Let q be even and let $\mathcal{F}_{q-1}$ be a partial flock of $q - 1$ conics of a quadratic cone φ in $PG(3, q)$. Then $\mathcal{F}_{q-1}$ may be uniquely extended to a flock of φ.*

PROOF OF THEOREMS A, B. Referring to B and using (2.5)(1) and C, the net defined by Fix $\mathcal{B}$ is derivable. To prove A, use B and the result of Johnson in [10] that derivable $(q, 2)$ Baer-Elation planes are Hall.

References

1. Biliotti, M., Jha, V., Johnson, N. L., Menichetti, G., *The collineation groups of Baer-Elation planes*, Atti Sem. Mat. Fis. Univ. Modena **36**, no.1 (1988), 23–35.
2. Foulser, D. A., *Subplanes of partial spreads in translation planes*, Bull. London Math. Soc. **4** (1972), 32–38.
3. Gevaert, H., Johnson, N. L., *Flocks of quadratic cones, generalized quadrangles and translation planes*, Geom. Ded. **27**, no.3 (1988), 301–317.
4. _____, *On maximal partial spreads in* $PG(3, q)$ *of cardinalities* $q^2 - q + 1$, $q^2 - q + 2$, Ars. Comb. **26** (1988), 191–196.
5. Gevaert, H., Johnson, N. L., Thas, J. A., *Spreads covered by reguli*, Simon Stevin **62** (1988), 51–62.
6. Huang, H., Johnson, N. L., 8 *semifield planes of order* 8^2, Discrete Math. **80** (1990), 69–79.
7. Jha, V., Johnson, N.L., *Baer-Elation planes*, Rend. Sem. Mat. Univ. Padova **78** (1987), 27–45.
8. _____, *Coexistence of elations and large Baer groups in translation planes*, J. London Math. Soc.(2) **32** (1985), 297–403.
9. _____, *Baer involutions planes admitting large elation groups*, Resultate d. Math. **11** (1987), 63–71.
10. Johnson, N. L., *On Derivable Baer-Elation Planes*, Note di Mat.(Lecce) **III** (1987), 19–27.
11. _____, *Translation planes admitting Baer groups and partial flocks of quadric sets*, Simon Stevin vol. 63, no. 3 (1989), 163–187.
12. _____, *Flocks of hyperbolic quadrics and translation planes admitting affine homoligies*, J. Geom. **34** (1989), 50–73.
13. Payne, S., Thas, J. A., *Conical flocks, partial flocks, derivation, and generalized quadrangles (preprint)*.
14. Thas, J.A., *Generalized quadrangles and flocks of cones*, European J. Combin. **81** (1987), 441–452.

DEPARTMENT OF MATHEMATICS, UNIVERSITY OF IOWA, IOWA CITY, IOWA 52242

Contemporary Mathematics
Volume **111**, 1990

The Finite Flag-Transitive Linear Spaces
with an Exceptional Automorphism Group

PETER B. KLEIDMAN

ABSTRACT. Assume that S is a finite linear space and that G is a flag-transitive group of automorphisms of S. If G is almost simple and the socle of G is an exceptional group of Lie type, then S must be the Ree Unital $U_R(q)$ and the socle of G must be $^2G_2(q)$. The proof of this result breaks down essentially into ten cases, corresponding to the ten families of exceptional groups $^2B_2$, $^3D_4$, E_6, $^2E_6$, E_7, E_8, F_4, $^2F_4$, G_2 and $^2G_2$. Here we offer a proof in three of the cases, namely $^2G_2$, E_7 and E_8. We make some remarks about the proof in the other seven cases. The proof of all ten cases will appear together in [5].

1. Introduction. This paper is a contribution to the program to classify the finite flag-transitive linear spaces. Currently, a six-person team, comprised of F. Buekenhout, A. Delandtsheer, J. Doyen, M. W. Liebeck, J. Saxl and myself, is working towards this classification. This subject has received a considerable degree of attention in the last few years, and we refer the reader to [4] for a survey of the various advances which have been obtained.

Let S be a finite linear space and assume that G is a flag-transitive group of automorphisms of S. Recall that a *linear space* is an incidence structure comprised of *points* and *lines* such that any two points are incident with exactly one line, every point is incident with at least two lines, and every line is incident with at least two points. A *flag* is a point-line pair, such that the point is incident with the line. Thus to say that G is *flag-transitive* means that if p_1, p_2 are points, ℓ_1, ℓ_2 are lines, and p_i is incident with ℓ_i for $i = 1, 2$, then there is an automorphism of S taking (p_1, ℓ_1) to (p_2, ℓ_2).

The starting point of the classification of the flag-transitive linear spaces is a result of Higman-McLaughlin [14], which asserts that G acts primitively on the points of S. Using this result, along with the O'Nan-Scott Theorem

1980 *Mathematics Subject Classification* (1985 *Revision*). Primary 51E30.
This paper is in final form and no version of it will be submitted for publication elsewhere.

(see [22] or [3], for example) it has been shown in [4, Section 5] and in [26, 27] that one of the following holds:

(I) G is almost simple;

(II) G is of affine type.

In (I), G is *almost simple* means that $G_0 \trianglelefteq G \leq Aut(G_0)$ for some non-abelian simple group G_0. In (II), G is of *affine type* means that G is isomorphic to the split extension of V by H, where V is vector space of size p^d for some prime p and some d, and H is a subgroup of $GL(V) \cong GL_d(p)$ with H irreducible on V; here the number of points in S is p^d.

Case (II) will ultimately present the most obstacles in the classification program, especially when d is small. Such difficulties have already been encountered by Kantor [15] in his attempt to classify the finite projective planes with a flag-transitive automorphism group. He elegantly shows that any such projective plane is Desarguesian, except when the automorphism group is of affine type with $d = 1$. Since every projective plane is a linear space, problems will no doubt arise in our analysis, too.

Part (I), on the other hand, is now for the most part under control. Let us assume here that G is almost simple and that G_0 is the socle soc(G), so that G_0 is a non-abelian simple group and $G_0 \trianglelefteq G \leq Aut(G_0)$. To study this situation, we invoke the classification of finite simple groups. Thus G_0 is either alternating, a classical group of Lie type, an exceptional group of Lie type, or one of the 26 sporadic groups.

Recently, it has been shown in [6] that G_0 cannot be sporadic, and that if G_0 is alternating, then $S = PG(3, 2)$ and $G \cong A_7$ or A_8. Thus to complete the analysis of part (I), we must consider the case where G_0 is either a classical or exceptional group of Lie type. My contribution to the program is to handle the case where G_0 is exceptional. Our main result is

THEOREM A. *Let S be a finite linear space and assume that G is a flag-transitive group of automorphisms of S. Suppose further that the socle soc(G) is a simple exceptional group of Lie type. Then S is the Ree unital $U_R(q)$ of order $q = 3^{2m+1} (m \geq 1)$ and ${}^2G_2(q) \trianglelefteq G \leq Aut({}^2G_2(q))$.*

The Ree Unital $U_R(q)$, which was discovered by Lüneburg [23], is the linear space whose points and lines are respectively the Sylow 3-subgroups and the involutions of the Ree group ${}^2G_2(q)$. A point and a line are incident if and only if the line normalizes the point. Observe that by Sylow's Theorem, ${}^2G_2(q)$ is transitive on points; and since $N_{{}^2G_2(q)}(P)/P$ is cyclic of order $q - 1$ for any point P, we know that $N_{{}^2G_2(q)}(P)$ is transitive on its involutions. Consequently ${}^2G_2(q)$ acts flag-transitively on $U_R(q)$. Moreover, so does any group G satisfying ${}^2G_2(q) \trianglelefteq G \leq Aut({}^2G_2(q))$. We have stipulated $m \geq 1$ in the statement of Theorem A in view of the isomorphism ${}^2G_2(3) \cong Aut(L_2(8)) \cong L_2(8).3$. Of course there is a Ree Unital $U_R(3)$, but its automorphism group has socle $L_2(8)$, which we regard as classical rather than exceptional.

The proof of Theorem A relies on the classification of finite simple groups, along with a wealth of information concerning the subgroup structure of the exceptional groups of Lie type. In particular, it relies on

(a) certain properties of their parabolic subgroups;

(b) the classification of their large subgroups (see [21]);

(c) the classification of their local maximal subgroups (see [9]);

(d) the determination of their conjugacy classes of semisimple elements (see [12]).

As there are ten families of exceptional groups of Lie type, the proof of Theorem A breaks down into ten seperate cases, accordingly. In this paper we do not treat all ten families; rather, we consider only the case where G_0 is $^2G_2(q)$, $E_7(q)$ or $E_8(q)$. The case $G_0 = {}^2G_2(q)$ is of course of special interest, for here (and only here) do examples actually arise. The case $G_0 = E_8(q)$ also has its appeal, as $E_8(q)$ is in some sense the 'biggest' exceptional group. We include $E_7(q)$, for the analysis here is for the most part more intricate than that in the other nine cases. The arguments presented in this paper certainly capture the flavor of the entire proof of Theorem A, which will appear in [5].

2. Preliminaries. We continue the notation introduced in §1, so that S is a finite linear space and G is a flag-transitive group of automorphisms of S. Assume further that the socle $G_0 = soc(G)$ is a simple exceptional group of Lie type over $\mathbf{F}_q$, where $q = p^a$ and p is prime. Thus G_0 is one of the simple groups $G_2(q)\,(q \geq 3)$, $F_4(q)$, $F_6(q)$, $E_7(q)$, $E_8(q)$, $^2B_2(q)$, $^2G_2(q)$, $^2F_4(q)'$, $^2E_6(q)$ or $^3D_4(q)$. Our goal is to classify all possible pairs (S, G).

Write v and b for the number of points and lines in S, respectively. Also let r be the number of lines through a point and k the number of points on a line (note that by flag-transitivity, r and k are independent of the point or line chosen). Now let M be the stabilizer in G of a point x in S, and let L be the stabilizer of a line ℓ through x. Thus $M \cap L$ is the stabilizer of a flag and we have

$$v = |G : M|,$$
$$b = |G : L|,$$
$$r = |M : M \cap L|,$$
$$k = |L : M \cap L|.$$

Now set

$$M_0 = M \cap G_0, \qquad L_0 = L \cap G_0,$$

and define

$$m = |G : G_0|,$$
$$r_0 = |M_0 : M_0 \cap L_0|.$$

In the following Lemma we record some fundamental information which is essential to the later arguments.

LEMMA 2.1.
(i) $vr = bk$.
(ii) M is maximal in G.
(iii) $G_0 \not\leq M$ and $|M : M_0| = m$.
(iv) $|M|^3 > |G|$.
(v) $m^2|M_0|^3 > |G_0|$.
(vi) $v = |G_0 : M_0|$.
(vii) $r \,|\, (|M|, v - 1)$.
(viii) $r^2 > v$.
(ix) $|Out(G_0)|^2 (|M_0|, |G_0 : M_0| - 1)^2 > |G_0 : M_0|$.
(x) $r_0 \,|\, r$ and $\frac{r}{r_0} \,|\, m$.
(xi) If u is a prime divisor of r_0, then M_0 contains a Sylow u-normalizer in G_0.
(xii) If $O_u(M) \neq 1$ for some prime u, then M contains a Sylow u-subgroup of G_0.
(xiii) $r \,|\, |M \cap M^g|$ for all $g \in G \setminus M$.

PROOF. Assertions (i), (ii), (iv), (vii), (viii), (xii) and (xiii) appear in [4, §3]. To prove (iii), note that G acts faithfully on the points of S, and so $G_0 \not\leq M$. Therefore $MG_0 = G$ by the maximality of M, whence $|M : M_0| = |G : G_0| = m$. Parts (v) and (vi) follow immediately from (iii) and (iv). Furthermore (ix) is a direct consequence of (iii), (vi), (vii) and (viii). Clearly the orbits of M_0 on the lines through x all have the same length, and hence (x) follows. To prove (xi), first observe that $u \,|\, r$ by (x). Therefore $u \,|\, v - 1$ by (vii), and hence M contains a Sylow u-subgroup of G. Therefore M_0 contains a Sylow u-subgroup U of G_0. Now take $g \in N_{G_0}(U)$. We seek to show that $g \in M_0$, so assume for a contradiction that $g \notin M_0$. Then g moves the point x to another point y, and thus U fixes the unique line ℓ' through x and y. Therefore $U \leq M_0 \cap L_0'$, where $L_0' = N_{G_0}(\ell')$, and consequently u does not divide $\left| M_0 : M_0 \cap L_0' \right|$. But we have already seen that $r_0 = \left| M_0 \cap L_0' \right|$, and this is a contradiction as $u \,|\, r_0$. The proof is now complete. □

Note that if all the maximal subgroups of G are known, then one can usually eliminate all possibilities for M using the results provided by Lemma 2.1 (in particular, Lemma 2.1.ix). This is essentially the strategy adopted in [11], where all pairs (S, G) are determined provided G is one of the groups low-rank $L_2(q)$, $L_3(q)$, $U_3(q)$ or $Sz(q) = {}^2B_2(q)$. Similarly, one can handle other low-rank cases where G_0 is $^2G_2(q)$, $G_2(q)$ or $^3D_4(q)$, for the maximal subgroups of these groups (and of their automorphism groups) are known (see [1, 10, 16, 17, 19, 24]). As for the high-rank exceptional groups, complete lists of the maximal subgroups are not as yet determined, and so additional information (see (a)-(d) above) is required.

Aschbacher (see [2]) has recently completed a detailed investigation of the subgroups of E_6. As a consequence of his theorems, it will be rather easy to show that no maximal subgroup of any automorphism group of $E_6(q)$ can satisfy the conditions in Lemma 2.1. Furthermore, it will not be long before Aschbacher's E_6 theorems yield strong results concerning the maximal subgroups of F_4 and $^2E_6$. In view of this recent progress, my original proof that G_0 cannot be E_6, $^2E_6$ or F_4 will most likely be superceded by a new proof relying on Aschbacher's work. This fact provides us with a fairly good excuse for omitting an analysis of these groups in this paper!

It will be convenient in §§ 4,5 to make use of the *cyclotomic polynomials*, defined by

$$\Phi_d(x) = \prod_{\substack{1 \le i \le d \\ (i,d)=1}} (x - \omega_d^i),$$

where ω_d is a primitive d^{th} root of unity in the complex numbers. We set $\Phi_d = \Phi_d(q)$, and when d is odd we define $\Phi_{-d} = \Phi_{2d}$. Thus for $d \in \{1, \ldots, 10, 12, 14, 15, 18, 20, 24, 30\}$, the polynomials Φ_d are given by

$$\Phi_1 = q - 1$$
$$\Phi_{-1} = \Phi_2 = q + 1$$
$$\Phi_3 = q^2 + q + 1$$
$$\Phi_4 = q^2 + 1$$
$$\Phi_5 = q^4 + q^3 + q^2 + q + 1$$
$$\Phi_{-3} = \Phi_6 = q^2 - q + 1$$
$$\Phi_7 = q^6 + q^5 + q^4 + q^3 + q^2 + q + 1$$
$$\Phi_8 = q^4 + 1$$
$$\Phi_9 = q^6 + q^3 + 1$$
$$\Phi_{-5} = \Phi_{10} = q^4 - q^3 + q^2 - q + 1$$
$$\Phi_{12} = q^4 - q^2 + 1$$
$$\Phi_{-7} = \Phi_{14} = q^6 - q^5 + q^4 - q^3 + q^2 - q + 1$$
$$\Phi_{15} = q^8 - q^7 + q^5 - q^4 + q^3 - q + 1$$

$$\Phi_{-9} = \Phi_{18} = q^6 - q^3 + 1$$
$$\Phi_{20} = q^8 - q^6 + q^4 - q^2 + 1$$
$$\Phi_{24} = q^8 - q^4 + 1$$
$$\Phi_{-15} = \Phi_{30} = q^8 + q^7 - q^5 - q^4 - q^3 + q + 1.$$

Let n and c be integers with $n, c \geq 2$. A prime which divides $n^c - 1$ but which does not divide $n^i - 1$ whenever $1 \leq i < c$ will be called a *primitive prime divisor of* $n^c - 1$, and we will write n_c for such a prime. A famous theorem of Zsigmondy shows that n_c exists for almost all pairs n, c.

PROPOSITION 2.2 (Zsigmondy [28]). *Assume that* n *and* c *are integers with* $n, c \geq 2$. *Then there is a primitive prime divisor of* $n^c - 1$, *unless* $(n, c) = (2, 6)$, *or* $c = 2$ *and* $n + 1$ *is a power of* 2.

Remarks. (i) If u is a prime divisor of Φ_c but u does not divide Φ_i for $1 \leq i < c$, then $u = q_c$.

(ii) Observe $c \mid n_c - 1$.

(iii) As a matter of convenience, when c is odd we set $n_{-c} = n_{2c}$, so that n_{-c} denotes any primitive prime divisor of $n^{2c} - 1$.

We conclude this section by giving our conventions for expressing the structures of groups. If X and Y are groups, then $X.Y$ denotes an extension of X by Y. Further, X^n denotes the direct product of n copies of X and $\frac{1}{n}X$ denotes a subgroup of index n in X. We write Z_n or simply n for the cyclic group of order n, and D_n denotes the dihedral group of order n. Furthermore, we write $[n]$ for an arbitrary group of order n. For any prime power s we write

$$L_m^+(s) = PSL_m(s) = L_m(s),$$
$$L_m^-(s) = PSU_m(s) = U_m(s),$$
$$E_6^+(s) = E_6(s),$$
$$E_6^-(s) = {}^2E_6(s).$$

The groups $E_6^\pm(s)$ and $E_7(s)$ always denote simple groups (as opposed to quasisimple groups with non-trivial center). Finally, the symbol ϵ denotes $+$ or $-$.

3. The case $G_0 \cong {}^2G_2(q)$. In this section assume that $G_0 \cong {}^2G_2(q)$, where $q = 3^a$, with a odd and $a \geq 3$. Thus $|G_0| = q^3(q^3 + 1)(q - 1)$ and $Out(G_0)$ is cyclic of order a. In particular, $m \mid a$, where (recall) $m = |G : G_0|$. Our goal is to show that S is the Ree Unital $U_R(q)$. We begin our analysis with the list of maximal subgroups of G. This list was first obtained for the simple group G_0 in [19, Theorem 1], and then independently in [17, Theorem C] for all groups with socle G_0.

PROPOSITION 3.1. *Assume that* $G_0 \cong {}^2G_2(q)$ *with* $q = 3^a \geq 27$, *and that* H *is a maximal subgroup of* G *not containing* G_0. *Then* $H \cap G_0$ *has one of the following structures*:

$$[q^3].Z_{q-1},$$
$$Z_{q\pm\sqrt{3q}+1}.Z_6,$$
$${}^2G_2(q_0), \qquad q = q_0^\alpha \text{ with } \alpha \text{ prime},$$
$$2 \times L_2(q),$$
$$(2^2 \times D_{(q+1)/2}).3.$$

THEOREM 3.2. *If* $G_0 \cong {}^2G_2(q)$ *with* $q = 3^a \geq 27$, *then* S *is the Ree Unital* $U_R(q)$.

PROOF. Since $m \mid \log_3(q)$, Lemma 2.1.v shows that M_0 is not $Z_{q\pm\sqrt{3q}+1}.Z_6$ or $(2^2 \times D_{(q+1)/2}).3$. Assume for the moment that $M_0 \cong {}^2G_2(q_0)$ with $q = q_0^\alpha$. Then Lemma 2.1.v forces $\alpha = 3$. In this case $v = q^2(q^2-q+1)(q^{2/3}+q^{1/2}+1) \geq q^{14/3}$. On the other hand, since ${}^2G_2(q^{1/3})$ does not contain a Sylow 3-subgroup of G_0 we know that $(r_0, 3) = 1$ by Lemma 2.1.xi, and hence $r_0 \leq (q+1)(q^{1/3}-1)$. But then $r \leq (q+1)(q^{1/3}-1)m < \sqrt{v}$, against Lemma 2.1.viii. And if $M_0 = 2 \times L_2(q)$, then $v - 1 = (q-1)(q^3+q+1)$, and hence $(v-1, |M_0|) = q-1$. Therefore by Lemma 2.1.vii $r \leq (v-1, |M|) \leq m(q-1) < \sqrt{v}$, which yields the same contradiction.

It therefore follows that $M_0 = [q^3].Z_{q-1}$, which is a Sylow 3-normalizer in G_0. We have $v = q^3 + 1 = 3^{3a} + 1$, and so $r = 3^e$ where $\frac{3a}{2} \leq e \leq 3a$. Thus $k = 3^{3a-e} + 1$ and so

(3.1) $$b = \frac{3^e(3^{3a}+1)}{3^{3a-e}+1}.$$

Recall L is the stabilizer in G of a line ℓ, so that $|G:L| = b$. It is clear that L does not contain G_0 since G acts faithfully on the lines of S. Hence $L_0 = L \cap G_0$ is a proper subgroup of G_0. Now let N_0 be a maximal subgroup of G_0 containing L_0. Then $|G_0 : N_0| \mid b$, and by running through the list in Proposition 2.1 and using (3.1), we find that the only possibility is $N_0 = 2 \times L_2(q)$. Assume for the moment that L_0 is solvable. Then knowledge of the maximal subgroups of $L_2(q)$ shows that L_0 embeds in either $2 \times [q].Z_{(q-1)/2}$, $2 \times D_{q\pm1}$ or $2 \times A_4$. If L_0 embeds in $2 \times [q].Z_{(q-1)/2}$, then it follows that $\frac{1}{2}q^2(q^3+1)$ divides b, which contradicts (3.1). Similarly one checks that L_0 cannot embed in $2 \times D_{q\pm1}$ or $2 \times A_4$, and hence L_0 must be non-solvable. Therefore L_0 embeds in $2 \times L_2(q_1)$, where $\mathbf{F}_{q_1}$ is some subfield of $\mathbf{F}_q$. Again, we must have $|G_0 : L_0|$ dividing b as in (3.1), and so it follows that $q_1 = q$. Consequently L_0 is either $L_2(q)$ or $2 \times L_2(q)$. Since the stabilizer of a point in S is a solvable group, the subgroup $L_2(q)$ of L_0 must act non-trivially on the line ℓ. Therefore $3^{3a-e} + 1 = k \geq q + 1$, as $q + 1$ is the smallest degree of a non-trivial action of $L_2(q)$. Hence $e \leq 2a$, and because $|G_0 : N_0| \mid b$ we know that $q^2 \mid 3^e$, which means $2a \leq e$. Consequently $e = 2a$, which is to say $r = q^2$. we now see that $k = q + 1$ and $b = q^2(q^2 - q + 1)$, whence $L_0 = 2 \times L_2(q)$.

We are now in a position to identify S as the Ree Unital. Define a map ϕ from S to $U_R(q)$ as follows. For a point $x' \in S$, we know that $N_{G_0}(x')$ is the normalizer of Sylow 3-subgroup in G_0, since $N_{G_0}(x')$ is G-conjugate to $N_{G_0}(x) = M_0$. We define $\phi(x')$ to be that Sylow 3-subgroup. Also, for a line ℓ', we know that $N_{G_0}(\ell')$ centralizes a unique involution G_0, and we define $\phi(\ell')$ to be that involution. Thus ϕ sets up a bijection between the points and lines of S with the points and lines of $U_R(q)$. To prove that ϕ is an isomorphism of linear spaces it remains to prove that ϕ preserves incidence. So suppose that x' and ℓ' are incident in S. Then (x', ℓ') forms a flag, and hence

$$(3.2) \qquad\qquad |N_{G_0}(x') \cap N_{G_0}(\ell')| = q(q - 1).$$

Now for all $g \in G_0$, we have $\phi(gx') = \phi(x')^g$ and $\phi(g\ell') = \phi(g\ell')^g$ (here G_0 acts on S on the left). Therefore (3.2) implies that $|H| = q(q-1)$, where $H = N_{G_0}(\phi(x')) \cap N_{G_0}(\phi(\ell'))$. Suppose for a contradiction that $\phi(x')$ and $\phi(y')$ are not incident in $U_R(q)$. Then H normalizes two distinct Sylow 3-subgroups of G_0, namely $\phi(x')$ and $\phi(x')^{\phi(\ell')}$. However, the normalizer of two distinct Sylow 3-subgroups has order $q - 1$, which contradicts the fact that $|H| = q(q - 1)$. Therefore $\phi(x')$ and $\phi(\ell')$ are in fact incident, and so ϕ is the desired isomorphism. $\square$

4. Proving $G_0 \not\cong E_8(q)$. In this section we shall demonstrate that $G_0 \not\cong E_8(q)$. Recall that $E_8(q)$ is a simple group of order

$$q^{120}(q^2 - 1)(q^8 - 1)(q^{12} - 1)(q^{14} - 1)(q^{18} - 1)(q^{20} - 1)(q^{24} - 1)(q^{30} - 1)$$
$$= q^{120}\Phi_1^8\Phi_2^8\Phi_3^4\Phi_4^4\Phi_5^2\Phi_6^4\Phi_7\Phi_8^2\Phi_9\Phi_{10}^2\Phi_{12}^2\Phi_{14}\Phi_{15}\Phi_{18}\Phi_{20}\Phi_{24}\Phi_{30},$$

and $Out(E_8(q))$ is cyclic of order $a = \log_p(q)$. Also, we write $W(E_8)$ for the Weyl group of E_8, so that $W(E_8) \cong 2.O_8^+(2)$.

We begin our analysis by presenting a theorem in [9], where the local maximal subgroups of $E_8(q)$ and its automorphism groups are determined. We say that a group X is *local* if $O_u(X) \neq 1$ for some prime u.

PROPOSITION 4.1. [9] *Assume that* $G_0 \cong E_8(q)$ *and that* H *is a maximal subgroup of* G. *Further suppose that* $H_0 = H \cap G_0$ *is local. Then either* H_0 *is a parabolic subgroup of* G_0 *or* H_0 *appears in Table* 1. *Conversely, each group in Table* 1 *is maximal in* $E_8(q)$. *In the right-hand column appears those primes* u *for which* H_0 *is a Sylow* u-*normalizer in* G_0.

The notation $\Phi_\epsilon^8 . W(E_8)$ denotes a group with a normal homocyclic subgroup $Z_{q-\epsilon} \times \cdots \times Z_{q-\epsilon}$ (8 times) and such that the quotient group is isomorphic to the Weyl group $W(E_8)$. Similarly, the entry $\Phi_{\epsilon 3}^4.[155520]$ denotes a group with a normal homocyclic subgroup $Z_{q^2+\epsilon q+1} \times \cdots \times Z_{q^2+\epsilon q+1}$ (4 times) such that the quotient group has order 155520.

THEOREM 4.2. *We have* $G_0 \cong E_8(q)$.

PROOF. Assume for a contradiction that $G_0 \cong E_8(q)$. Our starting point is a theorem of Liebeck and Saxl [21] which classifies the large subgroups of $E_8(q)$. The main theorem of [21] implies that one of the following holds. (a) M_0 is a parabolic subgroup of G_0; (b) M_0 is isomorphic to $d.P\Omega_{16}^+(q).d$, $d.(L_2(q) \times E_7(q)).d$ or $E_8(q^{1/2})$, where $d = (2, q-1)$; (c) $|M_0| < q^{110}$.

Table 1		
H_0	conditions on q	u
$2.P\Omega_{16}^+(q).2$	q odd	
$2.(L_2(q) \times E_7(q)).2$	q odd	
$2^2.P\Omega_8^+(q)^2.2^2.D_{12}$	q odd	
$2^2.P\Omega_8^+(q^2).D_{12}$	q odd	
$2^4.L_2(q)^8.2^4.2^3.L_3(2)$	q odd	
$2^{5+10}.L_5(2)$	q odd	
$3.L_9^\epsilon(q).S_3$	$9 \mid q - \epsilon$	
$3.(L_3^\epsilon(q) \times E_6^\epsilon(q)).S_3$	$3 \mid q - \epsilon$	
$L_2(2) \times E_7(2)$	$q = 2$	
$3^2.L_3^\epsilon(q)^4.3^2.2.S_4$	$3 \mid q - \epsilon$	
$5.L_5^\epsilon(q)^2.5.4$	$5 \mid q - \epsilon$	
$5.U_5(q^2).4$	$5 \mid q^2 + 1$	
$5^3.L_3(5)$	$9 \le q \equiv \pm 1 \pmod 5$, $q \le p^2$	
$\Phi_\epsilon^8.W(E_8)$	$(q, \epsilon) \neq (2, +), (3, +), (4, +)$	$u \mid \Phi_\epsilon$, $u \neq 2, 3, 5, 7$
$\Phi_{\epsilon 3}^4.[155520]$	$(q, \epsilon) \neq (2, -)$	$u \mid \Phi_{\epsilon 3}$, $u \neq 3$
$\Phi_4^4.[46080]$		$u \mid \Phi_4$, $u \neq 2, 5$
$\Phi_{\epsilon 5}^2.[600]$		$u \mid \Phi_{\epsilon 5}$, $u \neq 5$
$\Phi_{12}^2.[288]$		$u \mid \Phi_{12}$
$\Phi_{\epsilon 15}.[30]$		$u \mid \Phi_{\epsilon 15}$

Suppose first that (a) occurs. Using the notation of [7, §8.5], we have $M_0 = P_J = U_J L_J$, where $J \subseteq \Pi = \{p_1, \ldots, p_8\}$. Also let $H, N, \Phi, X_s (s \in \Phi), U, V$ be as given in [7, Chapters 5-7]. Thus we may identify N/H with $W(E_8)$. Let $w = w_1 \ldots w_8$ be a Coxeter element of $W(E_8)$, where w_i is the reflection corresponding to the fundamental root p_i (see [7, §10.3]). Then $|w| = 30$ and a direct calculation shows that w^{15} sends s to $-s$ for all roots $s \in \Phi$. Thus if $g \in N$ satisfies $gH/H = w^{15}$, then $X_s^g = X_{-s}$ for all root groups X_s, with $s \in \Phi$. Now g normalizes H and it is clear that if $s \in \Phi_J$ then $-s \in \Phi_J$. Consequently g normalizes L_J. Therefore $P_J^g = (U_J \cap U_J^g)L_J$. Moreover $U_J \leq U$ and hence $U_J^g \leq U^g = V$. However $U \cap V = 1$ (see [7, Lemma 7.1.2]), and so $|M_0 : M_0 \cap M_0^g| = |U_J|$, which is a power of p. Consequently $|M : M \cap M^g|$ divides mp^x for some x, and hence $r \mid mp^x$ by Lemma 2.1.xiii. We now claim that M_0 is a maximal parabolic subgroup of G_0. For write $M_0 = P_J \leq P_K < G_0$, where P_K is a maximal parabolic of G_0, so that $J \subseteq K \subset \Pi$ and $|K| = 7$. The Frattini argument shows $M = M_0 N_G(B)$, where B is the Borel subgroup $B = P_\varnothing \leq P_J$. But clearly $N_G(B)$ acts on the set of 2^8 parabolics containing B, and so $N_G(B)$ acts on the set of eight maximal parabolics containing B (see [7, Theorems 8.3.2, 8.3.4]). Using the Dynkin diagram associated with E_8, one checks that these eight parabolics have Levi factors of the following types: D_7, $A_1 \times A_6$, $A_1 \times A_2 \times A_4$, $A_4 \times A_3$, $D_5 \times A_2$, $E_6 \times A_1$, E_7 and A_7. Since these types are all distinct, $N_G(B)$ must fix each of these eight parabolics, and hence $N_G(B) \leq N_G(P_K)$. Therefore $M \leq N_G(P_K)$, and so by the maximality of M we have $M = N_G(P_K)$. Therefore $M_0 = N_{G_0}(P_K) = P_K$, as desired. Now one checks that the index in G_0 of each of these eight parabolics is congruent to $1 + q$ modulo q^2. In particular, $v - 1 \equiv q \pmod{q^2}$. We have already seen that $r \mid mp^x$, and since $r \mid v - 1$ by Lemma 2.1.vii, we conclude $r \leq mq$. But clearly this contravenes Lemma 2.1.viii, and so we have disposed of case (a).

Next suppose that (b) occurs, and take $M_0 \cong d.P\Omega_{16}^+(q).d$, so that $|M_0| = q^{56}\Phi_1^8\Phi_2^8\Phi_3^5\Phi_4^4\Phi_5\Phi_6^2\Phi_7\Phi_8^2\Phi_{10}\Phi_{12}$. Evidently $\Phi_3 \mid |G_0 : M_0|$, and so Φ_3 divides both v and $|M|$. Thus by Lemma 2.1.vii $(r, \Phi_3) = 1$. Similarly $(r, p\Phi_5\Phi_6\Phi_{10}\Phi_{12}) = 1$, and hence $(r, |M_0|) \leq q^{44}$. Therefore using lemma 2.1.iii,vii we have $r \leq mq^{44}$. On the other hand $v > q^{128}$, and so Lemma 2.1.viii is violated. In the same fashion we eliminate the case in which M_0 is isomorphic to $d.(L_2(q) \times E_7(q)).d$ or $E_8(q^{1/2})$.

It remains to consider situation (c). As $|M_0| < q^{110}$ and $|G_0| \geq \frac{1}{2}q^{248}$ we have $v > \frac{1}{2}q^{248-110}$, and hence by Lemma 2.1.viii

$$(4.1) \qquad\qquad\qquad r > \frac{1}{\sqrt{2}}q^{69}.$$

Suppose that u is a prime divisor of (r_0, Φ_4), so that M_0 contains a Sylow u-normalizer in G_0 by Lemma 2.1.xi. If $u \neq 2, 5$ then Table 1 implies that the Sylow u-normalizer is maximal in G_0. Thus M_0 is equal to the Sylow u-normalizer, which means $|M_0| = 46080\Phi_4^4$, which violates Lemma 2.1.v. This contradiction shows that $u \in \{2, 5\}$, and hence we have proved that (r_0, Φ_4) divides $(2, q - 1)(\Phi_4)_5$, where $(\Phi_4)_5$ denotes the 5-part of Φ_4 (e.g., if $q = 7$ then $(\Phi_4)_5 = 25$). In the same manner one can show that $(r_0, \Phi_{\pm 5})$ divides $(5, q \mp 1)$, that $(r_0, \Phi_{\pm 3})$ divides $(3, q \mp 1)$, and that $(r_0, \Phi_{12}\Phi_{15}\Phi_{30}) = 1$. Moreover, as M_0 is not a parabolic subgroup of G_0 we know that M_0 does not contain a Sylow p-subgroup of G_0, and hence by Lemma 2.1.xi we have $(r, p) = 1$. Consequently

$$r_0 \leq \Phi_1^8 \Phi_2^8 \Phi_7 \Phi_8 \Phi_9 \Phi_{14} \Phi_{18} \Phi_{20} \Phi_{24} (2, q-1)^4 (15, q^2-1)^4 (\Phi_4^4)_5.$$

But this inequality is incompatible with (4.1), and so the proof is complete. $\square$

5. Proving $G_0 \not\cong E_7(q)$. In this section we prove that $G_0 \not\cong E_7(q)$. Recall $E_7(q)$ has order

(5.1)
$$\frac{1}{d} q^{63}(q^2-1)(q^6-1)(q^8-1)(q^{10}-1)(q^{12}-1)(q^{14}-1)(q^{18}-1)$$
$$= \frac{1}{d} q^{63} \Phi_1^7 \Phi_2^7 \Phi_3^3 \Phi_4^2 \Phi_5 \Phi_6^3 \Phi_7 \Phi_8 \Phi_9 \Phi_{10} \Phi_{12} \Phi_{14} \Phi_{18},$$

where $d = (2, q-1)$, and $Out(E_7(q))$ has order ad, where $a = \log_p(q)$. Let us assume for a contradiction that $G_0 \cong E_7(q)$, and we begin the argument as for $E_8(q)$. Applying the theorem in [21] we see that either M_0 is a parabolic, M_0 is conjugate to one of three known subgroups of G_0, or $|M_0| < q^{64}$. Reasoning as for $E_8(q)$ (we omit the details here), we eliminate the first two possibilities, thereby reducing to the case in which

(5.2)
$$|M_0| < q^{64}.$$

As in §4 we require information concerning the local subgroups of G. Again we quote [9].

PROPOSITION 5.1. [9] *Assume $G_0 \cong E_7(q)$ and that H is a maximal subgroup of G. Further suppose that $H_0 = H \cap G_0$ is local. Then either H_0 is a parabolic or H_0 appears in Table 2.*

PROPOSITION 5.2. *The group M_0 is non-local.*

PROOF. Assume for a contradiction that M_0 is local. Using Proposition 5.1, along with (5.2) and Lemma 2.1.v, we see that M_0 must be one of the groups $3.(L_3^\epsilon(q) \times L_6^\epsilon(q)).S_3$, $L_2(3) \times F_4(3)$ or $2.L_8^\epsilon(q).[e]$. The first group is discarded with the help of Lemma 2.1.ix. Furthermore, note that as M_0 is u-local, so is M, and hence M contains a Sylow u-subgroup of G by Lemma 2.1.xii. But then M_0 must contain a Sylow u-subgroup of G_0, and this eliminates the latter two groups (recall $L_2(3)$ is 2-local). $\square$

Table 2	
H_0	conditions on q
$2.(L_2(q) \times P\Omega_{12}^+(q)).2$	q odd
$2.(L_8^\epsilon(q)).[e]$	q odd, $4 \mid q - \epsilon$, $e = \frac{1}{2}(8, q - \epsilon)$
$2^2.(L_2(q)^3 \times P\Omega_8^+(q)).2^2.S_3$	q odd
$(2^2 \times D_4(q)).S_3$	q odd
$L_2(3) \times F_4(3)$	$q = 3$
$2^3.L_2(q)^7.2^3.L_3(2)$	q odd
$3.(L_3^\epsilon(q) \times L_6^\epsilon(q)).S_3$	$3 \mid q - \epsilon$
$(L_2(2)^3 \times \Omega_8^+(2)).S_3$	$q = 2$
$L_2(2) \times \Omega_{12}^+(2)$	$q = 2$
$e_\epsilon.(E_6^\epsilon(q) \times (\frac{q-\epsilon}{de_\epsilon})).e_\epsilon.2$	$(q, \epsilon) \neq (2, +), (3, +), e_\epsilon = (3, q - \epsilon)$
$(\frac{1}{d}\Phi_\epsilon^7).W(E_7)$	$(q, \epsilon) \neq (2, +), (3, +), (4, +)$

As a consequence of Proposition 5.2, the socle S of M_0 satisfies

(5.3) $S = soc(M_0) = S_1 \times \cdots \times S_t$,

with each S_i a non-abelian simple group (recall that the *socle* of a group is the subgroup generated by its minimal normal subgroups). Note that $S \trianglelefteq M$, and hence $M = N_G(S)$ by the maximality of M. Consequently $M_0 = N_{G_0}(S)$. Moreover $C_{M_0}(S) = 1$ (since each S_i is non-abelian), and so it now follows that

(5.4) $C_{G_0}(S) = 1.$

The purpose of this next result is to exhibit a relatively large prime divisor of r_0. If u is a prime and X a group, then $m_u(X)$ denotes the maximal rank of an elementary abelian u-subgroup of X.

LEMMA 5.3. *There is a prime divisor u of r_0 such that*
(i) *u divides Φ_i for some $i \in \{\pm 5, \pm 7, \pm 9, 8, 12\}$;*
(ii) *u is a primitive prime divisor of $q^i - 1$;*
(iii) *$m_u(G_0) = 1$;*
(iv) *$u \geq 11$.*

PROOF. Assume for a contradiction that all prime divisors of r_0 divide $\Phi_1 \Phi_2 \Phi_3 \Phi_4 \Phi_6$. Then $r_0 \leq h\Phi_1^7 \Phi_2^7 \Phi_3^3 \Phi_4^2 \Phi_6^3$, where h is the π-part of $\Phi_5 \Phi_7 \Phi_8 \Phi_9 \Phi_{10} \Phi_{12} \Phi_{14} \Phi_{18}$, and π is the set of primes dividing $\Phi_1 \Phi_2 \Phi_3 \Phi_4 \Phi_6$. Thus $h = (210, q^2 - 1)$. On the other hand, (5.2) implies that $v > |G_0|/q^{64}$. And since $(mr_0)^2 > v$ by Lemma 2.1.viii,x, we obtain

$$m^2 h^2 \Phi_1^7 \Phi_2^7 \Phi_3^3 \Phi_4^2 \Phi_6^3 > \Phi_5 \Phi_7 \Phi_8 \Phi_9 \Phi_{10} \Phi_{12} \Phi_{14} \Phi_{18},$$

which is false. Thus there is a prime divisor u of r_0 such that u does not divide $\Phi_1\Phi_2\Phi_3\Phi_4\Phi_6$. Since M_0 is not a parabolic subgroup, we know that M_0 does not contain a Sylow p-subgroup of G_0, and hence $u \neq p$ by Lemma 2.1.xi. Consequently $u \mid \Phi_i$ for some $i \in \{\pm 5, \pm 7, \pm 9, 8, 12\}$. Moreover, by the Remarks after Proposition 2.2 we know that u is a primitive divisor of $q^i - 1$, and hence $i \mid u - 1$. In particular, $u \geq 11$. Finally, it follows from [13, 10-1] that $m_u(G_0) = 1$. Thus u satisfies (i)-(iv), as required.   □

For the rest of this discussion we shall assume that u is the prime provided by Lemma 5.3, and we let U be a Sylow u-subgroup of M_0. By Lemma 2.1.xi we know that U is also a Sylow u-subgroup of G_0, and in fact

(5.5) $$N_{G_0}(U) \leq M_0.$$

It will be crucial to understand the Sylow u-structure of G_0, and the relevant information is recorded in the following Proposition.

PROPOSITION 5.4. *The Sylow u-normalizer and Sylow u-centralizer is given in Table 3.*

Table 3		
u	$C_{G_0}(U)$	$N_{G_0}(U)/C_{G_0}(U)$
$u \mid \Phi_{\epsilon 5}, u \neq 5$	$\left(SL_3^\epsilon(q) \times (\frac{q^5 - \epsilon}{q - \epsilon})\right) \cdot (\frac{q - \epsilon}{d})$	10
$u \mid \Phi_{\epsilon 7}, u \neq 7$	$\frac{q^7 - \epsilon}{d}$	14
$u \mid \Phi_8, u \neq 2$	$d.(L_2(q) \times L_2(q^2) \times (\frac{q^4 + 1}{d})).d$	8
$u \mid \Phi_{\epsilon 9}, u \neq 3$	$(q^6 + \epsilon q^3 + 1) \times (\frac{q - \epsilon}{d})$	18
$u \mid \Phi_{12}$	$L_2(q^3) \times (q^4 - q^2 + 1)$	12

These next few Lemmas are devoted to showing that U is in fact a subgroup of one of the groups S_i. We start by quoting some facts concerning the ranks of the elementary abelian subgroups of $E_7(q)$ and the degree of its minimal non-trivial representations. For any group X and any prime c, write $R_c(X)$ for the smallest degree of a non-trivial c-modular projective representation of X—that is, $R_c(X)$ is the smallest integer n such that there is non-trivial homomorphism from X to $PGL_n(\overline{\mathbf{F}}_c)$, where $\overline{\mathbf{F}}_c$ is the algebraic closure of $\mathbf{F}_c$.

LEMMA 5.5. *Let c be prime.*
(i) *If $c \neq p$, then $m_c(G_0) \leq 8$.*
(ii) *If c is odd and $c \neq p$, then $m_c(G_0) \leq 7$.*
(iii) $R_p(G_0) = 56$.

PROOF. Parts (i) and (ii) are proved in [8] and (iii) is proved in [20, 2.10].   □

For this next result we introduce the notation $A_i = Aut(S_i)$. We assume the reader is familiar with the structure of the outer automorphism groups of the simple groups of Lie type. In particular, we quote the fact that these outer automorphism groups are generated by diagonal, field and graph automorphisms. An exposition can be found in [7, Chapter 12] and in [13, §7]. Furthermore, we will make use of the theorem in [18], which gives lower bounds for $R_p(X)$ when X is a simple group of Lie type in characteristic coprime to p.

LEMMA 5.6.

(i) If $S_1 \cong L_n^\epsilon(s)$ for some prime power s, then $n \le 9$.

(ii) Assume that $g \in A_i \setminus S_i$ and that g has order u modulo S_i. Then $N_{A_i}(\langle g \rangle) = C_{A_i}(g)$.

PROOF. (i) Suppose for a contradiction that $n \ge 10$. If $(p, s) = 1$, then it follows from [18] that $R_p(S_1) \ge 341$, which violates Lemma 5.5.iii. Therefore $p \mid s$. Now if $\epsilon = +$, then S_1 has Lie rank at least 9, and this runs contrary to [21, (1.7A)]. This leaves the case $S_1 \cong U_n(s)$. But here, $m_c(S_1) \ge 9$ where c is a prime divisor of $s+1$, and this contradicts Lemma 5.5.i.

(ii) The outer automorphism group of an alternating or sporadic simple group has order at most 4, and so S_1 must be of Lie type. Suppose now that S_1 has an outer diagonal automorphism of order u. Then $S_1 \cong L_n^\pm(s)$ for some prime power s and some integer n with $u \mid n$. But this contradicts part (i). Therefore, u does not divide the order of the group of outer diagonal automorphisms. However, modulo the group of inner and diagonal automorphisms, the full automorphism group is abelian. Therefore (ii) holds. □

LEMMA 5.7. We have $U \le S_i$ for some i.

PROOF. It suffices to show that u does not divide $|M_0 : S|$; for then $U \le S$, and then the result will follow as $m_u(G_0) = 1$. So assume for a contradiction that $u \mid |M_0 : S|$, and let $g \in M_0$ be a u-element which has order u modulo S. According to Lemma 5.6.ii, $N_{A_i}(\langle g \rangle) = C_{A_i}(g)$ for each i. On the other hand, the right-hand column of Table 3 shows that $N_{G_0}(U)$ conjugates g to at least 8 distinct powers of itself, and hence it follows that M_0 has an orbit of size at least 8 on the set $\{S_1, \dots, S_t\}$. But now if S_j is in this orbit and c is an odd prime divisor of $|S_j|$ distinct from p, then $m_c(G_0) \ge 8$, which violates Lemma 5.5.ii. This contradiction completes the proof. □

PROPOSITION 5.8. The group M_0 is almost simple.

PROOF. We know that $U \le S_j$, for some j, and we lose no generality in assuming $U \le S_1$. Now let i be as in Lemma 5.3 and suppose for the moment that $i = 7$. Then according to Table 3, the centralizer of a Sylow

u-subgroup is solvable, and so it must be the case that $t = 1$, as desired. The same reasoning applies if we assume that $i = 9, 14, 18$. Thus for the remainder of this proof we can assume that $i \in \{5, 8, 10, 12\}$, and we consider these cases separately. We shall also assume for a contradiction that $t \geq 2$, and we set $S_* = S_2 \times \cdots \times S_t$. For any group we write X^∞ for the last term of the derived series of X, and hence

$$(5.6) \qquad\qquad S_* \leq C_{G_0}(U)^\infty.$$

Also notice that S_1 is characteristic in M_0, since it is the only component of S whose order is divisible by u. Therefore $S_i \trianglelefteq M$, and so by maximality we have $M = N_G(S_1)$. Similarly $M = N_G(S_*)$, and hence

$$(5.7) \qquad\qquad M_0 = N_{G_0}(S_1) = N_{G_0}(S_*).$$

We now consider the various possibilities for i in turn.

Case $i = 8$. The group G_0 contains a commuting subgroups $X_1 = d.P\Omega_{12}^+(q)$ and $X_2 = SL_2(q)$, and X_1 contains a Sylow u-subgroup of G_0. Thus by Sylow's Theorem, we may assume that $U \leq X_1 < X_1 X_2$. Let us write V for the natural projective module for $X_1 / Z(X_1) = P\Omega_{12}^+(q)$, so that V is a 12-dimensional $\mathbf{F}_q$-vector space which supports a non-degenerate hyperbolic quadratic form. Evidently U fixes a non-degenerate 8-dimensional elliptic subspace W of V. Hence U is centralized by a group X_3, where $X_3 \leq X_1$ and the image of X_3 in $X_1 / Z(X_1)$ is a group $\Omega_4^-(q) \cong L_2(q^2)$ fixing the 4-dimensional elliptic space $W^\perp$. It now follows from Table 3 that $C_{G_0}(U)^\infty \leq X_2 X_3$, and hence by (5.6) we have $S_* \leq X_2 X_3$. And since $X_2 X_3 \leq C_{G_0}(U) \leq M_0$ by (5.5), we conclude $S_* \leq X_2 X_3 \leq M_0$. And because $S_* \trianglelefteq M_0$, we know that S_* is either X_2, X_3, or $X_2 X_3$. If $S_* = X_2$, then by (5.7) we have $X_1 \leq C_{G_0}(S_*)^\infty \leq M_0$, which contravenes (5.1). And if $S_* = X_2 X_3$, then X_2 is characteristic in S_*, and hence X_2 is characteristic in M_0. But then $X_2 \trianglelefteq M$, and hence $M = N_G(X_2)$, by maximality; this forces $X_1 \leq M_0$, which yields the same contradiction as before. We are left with the case in which $S_* = X_3$. At this stage we appeal to the determination of the centralizers of the semi-simple elements (that is, p'-elements) of G_0 as given in [12]. Let $x \in X_3$ be an element of odd order dividing $q^2 + 1$. Then $C_{G_0}(x)$ must appear in one of the entries in [12, Table 1], and we now determine which one. On the one hand, $|x|$ must divide the number given in the column headed $|(S^g)_\sigma|$. Furthermore, x is centralized by a group X_4, where $d.\Omega_8^-(q) \cong X_4 < X_1$ and the image of X_4 in $X_1 / Z(X_1)$ fixes the 8-space W. Therefore x is centralized by $X_2 X_4 \cong d.(L_2(q) \times \Omega_8^-(q))$, which is of type $A_1(q)^2 D_4(q^2)$ in the notation of [12]. Thus inspection of [12, Table 1] shows that x must correspond to the fourth entry appearing under $D_4 + A_1$ on the right-hand side of [12, p. 200]. It therefore follows that $C_{G_0}(x)^\infty = X_2 X_4$ (except when $q \leq 3$, in which case X_2 is solvable and $C_{G_0}(x)^\infty = X_4$). Thus $S_1 \leq X_2 X_4$, and since S_1 is simple and u does not divide $|X_2|$, we deduce $S_1 \leq X_4$. But then $X_2 \leq C_{G_0}(S_1 S_*) = C_{G_0}(S)$, against (5.4).

Case $i = \epsilon 5$. The group G_0 contains a covering group of $L_8^\epsilon(q)$, which in turn has a subgroup $Y_1 \times Y_2$, with $Y_1 \cong SL_5^\epsilon(q)$ and $Y_2 \cong SL_3^\epsilon(q)$. Since Y_1 contains a Sylow u-subgroup of G_0, we may write $U \leq Y_1$. And because $C_{G_0}(U)$ is non-solvable, it follows from Table 3 that $(q, \epsilon) \neq (2, -)$ and hence $C_{G_0}(U)^\infty = Y_2$. Therefore $S_* \leq Y_2$. On the other hand, we have $Y_2 \leq C_{G_0}(U) \leq M_0$, and so $S_* \leq Y_2 \leq M_0$. Since Y_2 is quasisimple and $S_* \trianglelefteq Y_2$, it now follows that $S_* = Y_2$. Hence $Y_1 \leq M_0$ by (5.7). Now we argue as in the previous case using [12]. Namely, if $x \in Y_2$ satisfies $|x| \mid q^3 - \epsilon$ and $(3, |x|) = 1$, then x must correspond to the entry $(A_4(q), q^3 - 1)$ or $(^2A_4(q^2), q^3 + 1)$ (see [12, p. 199]). In particular, $S_1 \leq C_{G_0}(x)^\infty = Y_1$. However, we have already seen that $Y_1 \leq M_0$, and so $S_1 \trianglelefteq Y_1$. We now conclude $S = Y_1 \times Y_2$. But then $|M_0| \leq |Aut(Y_1)| \times |Aut(Y_2)| \leq 2aq^{24} \times 2aq^8 = 4a^2 q^{32}$, which contradicts Lemma 2.1.v.

Case $i = 12$. The argument here is similar to that in the previous case, so we offer only a sketch. The group G_0 contains commuting subgroups Z_1, Z_2 with $Z_1 \cong {}^3D_4(q)$ and $Z_2 \cong L_2(q^3)$. Thus we may take $U \leq Z_1$, and the usual argument now implies that $S_* \leq Z_2 \leq M_0$. Therefore $S_* = Z_2$ and it follows from [12] that $C_{G_0}(Z_2)^\infty = Z_1$, and so $S = Z_1 \times Z_2$. Once again, this violates Lemma 2.1.v. $\quad\square$

We now know that $S = S_1$ is simple, and to determine all possibilities for S we invoke the classification of simple groups and use various facts about the representations of the simple groups. Our analysis breaks up into four parts: the case where S is alternating, of Lie type in characteristic prime to p, sporadic, and of Lie type in characteristic p.

PROPOSITION 5.9. *The group S is not alternating.*

PROOF. Assume for a contradiction that $S \cong A_n$ for some $n \geq 5$. Following the proof of (3B) Step 1 in [21] we find that $n \leq 18$. Thus Lemma 2.1.v forces $q = 2$. But then 125 does not divide $|G_0|$, and so $n \leq 14$, which contravenes Lemma 2.1.v again. $\quad\square$

PROPOSITION 5.10. *The group S is not of Lie type in characteristic prime to p.*

PROOF. Assume for a contradiction that S is of Lie type in characteristic prime to p. Then $R_p(S) \leq R_p(G_0) \leq 56$, and so according to the theorem in [18], S is one of:

$L_2(c)$ with $c \leq 113$ and $c \neq 32, 64$

$L_3^{\pm}(c)$ with $c \leq 7$

$U_3(8)$, $L_4^{\pm}(2)$, $L_4^{\pm}(3)$, $U_4(4)$

$L_5^{\pm}(2)$, $L_6^{\pm}(2)$

$PSp_4(c)'$ with $c \leq 9$ and $c \neq 8$

$PSp_6(2)$, $PSp_6(3)$

$PSp_8(2)$, $PSp_8(3)$

$\Omega_8^+(2)$, $\Omega_8^-(2)$, $\Omega_7(3)$

$F_4(2)$, $G_2(3)$

${}^3D_4(2)$, ${}^2F_4(2)'$, ${}^2B_2(8)$.

We now obtain a contradiction using Lemma 2.1.v and Lagrange's Theorem. For example, if $S \cong PSp_8(3)$, then $|M_0| \leq |Aut(PSp_8(3))| = 2^{15}.3^{16}.5^2.7.13.41$, and so Lemma 2.1.v shows that $q \leq 3$. However, our assumption on the characteristic means $(3, q) = 1$, and so $q = 2$. But 41 does not divide $|E_7(2)|$. All the other possibilties are just as easy to treat, and these are left to the reader. $\square$

PROPOSITION 5.11. *The group S is not sporadic.*

PROOF. For a contradiction assume that S is sporadic. Then $R_p(S) \leq 56$ and hence it follows from [20, §5] that S cannot be J_4, Fi_{23}, Fi_{24}', the Baby Monster or the Monster. As for the remaining 21 sporadic groups, $|Aut(S)|$ is maximized when $S = Co_1$, and so $|Aut(S)| < 2^{62}$. Therefore, Lemma 2.1.v forces $G_0 = E_7(2)$. However, using Lagrange's Theorem and Lemma 2.1.v again, we eliminate all possibilites for the S, apart from Fi_{22}. However, if $S \cong Fi_{22}$, then M contains no Sylow 2-, 3-, 7-subgroup of G, and so $r \leq 5.11.13 < \sqrt{v}$, a contradiction. $\square$

We complete our analysis by proving

PROPOSITION 5.12. *The group S is not of Lie type in characteristic p.*

PROOF. Assume for a contradiction that S is of Lie type in characteristic p. First of all, it follows from [21, (1.7A)] that the Lie rank of S is at most 7, and also S is not of type U_n for any $n \geq 10$ by Lemma 5.6.i. Thus we may write $S \cong X(p^b)$, where X is one of the symbols L_n ($2 \leq n \leq 8$), U_n ($3 \leq n \leq 9$), Ω_{2n+1} ($3 \leq n \leq 7$), PSp_{2n} ($2 \leq n \leq 7$), $P\Omega_{2n}^{\pm}$ ($4 \leq n \leq 7$), $P\Omega_{16}^-$, E_n ($n = 6, 7$), ${}^2E_6$, F_4, ${}^2F_4$, G_2, ${}^2G_2$, ${}^2B_2$ or ${}^3D_4$. In view of (5.2) we have $v > q^{68}$, and so

(5.8) $r \geq q^{34}$.

Recall that $(p, r) = 1$ since M does not contain a Sylow p-subgroup of G_0, and hence $r \leq mr_0 \leq m|M_0|_{p'}$, which implies $m|Aut(S)|_{p'} \geq q^{34}$ (here $x_{p'}$ denotes the p'-part of the integer x). As $m \leq |Out(G_0)| = da = (2, q - 1) \log_p(q) < q$, we deduce

$$(5.9) \qquad\qquad |Aut(S)|_{p'} > q^{33}.$$

Along with (5.9) the following result will also be quite useful. Recall the discussion of primitive prime divisors given in §2.

LEMMA 5.13. *Assume that* $p_{cb} \mid |S|$ *for some integer* c, *where* $cb \geq 2$ *and* p_{cb} *is a primitive prime divisor of* $p^{cb} - 1$.
(i) $cb \mid ia$ *for some* $i \in I = \{8, 10, 12, 14, 18\}$.
(ii) *Either* $cb = ja$ *for some* $j \in J = \{6, 7, 8, 9, 10, 12, 14, 18\}$, *or* $cb \leq 5a$.

PROOF. Obviously p_{cb} divides $q^i - 1$ for some $i \in I$ (see (5.1)). Therefore $p_{cb} \mid p^{ia} - 1$, and it is now clear that $cb \mid ia$, as claimed. If $cb \neq ja$ for all $j \in J$, then $cb \leq \max \{\frac{18}{4}a, \frac{14}{3}a, \frac{12}{3}a, \frac{10}{2}a, \frac{8}{2}a\} = 5a$, proving (ii). $\square$

We resume the proof of Proposition 5.12 by considering the various possibilities for S.

Case $S = L_2(p^b)$. Obviously $b \neq 1$ by Lemma 2.1.v, and also $S \neq L_2(8)$. Therefore by Proposition 2.2 there exists p_{2b}, and by Lemma 5.13.i we have $2b \leq 18a$. Hence $b \leq 9a$. But then $|Aut(S)| = bp^b(p^{2b} - 1) < bp^{3b} \leq 9aq^{27} < q^{32}$, contrary to (5.9).

Case $S = L_3(p^b)$. Here $S \neq L_3(4)$, and so there exists p_{3b}. Hence $3b \leq 18a$, forcing $b \leq 6a$. Therefore $|Aut(S)|_{p'} = (2bp^{6b}(p^{3b} - 1)(p^{2b} - 1))_{p'} \leq (2, q - 1)bp^{5b} \leq (2, q - 1)6aq^{30} < q^{33}$, violating (5.9).

Case $S = L_4(p^b)$. As before we obtain $4b \leq 18a$. Suppose for the moment that $4b = 18a$. Since p_{3b} divides $|S|$, Lemma 5.13.i implies that $3b$ divides ia for some $i \in I$. However $3b = \frac{27}{2}a$, which divides no such ia. Similarly $4b \neq 14a$. Therefore $4b \leq 12a$ and so the usual reasoning yields $|Aut(S)|_{p'} < 6aq^{27} < q^{33}$, a contradiction.

Case $S = L_5(p^b)$. We have $p_{5b} \mid |S|$, and if $5b \leq 10a$, then $|Aut(S)|_{p'} < q^{33}$, against (5.9). Therefore by Lemma 5.13 we see that $5b = ja$ for $j = 12, 14$ or 18. But $p_{4b} \mid |S|$ and we find that $4b = \frac{4j}{5}a$ does not divide ia for all $i \in I$. This is a contradiction.

The remaining possibilities for S are treated in an entirely similar fashion, and we omit most of the details. We include a few more cases, some of which require some arguments not illustrated so far.

Case $S = {}^2F_4(p^b)'$ *and* ${}^3D_4(p^b)$. Here $12b \leq 18a$, and hence $b \leq \frac{3}{2}a$. Therefore $|Aut(S)|_{p'} \leq 3bp^{16b} \leq \frac{9}{2}aq^{24}$, contradicting (5.9).

Case $S = L_8^\epsilon(p^b)$. Here p_{8b} divides $|S|$, and if $8b \leq 5a$ then $|Aut(S)|_{p'}$ $\leq 2bp^{35b} \leq \frac{5a}{4}q^{22}$, against (5.9). Therefore $8b = ja$ for some $j \in J$. Now $p_{\epsilon 7b}$ also divides $|S|$, and so $7b \mid ia$ for some $i \in I$. But $7b = \frac{7ja}{8}$, and hence $7j \mid 8i$, forcing $i = 14$ and $j = 8$. Thus $S \cong L_8^\epsilon(q)$. Now let u be an odd prime divisor of Φ_8. Then the Sylow u-normalizer in $L_8^\epsilon(q)$ is solvable, and hence it is solvable in M_0. However according to Table 3, the Sylow u-normalizer in G_0 involves $L_2(q^2)$, which is non-solvable. Therefore M_0 does not contain a Sylow u-normalizer in G_0, and so $(r, \Phi_8) = (2, q - 1)$. Now $|PGL_8^\epsilon(q)| = q^{28} \prod_{c=2}^8 (q^c - \epsilon^c) = q^{28}\Phi_\epsilon^7\Phi_{-\epsilon}^4\Phi_{3\epsilon}^2\Phi_{-3\epsilon}\Phi_4^2\Phi_{\epsilon 5}\Phi_{\epsilon 7}\Phi_8$, and hence

$$r \leq mr_0 \leq m \mid Aut(S) : PGL_8^\epsilon(q) \mid (2, q - 1)\Phi_\epsilon^7\Phi_{-\epsilon}^4\Phi_{\epsilon 3}^2\Phi_{-\epsilon 3}\Phi_4^2\Phi_{\epsilon 5}\Phi_{\epsilon 7}$$
$$< 2(2, q - 1)^2 a^2 q^{31} < q^{34},$$

which violates (5.8).

Case $S = PSp_6(p^b)$ *and* $S = \Omega_7(p^b)$. If $p^b = 2$ then Lemma 2.1.v fails, and hence p_{6b} divides $|S|$. If $6b \leq 14a$, then $|Aut(S)|_{p'} \leq bp^{12b} \leq \frac{7}{3}aq^{28}$, against (5.9). Consequently $6b = 18a$, and hence $b = 3a$. However it is shown in [21, (4B)] that neither $PSp_6(q^3)$ nor $\Omega_7(q^3)$ is contained in G_0, and so this case is finished.

Case $S = E_7(p^b)$. Here p_{18b} divides $|S|$, and if $18b < 5a$, then $|Aut(S)|$ $\leq bp^{133b} < \frac{5a}{18}q^{37} < q^{38}$, which violates Lemma 2.1.v. Therefore by Lemma 5.13 we have $18b = ja$ for some $j \in J$. Since p_{14b} also divides $|S|$, we know that $14b = \frac{7ja}{9}$ must divide ia for some $i \in I$. Thus $i = 14$ and $j \mid 18$. Therefore j is 6 or 9. However $j \neq 9$ in view of (5.2), and so $S \cong E_7(q^{1/3})$. But then $|Aut(S)|_{p'} \leq bp^{70} = \frac{a}{3}q^{70/3} < q^{33}$, contrary to (5.9).

Case $S = P\Omega_{12}^\pm(q)$. We have $p_{10b}||S|$, and if $10b \leq 5a$ then Lemma 2.1.v fails. Therefore $10b = ja$ for some $j \in J$. Since p_{8b} also divides $|S|$, we see as before that $4j \mid 5i$ for some $i \in I$, and hence i is 8 or 12. Thus $j = 10$, which is to say $a = b$. But this contradicts (5.2).

The reader may easily eliminate the remaining possibilities for S. $\quad\square$

Combining the results in this section we obtain

THEOREM 5.13. *We have* $G_0 \not\cong E_7(q)$.

REFERENCES

1. M. Aschbacher, *Chevalley groups of type* G_2 *as the group of a trilinear form*, J. Algebra **109** (1987), 193–259.
2. _____, *The 27-dimensional module for* E_6. *I*, Invent. Math. **89** (1987), 159–195.
3. F. Buekenhout, *On a theorem of O'Nan and Scott*, Bull. Soc. Math. Belg. (B) **40** (1988) 1–9.
4. F. Buekenhout, A. Delandtsheer and J. Doyen, *Finite linear spaces with flag-transitive groups*, J. Combinatorial Theory (A) **49** (1988) 268–293.

5. F. Buekenhout, A. Delandtsheer, J. Doyen, P.B. Kleidman, M.W. Liebeck and J. Saxl, *A classification of the finite flag-transitive linear spaces,* in preparation.

6. F. Buekenhout, A. Delandtsheer and J. Doyen, *Finite linear spaces with a flag-transitive sporadic or alternating group,* preprint.

7. R.W. Carter , *"Simple Groups of Lie Type",* Wiley, New York-London, 1972.

8. A.M. Cohen and G.M. Seitz, *The r-rank of the groups of exceptional Lie type,* Proc. Koninklijke Nederlandse Akademie van. Wetenschappen (A) **90** (1987) 251–259.

9. A.M. Cohen , M.W. Liebeck, J. Saxl and G.M. Seitz, *The local maximal subgroups of the exceptional groups of Lie type,* (preprint).

10. B.N. Cooperstein, *Maximal subgroups of $G_2(2^n)$ ',* J. Algebra **70** (1981), 23–36.

11. A. Delandtsheer, *Flag-transitive finite simple groups,* Arch. Math. **47** (1986), 395–400.

12. D.I. Deriziotis, *The centralizers of semisimple elements of the Chevalley groups E_7 and E_8,* Tokyo J. Math. **6** (1983), 191–216.

13. D. Gorenstein and R. Lyons, *The local structure of finite groups of characteristic 2 type,* Mem. Amer. Math. Soc. **42** (1983).

14. D. G. Higman and J.E. McLaughlin, *Geometric ABA-groups,* Illinois J. Math. **5** (1961), 382–397.

15. W.M. Kantor, *Primitive permutation groups of odd degree, and an application to finite projective planes,* J. Algebra **106** (1987), 15–45.

16. P.B. Kleidman, *The maximal subgroups of the Steinberg triality groups $^3D_4(q)$ and of their automorphism groups,* J. Algebra **115** (1988), 182–199.

17. ____, *The maximal subgroups of the Chevalley groups $G_2(q)$ with q odd, of the Ree groups $^2G_2(q)$, and of their automorphism groups,* J. Algebra. **117** (1988), 30–71.

18. V. Landazuri and G.M. Seitz, *On the minimal degrees of projective representations of the finite Chevalley groups,* J. Algebra **32** (1974), 418–443.

19. V.M. Levchuk and Ya. N. Nuzhin, *Structure of Ree groups,* Algebra i Logika **24** (1985), 26–41.

20. M.W. Liebeck, *The affine permutation groups of rank three,* Proc. London Math. Soc. **54** (1987), 477–516.

21. M.W. Liebeck and J. Saxl, *On the orders of maximal subgroups of the finite exceptional groups of Lie type,* Proc. London Math. Soc **55** (1987), 299–330.

22. M.W. Liebeck, C.E. Praeger and J. Saxl, *On the O'Nan-Scott reduction theorem for finite primitive permutation groups,* J. Australian Math. Soc. (A) **44**(1988), 389–396.

23. H. Lüneburg, *Some remarks concerning the Ree groups of type (G_2)* J. Algebra **3** (1966), 256–259.

24. E.T. Migliori, *The determination of the maximal subgroups of $G_2(q)$, q odd,* U.C.S.C. Thesis, 1982.

25. M. Suzuki, *On a class of doubly transitive groups,* Ann. of Math. **75** (1962), 105–145.

26. P.H. Zieschang, *Fahnentransitive Automorphismengruppen von Bloclplänen,* Geom. Dedicata **18** (1985), 173–180.

27. ____, *Flag transitive automorphism groups of finite linear spaces,* preprint.

28. K. Zsigmondy, *Zur Theorie der Potenzreste,* Monatsh. für Math. u. Phys. **3** (1892),265–284.

DEPARTMENT OF MATHEMATICS, CALIFORNIA INSTITUTE OF TECHNOLOGY, PASADENA, CALIFORNIA 91125

Contemporary Mathematics
Volume 111, 1990

Constructing 6-(14,7,4) Designs

DONALD L. KREHER AND STANISŁAW P. RADZISZOWSKI

ABSTRACT. A summary of the algebraic and computational techniques used in the construction of two non-isomorphic simple 6-(14,7,4) designs and four non-isomorphic simple 5-(13,6,4) designs is presented. With the exception of the 6-(33,8,36) designs discovered by Magliveras and Leavitt, and the 6-(20,9,112) designs discovered by Kramer, Leavitt and Magliveras, this is the only other small parameter situation in which a simple 6-design is known to exist.

1. Introduction

The results and ideas in this paper are not new but instead are spread over three of our papers [KR1], [KR2], [KR4]. Thus the authors were at first reluctant to enunciate them again. However, due to the encouragement of colleagues and the apparent importance of the results, this paper was presented and received favorably at the 307th AMS meeting held in Lincoln, Nebraska. This is the only paper in which the entire details of our method appear.

2. Notation and background

The construction of the 6-(14,7,4) designs had three essential components.

   I: Incidence Matrices
  II: Basis Reduction
 III: Extension

Each of these components posed difficult and interesting problems both computational and mathematical. They will be discussed in the sections that follow. First recall that a *t-design*, or $t - (v, k, \lambda)$ *design* is a pair $(X, \mathscr{B})$ with a *v-set X of points* and a family $\mathscr{B}$ of k-subsets of X called *blocks* such that any t points are contained in exactly λ blocks. A $t - (v, k, \lambda)$ design $(X, \mathscr{B})$ is *simple* if no block in $\mathscr{B}$ is repeated. A group $G \leq Sym(X)$ is

---

1980 *Mathematics Subject Classification* (1985 *Revision*). Primary 51E05,68R05.
Research supported by National Science Foundation grant No. CCR-8711229.
This paper is in final form and no version of it will be submitted for publication elsewhere.

an automorphism group of a $t - (v, k, \lambda)$ design $(X, \mathscr{B})$ if every $g \in G$ preserves $\mathscr{B}$. For example a $2 - (7, 3, 1)$ design $(X, \mathscr{B})$ is given by $X = \{1,2,3,4,5,6,7\}$ and $\mathscr{B} = \{124,235,346,457,156,267,137\}$. The points on the 6 lines and 1 circle in Figure 1 are the 7 blocks of this design. Thus easy observation shows that it has $G = <(1\ 4\ 5)(2\ 7\ 6), (2\ 6)(4\ 5) > \simeq S_3$ as an automorphism group.

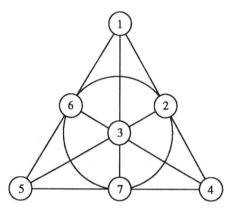

FIGURE 1. The $2 - (7, 3, 1)$ design.

The set of *all* automorphisms of a $t - (v, k, \lambda)$ design $(X, \mathscr{B})$ is said to be the *full automorphism group* and is denoted by $Aut(\mathscr{B})$. Indeed, as it is well known, for the 2-(7,3,1) design given above $Aut(\mathscr{B})$ is isomorphic to $PSL_2(7)$ and thus it has in fact 168 automorphisms.

3. Incidence matrices

Incidence matrices have been investigated by a number of researchers but probably their intimate connection to $t - (v, k, \lambda)$ designs with a given automorphism group was first given by Earl Kramer and Dale Mesner in 1976 [**KM**]. Their observation was:

A $t - (v, k, \lambda)$ design exists with $G \leq Sym(X)$ as an automorphism group if and only if there is an integer solution U to the matrix equation

$$(1) \qquad A_{tk} U = \lambda \cdot J_{N_t}$$

where:

    (a) *The N_t rows of A_{tk} are indexed by $\Delta_1, \Delta_2, \ldots, \Delta_{N_t}$ the G-orbits of t-subsets of X;*

    (b) *The N_k columns of A_{tk} are indexed by $\Gamma_1, \Gamma_2, \ldots, \Gamma_{N_k}$ the G-orbits of k-subsets of X;*

    (c) $A_{tk}[\Delta_i, \Gamma_j] = |\{K \in \Gamma_j : K \supseteq T, T \in \Delta_i \text{ fixed}\}|$;

    (d) $J_{N_t} = [1, 1, 1, \ldots, 1]^T$.

We call the matrix A_{tk} defined above an incidence matr x, and when it is important to keep track of the automorphism group we write $A_{tk}(G|X)$ for

A_{tk} so that no confusion arises. An example appears in Figure 2. A solution to equation (1) in this case is $U = [0, 1, 0, 0, 0, 0, 1, 0, 0, 1]^T$ and the correspondingly chosen 3-subsets form a 2-(7,3,1).

123	125	127
347	147	467
136	146	167

356	124	456	126	256	134	137		236	
357	457	157	247	257	345	346		237	
234	156	245	567	246	135	235	145	367	267

{12 47 16 56 57 24}	1	1	1	1	1	0	0	0	0	0
{13 34 35}	2	0	0	0	0	2	1	0	0	0
{14 45 15}	0	1	2	0	0	1	0	1	0	0
{17 46 25}	0	0	2	0	2	0	1	0	0	0
{23 37 36}	2	0	0	0	0	0	1	0	2	0
{26 27 67}	0	0	0	1	2	0	0	0	1	1

FIGURE 2. $A_{23}(G \mid X)$, $G = <(1\ 4\ 5)(2\ 7\ 6), (2\ 6)(4\ 5)>$

This observation led directly to the discovery of many previously unknown designs and we mention a few.

1975: Two 5-(17,8,80) designs. The first examples of 5-designs on an odd number of points, Kramer [K1].

1982: A large number of 5-(33,6,12), 5-(33,7,42) and 5-(33,7,126) designs. The second, third and fourth sets of examples of 5-designs on an odd number of points, Magliveras and Leavitt [ML].

1982: Thirteen 6-(33,8,36) designs. The first examples of 6-designs, Magliveras and Leavitt [ML].

1984: Two 6-(20,9,112) designs. The second set of examples of 6-designs, Kramer, Leavitt and Magliveras [KLM].

1986: Four 5-(13,6,4) designs. The fifth set of examples of 5-designs on an odd number of points, Kreher and Radziszowski [KR2].

1986: Two 6-(14,7,4) designs. The third set of examples of 6-designs, and the smallest possible 6-designs that can exist. Kreher and Radziszowski [KR2].

We also point out that an initial investigation of the algebraic properties of the A_{tk} matrices was done in [K2, K3] where some new combinatorial identities were found.

Now given this mathematical motivation our first computational problem is:

How can the matrix A_{tk} be computed efficiently?

3.1. Computing A_{tk}. If $G \leq Sym(X)$, then for any integer t, $0 \leq t \leq v$ the number of orbits of t-subsets can be computed by using the well known Cauchy-Frobenius-Burnside Lemma

$$N_t = \sum_{g \in G} \chi(g)$$

where $\chi(g)$ is the number of t-subsets fixed by g. It is easy to determine the value of $\chi(g)$ from the cycle representation of g on the point set X. If $G \leq Sym(X)$ and $0 < t < k < v$, where $v = |X|$, then given orbit representatives $\{T_i : 1 \leq i \leq N_t\}$ of t-subsets and orbit representatives $\{K_i : 1 \leq i \leq N_k\}$ of k-subsets it is easy to see that the A_{tk} matrix of G acting on X can be efficiently generated by making one pass over the group elements. For example algorithm Mat computes A_{tk} in time $O(|G| \cdot N_t \cdot N_k)$.

Algorithm Mat
Let A be an N_t by N_k array with each entry set to 0.
Let $stab$ be a 1-dimensional array of N_k entries all set to 0.
for each (g, j) where $g \in G$ and $1 \leq j \leq N_k$ do
   if $K_j = K_j^g$ then $stab[j] = stab[j] + 1$;
   for each i, $1 \leq i \leq N_t$ do
      if $T_i \subseteq K_j^g$ then $A[i, j] = A[i, j] + 1$;
   for each (i, j) where $1 \leq i \leq N_t$ and $1 \leq j \leq N_k$ do
      $A[i, j] = A[i, j]/stab[j]$.

Thus the problem of constructing an A_{tk} matrix is reduced to being able to compute a complete list of orbit representatives $Reps(t) = \{T_i : 1 \leq i \leq N_t\}$ of t-subsets and a complete list of orbit representatives $Reps(k) = \{K_i : 1 \leq i \leq N_k\}$ of k-subsets.

In our next algorithm "$\leq$" is a total linear ordering on the set of all subsets of X. In particular if we take $X = \{0, 1, 2, ..., v - 1\}$, then $code : P(X) \rightarrow \mathbb{Z}$ given by

$$code(S) = \sum_{i \in S} 2^i$$

gives such a linear ordering. That is

$$S \leq T \text{ if and only if } code(S) \leq code(T).$$

This is particularly effective when $|X| \leq$ "the machine word size" (e.g. 32) since one can take advantage of bit operations for doing set operations. If for some t a complete list of orbit representatives $Reps(t)$ of t-subsets is known we define $XReps(t+1)$ to be $\{S \cup \{x\} : S \in Reps(t) \text{ and } x \in X - S\}$.

It is easy to see for any complete list of orbit representatives, $Reps(t + 1)$, of $(t + 1)$-sets that for all $S \in Reps(t + 1)$ there is a $g \in G$ such that S^g, the image of S under g, is an element of $XReps(t + 1)$. This observation leads to algorithm **Reps**.

Let L be an empty list to which the following two operations apply: IN-SERT S, which means insert S on list L and DELETE S, which means delete S from list L.

Algorithm Reps
    for each $S \in Reps(t)$ do
        for each $x \in X - S$ do
            INSERT $\{S \cup \{x\}\}$;
    $n \leftarrow N_t \cdot (v - t)$;
    repeat
        for each S *on* L do
            for each $g \in G$ do
                $K \leftarrow S^g$;
                if $K < S$ then
                      DELETE S
                      if K is on L then
                          $n \leftarrow n - 1$
                      else
                          INSERT K
                  $S \leftarrow K$
    until $(n = N_{t+1})$;
    $Reps(t + 1) = \{S : S$ is on list L$\}$.

This algorithm is the (asymptotically) best algorithm we know of for computing $Reps(t + 1)$ from $Reps(t)$. It is easy to modify **Reps** so that the output, $Reps(t + 1)$, has the property that whenever $S \in Reps(t + 1)$ and $g \in G$, then $S \leq S^g$. This canonical form for $Reps(t + 1)$ is often useful in other applications. Since $Reps(0) = \{\varnothing\}$ it is easy to construct from algorithm **Reps** an algorithm to compute a complete list of orbit representatives of t-sets for all $0 \leq t \leq k$. An algorithm similar in spirit but different in implementation is used by Magliveras [M]. A different algorithm has also been constructed by Leo Chouinard [C]. Although his algorithm uses ideas similar to ours he makes a clever use of data structures to compute the orbit representatives and the resulting algorithm appears to run faster in practice. Its asymptotic running time has not yet been determined. A probabilistic algorithm has also been successfully used by Kramer, Leavitt and Magliveras [KLM, ML].

Now that an efficient method has been developed to obtain the matrix A_{tk} we ask the second and more difficult question.

How can equation (1) be effectively solved?

4. Solving equation (1)

Logically our approach to solving the integer linear equation (1) does not rely on the fact that the matrix A_{tk} is special, however the performance of our algorithm on general matrices has not been investigated. The related general decision problem is known to be NP-complete and thus we do not present an efficient deterministic algorithm. Instead we use the powerful tool of basis reduction [**LLL**] with some heuristics and do not guarantee that a solution, if it exists, will always be found. On the other hand we know of no situation for $t - (v, k, \lambda)$ designs and automorphism group G with $N_k < 200$ in which a solution is known to exist but our algorithm is unable to find it.

Let X be a v-set, $G \leq Sym(X)$ and consider the $(m + n)$ by $(n + 1)$ matrix B below:

$$(2) \qquad\qquad B = \begin{bmatrix} I_n & 0 \\ A_{tk} & -\lambda J_m \end{bmatrix},$$

where A_{tk} is the m by n Kramer-Mesner matrix described in (1) I_n is the n by n identity matrix, and $J_m = [1, 1, 1, \ldots 1]^T$. Let L be the $n + 1$ dimensional integer lattice spanned by the columns of B. That is

$$L = \{R \in \mathbb{Z}^{m+n} : R = B \cdot S, \text{ for some } S \in \mathbb{Z}^{n+1}\}.$$

Let E_m be the m-dimensional zero vector. Then the following proposition is clear.

PROPOSITION 1. $A_{tk} U = d \cdot \lambda J_m$ for some integer d if and only if $[U, E_m]^T \in L$.

Thus to find a (0,1)-solution U to $A_{tk} U = \lambda J_m$ we need only look for a linear combination $U = [U, E_m]^T$ of the columns of B such that U is a (0,1)-vector. If $U \neq J_m$, then we will have found a nontrivial $t - (v, k, d \cdot \lambda)$ design for some positive integer d. Note that since the complement of a design is a design, then we may assume $\|U\|^2 \leq n/2$. That is, U is a "short" vector in L. Our algorithm will try to find for L a new basis all of whose vectors are as short as we can make them. It has been our experience that the vector U tends to appear as one of the vectors in such a new basis.

4.1. Basis reduction. Let n be a positive integer. A subset L of the r-dimensional real vector space $R^r, r \geq n$ is called a *lattice* if and only if there is a basis $B = \{b_1, b_2, \ldots, b_n\}$ of an n-dimensional subspace of R^r such that every member of L is an integer linear combination of the vectors in B. Recall that given a basis $B = \{b_1, b_2, \ldots, b_n\}$ of an n-dimensional subspace of R^r, an orthogonal basis $B^* = \{b_1^*, b_2^*, \ldots, b_n^*\}$ of it may be obtained inductively via the Gram-Schmidt process of orthogonalization as follows:

(3) $$b_i^* = b_i - \sum_{j=1}^{i-1} \mu_{ij} b_j^*, \quad \text{for } 1 \le i \le n,$$

(4) $$\mu_{ij} = (b_i, b_j^*)/(b_j^*, b_j^*), \quad \text{for } 1 \le j \le i \le n,$$

where $(\cdot, \cdot)$ denotes the ordinary inner product on R^r. An ordered basis $B = [b_1, b_2, \ldots, b_n]$ for a lattice L will be said to be *y-reduced* (*or reduced*) if the following two conditions hold:

(i) $|\mu_{ij}| \le 1/2$ for $1 \le j < i \le n$,
(ii) $\|b_i^* + \mu_{i,i-1} b_{i-1}^*\|^2 \ge y \cdot \|b_{i-1}^*\|^2$ for $1 < i \le n$,

where y, $\frac{1}{4} < y < 1$ is a constant and $\|\cdot\|$ denotes ordinary Euclidean length. Lenstra et al. [LLL] describe an algorithm which, when presented with y, $\frac{1}{4} < y < 1$ and an ordered basis $B = [b_1, b_2, \ldots, b_n]$ for a lattice L as input, produces a reduced ordered basis $B' = [b_1', b_2', \ldots, b_n']$ as output. The L^3 algorithm consists of applying a finite number of two kinds of linear transformations. These are:

T1: Interchange vectors b_i and b_{i-1} if $\|b_i^* + \mu_{ii-1} b_{i-1}^*\|^2 \ge y \|b_{i-1}^*\|^2$ does not hold, for some $1 < i \le n$, and the global constant $y \in (\frac{1}{4}, 1)$.

T2: Replace b_i by $b_i - r b_j$, where $r = round(\mu_{ij})$ is the nearest integer to μ_{ij}, and $|\mu_{ij}| > \frac{1}{2}$, for some $1 \le j < i \le n$.

The efficient implementation of the sequence of transformations T1 and T2 relies mainly on the fact that old values of μ_{ij} and $\|b_i^*\|^2$ can be easily updated after each transformation without using the full process of orthogonalization. The L^3 algorithm performs the transformations T1 and T2 using a strategy somewhat resembling the bubble-sort. However as H.W. Lenstra [LLL] remarks, *any* sequence of these transformations will lead to the reduced basis.

The L^3 algorithm terminates when neither T1 nor T2 can be applied and such a situation implies that conditions (i) and (ii) are satisfied. The resulting reduced basis B' is an integer approximation to the basis B^* defined by the Gram-Schmidt orthogonalization process and as a consequence contains a "short" vector, as can be seen in the following proposition of Lenstra et al. [LLL, Proposition 1.11]:

PROPOSITION 2. *Let $B' = [b_1', b_2', \ldots, b_n']$ be a reduced basis of a lattice L. Then*

$$\|b_1'\|^2 \le 2^{n-1} \cdot \min\{\|b\|^2 : b \in L \text{ and } b \ne 0\}.$$

They also give the following polynomial worst-case running time for its performance [LLL, Proposition 1.26].

PROPOSITION 3. *Let* $B = [b_1, b_2, \ldots, b_n]$ *be an ordered basis for an integer lattice* L *and let* Max *be an integer such that* $\|b_i\|^2 \le$ Max *for* $1 \le i \le n$. *Then the* $\mathbf{L}^3$ *algorithm produces a reduced basis* $B' = [b_1', b_2', \ldots, b_n']$ *for* L *using at most* $O(n^4 \log_2 \text{Max})$ *arithmetic operations, and the integers on which these operations are performed have length at most* $O(n \log -2 \text{Max})$.

In summary, the effect of the $\mathbf{L}^3$ algorithm is such that when given a basis B of an n-dimensional lattice $L \subseteq \mathcal{Z}^r$ it produces a *reduced* basis B' of L, and

   (i) $\mathbf{L}^3$ uses at most $O(n^4 \log \text{Max})$ arithmetic operations,
   (ii) B' is *almost* orthogonal (integer approximation to Gram-Schmidt orthogonal basis),
   (iii) B' contain a "short" vector.

Furthermore, we point out that although it is only proven [**LLL**] that B' contains a vector not longer than $2^{(n-1)/2} \cdot s$, where s is the length of a shortest nonzero vector in L, in practice the $\mathbf{L}^3$ algorithm finds much shorter vectors [**LO**].

When the number of rows in the A_{tk} matrix is $m = 1$, then (1) reduces to the knapsack or subset-sum problem. The application of the $\mathbf{L}^3$ algorithm to solve the subset sum problem was first studied in 1985 by J. C. Lagarias and A. M. Odlyzko [**LO**]. Our improvements in this direction appear in [**KR3**].

When the number of rows in the A_{tk} matrix is greater than 1, then our experience has been that $\mathbf{L}^3$ by itself often doesn't find t-designs. Thus further reduction methods are necessary.

4.2. Weight reduction. If B is the $(n+1)$-dimensional reduced basis produced by the $\mathbf{L}^3$ algorithm applied to (2), then there often exist, as extensive experiments showed, pairs of indices i and j, $1 \le i, j \le n+1$, $i \ne j$, and $\epsilon \in \{+1, -1\}$ such that

(5)   if $v = b_i + \epsilon b_j$, then $\|\mathbf{v}\| < \max\{\|b_i\|, \|b_j\|\}$.

A pair (i, j), $i \ne j$, satisfies the last condition for some $\epsilon \in \{+1, -1\}$ if and only if $\min\{\|b_i\|^2, \|b_j\|^2\} < 2 \cdot |(b_i, b_j)|$. In such a case we take ϵ to have a different sign from (b_i, b_j) and substitute v for the longer of b_i and b_j, obtaining a new basis with decreased *total weight*

$$w(B) = \sum_{p=1}^{n+1} \|b_p\|^2.$$

To facilitate the process of finding successive pairs of indices (i, j) satisfying (5) and decreasing $w(B)$ the algorithm keeps an array containing (b_i, b_j) and $\|b_i\|^2$ for $1 \le i, j \le n+1$. Whenever a pair of indices (i, j) is found satisfying (5) leading to substitution of some b_i by v these arrays must be updated. It is not necessary however to recalculate $\|v\|^2$ and (v, b_k) for $k \neq i$ from the definitions. Instead we keep track of the integers $\|b_i\|^2$ and $inn_{ij} = (b_i, b_j)$, for $1 \leq j < i < n+1$. These are then sufficient since we have the formulas:

$$\|v\|^2 = \|b_i\|^2 + \|b_j\|^2 - 2|inn_{ij}|, \text{ and}$$

$$(v, b_k) = inn_{ik} + \epsilon \cdot inn_{jk}, \text{ for } 1 \leq k \leq n+1, \ k \neq i \text{ and } k \neq j.$$

A simple algorithm for finding all such pairs can be designed and implemented in time $O(n^2)$ for each reduction, producing as output a basis with smaller weight. This algorithm, let us call it *Weight-Reduction*, is a useful complement to the $\mathbf{L}^3$ algorithm. When used as follows, the algorithms $\mathbf{L}^3$ and *Weight-Reduction* jointly tend to produce much shorter vectors than using $\mathbf{L}^3$ or *Weight-Reduction* alone:

$B \leftarrow \mathbf{L}^3(B)$;
repeat
    Weight-Reduction;
    sort basis with respect to $\|b_i\|^2$;
    $B \leftarrow \mathbf{L}^3(B)$;
until $(w(B)$ does not decrease$)$;
Weight-Reduction.

The $\mathbf{L}^3$ algorithm can remove the vector b_1 from the basis B only by replacing it with a shorter vector, since for $i = 2$ if the transformation T1 can be applied then $\|b_i\|^2 < \|b_{i-1}\|^2$ (note that this is not true when $i > 2$). Hence sorting the basis with respect to $\|b_i\|^2$ guarantees that the shortest vector in the basis B will not disappear in the next iteration, unless a new shorter vector is found.

Following the above approach one can try in general to find a k-tuple of distinct vectors $b_{i_1}, \ldots, b_{i_k}$, for some $k \geq 2$, in the basis B, such that the vector

$$v = \sum_{p=1}^{k} \epsilon_p b_{i_p}, \text{ for some choice of } \epsilon_p = \pm 1, \ 1 \leq p \leq k,$$

is shorter than b_{i_k}, where b_{i_k} is the longest vector in the k-tuple. In the latter case the weight of basis B can be decreased by substituting v for b_{i_k}. Note that $\|v\| < \|b_{i_k}\|$ if and only if

(6) $$\|v\|^2 = \sum_{j=1}^{k} \|b_{i_j}\|^2 + \sum_{h \neq j} \epsilon_{i_h} \epsilon_{i_j} (b_{i_h}, b_{i_j}) < \|b_{i_k}\|^2,$$

and a necessary condition for (6) is

$$\sum_{j=1}^{k-1} \|b_{i_j}\|^2 < \sum_{h \neq j} |(b_{i_h}, b_{i_j})|.$$

Consequently, our approach is to search for such k-tuples of vectors by considering the complete graph whose vertices are the basis vectors b_i and whose edges are labeled by edge weights $|(b_i, b_j)|$. The endpoints of edges with large weight are "less" orthogonal, and hence they are good candidates for the desired k-tuple. We can try to construct it by finding subgraphs with large edge weight.

Obviously, the complete analysis of all subgraphs would be too expensive. However, we are satisfied with a heuristic search for just a few of relatively small size. As before they are used to decrease the weight of basis B. This technique leads to a generalization of the *Weight-Reduction* algorithm and improves further the behavior of the $\mathbf{L}^3$ algorithm. We have implemented this strategy for $k = 3$ and $k = 4$.

4.3. Size reduction. Recall that $[U, E_m]^T \in \mathbf{L}$ if and only if there exist integers $a_1, a_2, \ldots, a_{n+1}$ such that

$$\begin{bmatrix} U \\ E_m \end{bmatrix} = B \begin{bmatrix} a_1 \\ \vdots \\ a_{n+1} \end{bmatrix}.$$

Whence, it follows that:

(*) *If there is exactly one j such that $b_{hj} \neq 0$ for some h, $n < h \leq n+m$, then $a_j = 0$.*

In this case we let B' be B with row h and column j removed and L' be the lattice spanned by B'. Then the $(n + m - 1)$-dimensional vector $[U, E_{m-1}]^T \in L'$ if and only if the $(n+m)$-dimensional vector $[U, E_m]^T \in L$.

To achieve situation (*) for row h, $n < h \leq n+m$, we perform the following two operations

(1) Multiply row h by $c = \max_i \|b_i\|^2$.

(2) Apply *Weight-Reduction* and/or $\mathbf{L}^3$.

This almost always produces such a situation. If this procedure is successfully iterated for each h, $h = n + m, n + m - 1, \ldots, n + 1$, then the resulting basis B' will consist of $n - m + 1$, n-dimensional vectors. Furthermore

$$U \in L' \text{ if and only if } A_{tk}U = d \cdot \lambda J_m,$$

for some integer d (see Proposition 1). Thus the result of these iterations, let us call them collectively *Size-Reduction,* is a basis of shc t vectors for the integer solution space to the matrix equation $A_{tk}U = d \cdot \lambda J_t$. Consequently, to discover a $t - (v, k, d \cdot \lambda)$ design we need only search t e lattice spanned by B' for a $(0,1)$-vector. Finally, our complete algorithm is given in Figure

3.

    Algorithm MSV (Matrix Short Vector)
      input basis B of the form in (2);
      $B \leftarrow \mathbf{L}^3(B)$;
      $B \leftarrow$ *Size-Reduction*, (B);
      repeat
          Weight-Reduction;
          sort basis with respect to $\|b_i\|^2$;
          $B \leftarrow \mathbf{L}^3(B)$;
      until $(weight(B) = \sum \|b_i\|^2$ does not decrease);
      Weight-Reduction.
    Check for solution after each *Weight-Reduction* and $\mathbf{L}^3$.

FIGURE 3. The MSV algorithm.

5. Extension

If $(X, \mathscr{B})$ is a $t-(v, k, \lambda)$ design and $x \in X$ then $(\Delta, \mathscr{B}_x)$ where $\Delta = X-\{x\}$ and $\mathscr{B}_x = \{K-\{x\} : x \in K$ and $K \in \mathscr{B}\}$ is a $(t-1)-(v-1, k-1, \lambda)$ design and is said to be the *derived design with respect to x*. Similarly, if there is a point $\infty \notin X$ and a $(t+1)-(v+1, k+1, \lambda)$ design $(Y, \mathscr{D})$ whose derived design with respect to $\infty \in Y$ is $(X, \mathscr{B})$, we say that $(Y, \mathscr{D})$ is an *extension of* $(X, \mathscr{B})$ *by* ∞.

Let G be an automorphism group of a $t - (v, k, \lambda)$ design $(X, \mathscr{B})$. Then there is a $(0,1)$-vector U such that $A_{tk}(G \mid X) \cdot U = \lambda J_{N_t}$. Further suppose that $\infty \notin X$, set $Y = X \cup \{\infty\}$ and let $H \leq Sym(Y)$ be defined by $h \in H$ if $h(\infty) = \infty$ and $h \mid X \in G$. Then it is easy to see, rearranging orbits if necessary, that

$$A_{t+1,k+1}(H \mid Y) = \begin{bmatrix} A_{tk}(G \mid X) & 0 \\ A_{t+1,k}(G \mid X) & A_{t+1,k+1}(G \mid X) \end{bmatrix},$$

where 0 is the matrix of all zeros. Consequently, an $(X, \mathscr{B})$ has an extension to a $(t+1)-(v+1, k+1, \lambda)$ design $(Y, \mathscr{D})$ if and only if there is a $(0,1)$-vector V such that

$$A_{t+1,k+1}(H \mid Y) \begin{bmatrix} U \\ V \end{bmatrix} = \begin{bmatrix} \lambda J_{N_t} \\ \lambda J_{N_{t+1}} \end{bmatrix}.$$

In particular V is a solution to

(7) $$A_{t+1,k+1}(G \mid X) \cdot V = \lambda J_{N_{t+1}} - A_{t+1,k}(G \mid X) \cdot U.$$

Such a solution can be found by using the basis reduction method described in previous sections. However, if t is odd and the parameters of the design satisfy $v = 2k + 1$ and $|\mathscr{B}| = \frac{1}{2}\binom{v}{k}$, then by Theorem A of Alltop in [A] a solution is immediate. We would also like to point out that if $k + t + 2 = v$, then by Lemma 10 of [K2] $A_{t+1,k+1}(G \mid X)$ has an inverse and thus a $(0,1)$-solution to (7), if it exists, would represent a unique extension.

6. Cyclic 5-(13,6,4) designs

If $X = \{0, 1, 2, \ldots, 12\}$ and $G = <(0\ 1\ 2\ 3\ 4\ 5\ 6\ 7\ 8\ 9\ 10\ 11\ 12)>$, then G is cyclic and the A_{56} matrix belonging to G has 99 rows and 132 columns. A direct search applied to the reduced basis obtained from the basis reduction algorithm after doing size reduction found that there are exactly two non-isomorphic solutions. A complete description of these designs can be found in [KR2]. Also in the same paper it was established that these two designs both have full automorphism group cyclic of order 13. Furthermore, between them they partition all of the 6-subsets.

If $(X, \mathscr{B})$ is a $t-(v, k, \lambda)$ design then $(X, \bar{\mathscr{B}})$, where $\bar{\mathscr{B}} = \binom{X}{k} - \mathscr{B}$, is a $t-(v, k, \binom{v-t}{k-t} - \lambda)$ design and is called the *complementary design of* $(X, \mathscr{B})$. Using Alltop's Theorem [A] any 5-(13,6,4) design $(X, \mathscr{B})$ can be extended to a 6-(14,7,4) design $(X \cup \{\infty\}, \mathscr{B}' \cup \mathscr{B}'')$ where

$$\mathscr{B}' = \{K \cup \{\infty\} : K \in \mathscr{B}\},$$
$$\mathscr{B}'' = \{X - K : K \in \bar{\mathscr{B}}\}.$$

We note that the complementary design of a 6-(14,7,4) design is also a 6-(14,7,4) design. These two 6-designs each have full automorphism group cyclic of order 13 and they partition all of the 7-subsets.

Using the results in Section 5 it is easy to see that the extension from any 5-(13,6,4) design to a 6-(14,7,4) design is unique. So we have the following theorem.

THEOREM 4. *If* $(X, \mathscr{B})$ *is a 5-(13,6,4) design it has a unique extension to a 6-(14,7,4) design.*

Recall that if $G \le Sym(X)$ fixes a subset $\Delta \subseteq X$, then each permutation $g \in G$ induces a permutation g^{Δ} on Δ and we denote the totality of g^{Δ}'s formed, by G^{Δ}. Note that G_x, the stabilizer in G of x, fixes the set $\Delta = X - \{x\}$.

THEOREM 5. *Let* $(X, \mathscr{B})$ *be a 6-(14,7,4) design with full automorphism group G. Then for each* $x \in X$, *the derived design with respect to x has full automorphism group* G_x^{Δ} *where* $\Delta = X - \{x\}$.

PROOF. Let $(X, \mathscr{B})$ be a 6-(14,7,4) design. Fix $x \in X$, set $\Delta = X - \{x\}$ and let H be the full automorphism group of the derived design $(\Delta, \mathscr{B}_x)$ of $(X, \mathscr{B})$ with respect to x. It is easy to see that $G_x^{\Delta} \subseteq H$. However, we note by Theorem 4 that $(X, \mathscr{B})$ is the unique extension of $(\Delta, \mathscr{B}_x)$ so it must be obtained by Alltop's construction [A]. That is

$$\mathscr{B} = \{K \cup \{x\} : K \in \mathscr{B}_x\} \cup \{\Delta - K : K \in \bar{\mathscr{B}}_x\}.$$

Observe that H is also the full automorphism group of the complementary design $\bar{B}_x$. Whence, if $\alpha \in H$, then $(\Delta - K)^{\alpha} = \Delta - (K^{\alpha}) \in \mathscr{B}$. Consequently the extension of α to $\hat{\alpha} : X \to X$ given by $y^{\hat{\alpha}} = y^{\alpha}$ if $y \neq x$ and $x^{\hat{\alpha}} = x$ must preserve $\mathscr{B}$. Thus $H \subseteq G_x^{\Delta}$ and hence the two are equal.

THEOREM 6. *Let* $(X, \mathscr{B})$ *be a* 6-(14,7,4) *design and let* $(X, \bar{\mathscr{B}})$ *be its complementary design. Then for each* $x \in X$ *any isomorphism* $\alpha : (X - \{x\}, \mathscr{B}_x) \to (X - \{x\}, \bar{\mathscr{B}}_x)$ *lifts to an isomorphism* $\hat{\alpha} : (X, \mathscr{B}) \to (X, \bar{\mathscr{B}})$.

PROOF. Fix $x \in X$, set $\Delta = X - \{x\}$ and suppose $\alpha : (\Delta, \mathscr{B}_x) \to (\Delta, \bar{\mathscr{B}}_x)$ is an isomorphism. Define $\hat{\alpha} : X \to X$ by $y^{\hat{\alpha}} = y^{\alpha}$ if $y \neq x$ and $x^{\hat{\alpha}} = x$. Then $\hat{\alpha}$ is a isomorphism from $(X, \mathscr{B})$ to $(X, \bar{\mathscr{B}})$ since by Alltop's construction and Theorem 4 we have

$$\mathscr{B} = \{K \cup \{x\} : K \in \mathscr{B}_x\} \cup \{\Delta - K : K \in \mathscr{B}_x\},$$
$$\bar{\mathscr{B}} = \{K \cup \{x\} : K \in \bar{\mathscr{B}}_x\} \cup \{\Delta - K : K \in \bar{\mathscr{B}}_x\}.$$

A t-design is *rigid* if it has no nontrivial automorphism. The lack of automorphisms make them often difficult to find.

COROLLARY 7. *The* 6-*subsets of a* 13-*set can be partitioned into two nonisomorphic rigid* 5-(13,6,4) *designs.*

PROOF. Let $\Delta = \{0, 1, 2, \ldots, 12\}$, $X = \Delta \cup \{\infty\}$ and consider the two complementary 6-(14,7,4) designs $(X, \mathscr{B})$ and $(X, \bar{\mathscr{B}})$ found in [KR2]. These two designs are nonisomorphic with full automorphism group cyclic of order 13 generated by the permutation $\alpha = (0\ 1\ 2\ 3\ 4\ 5\ 6\ 7\ 8\ 9\ 10\ 11\ 12)(\infty)$. Now fix $x \in \Delta$ and apply Theorems 5 and 6. The result follows.

This section suggests the following problems.

PROBLEM 1. *Does there exist a way to partition the* 6-*subsets of a* 13-*set into two isomorphic* 5-(13,6,4) *designs?*

PROBLEM 2. *Describe in a compact way the* 6-(14,7,4) *designs found in* [KR2] *without listing all of the orbit representatives.*

7. Concluding remarks

The methods described above were also successful in finding some other designs. These designs are listed in Table 1. Other small configurations when needed, even though their parameter situation had been settled, were also found quite easily with this method. We believe that basis reduction and similar tools will become a valuable aid to the combinatorialist and are currently conducting research to enhance the productivity of our algorithms.

TABLE 1

Parameters	Group	Remarks	Reference
$5 - (28, 6, \lambda)$	$PSL_2(27)$	New designs for all λ, $2 \le \lambda \le 21$	[KR4]
$4 - (12, 6, 10)$	C_{11}	New design	[KR3]
$4 - (15, 5, 5)$	C_{13}	New design	[K4]
$3 - (20, 5, \lambda)$	S_6	New designs for $\lambda \in \{18, 28, 48, 58\}$	[KdeC]
$3 - (20, 5, \lambda)$	A_6	New designs for $\lambda \in \{24, 54\}$	[KdeC]
$3 - (20, 5, \lambda)$	$H \le AF(19)$, $|H| = 114$	New designs for $\lambda \in \{12, 42, 22, 52, 34, 64\}$	[KdeC]

Note added in proof: Several thousand new designs were recently found by Kreher, Chee, DeCaen, Colburn, and Kramer, using the methods described in this paper. (Cf. *Some new simple t-designs*, to appear in the Journal of Combinatorial Mathematics and Combinatorial Computing.)

REFERENCES

[A]   W. O. Alltop, *Extending t-designs*, J. Combinatorial Theory Series A **18** (1975), 177–186.
[C]   L. G. Chouinard II, Personal communication..
[K1]   E. S. Kramer, *Some t-Designs for $t \ge 4$ and $t = 17$, 18*, Congressus Numerantium XIV, Proceedings of the Sixth Southeastern Conference on Combinatorics, Graph Theory and Computing (1975), 443–460.
[K2]   D. L. Kreher, *An Incidence Algebra for t-Designs with Automorphisms,*, Journal of Combinatorial Theory A **42, No. 2** (1986), 239–251.
[K3]   _____, *A Generalization of Connor's Inequality to t-Designs with Automorphisms*, Journal of Combinatorial Theory A **50** (1989), 259–268.
[K4]   _____, *A New 4-(15,5,4) design (to appear)*.
[KdeC] D. L. Kreher and D. de Caen, *On 3-(20,5,lambda) designs*, Rochester Institute of Technology, Computer Science Report: 001-NOV-1988.
[KLM] E. S. Kramer, D. W. Leavitt and S.S. Magliveras, *Construction procedures for t-Designs and the existence of new simple 6-designs*, Annals of Discrete Mathematics **26** (1985), 247–274.
[KM]   E. S. Kramer and D. Mesner, *t-designs on Hypergraphs*, Discr. Math. **15** (1976), 263–296.
[KR1]   D. L. Kreher and S. P. Radziszowski, *Finding Simple t-Designs by Using Basis Reduction*, Congressus Numerantium, Proceedings of the 17-th Southeastern Conference on Combinatorics, Graph Theory and Computing **55** (1986), 235–244.
[KR2]   _____, *The Existence of Simple 6-(14,7,4) designs*, Journal of Combinatorial Theory Series A, **43 No. 2** (1986), 237–244.
[KR3]   _____, *Solving Subset Sum Problems with the L^3 Algorithm*, Journal of Combinatorial Mathematics and Combinatorial Computing 3 (1988), 49–63.
[KR4]   _____, *New t-designs found by basis reduction*, Congressus Numerantium, Proceedings of the 18-th Southeastern Conference on Combinatorics Graph Theory and Computing **59** (1987), 155–164.
[LLL]   A. K. Lenstra, H. W. Lenstra and L. Lovász, *Factoring Polynomials with Rational Coefficients,*, Mathematische Annalen **261** (1982), 515–534.
[LO]   J. C. Lagarias and A. M. Odlyzko, *Solving Low-Density Subset Sum Problems*, Journal of the ACM **32, No. 1** (1985), 229–246.

[M]   S. S. Magliveras, Private communication..

[ML]   S. S. Magliveras and D. W. Leavitt, *Simple* 6-(33,8,36) *Designs from* $P\Gamma L_2(32)$, Computational Group Theory, Proceedings of the London Mathematical Society Symposium on Computational Group Theory, Academic Press, 1984, pp. 337–352.

SCHOOL OF COMPUTER SCIENCE, ROCHESTER INSTITUTE OF TECHNOLOGY, ROCHESTER, NEW YORK 14623

Current address, D. L. Kreher: Department of Mathematics, University of Wyoming, P. O. Box 3036 University Station, Laramie, Wyoming 82071

Contemporary Mathematics
Volume **111**, 1990

On the Classification of Finite C_n-Geometries with Thick Lines

ANTONIO PASINI

ABSTRACT. We prove that all finite C_n-geometries ($n \geq 4$) with thick lines and known parameters are buildings.

1. Introduction and main results. The reader is referred to [20] and [4] for all basic definitions concerning geometries of type M (GABs in [4]) and to [8] and [5] (or [7], or [9]) for Hecke algebras of geometries of type M. A geometry of type M is of type C_n (is a C_n-geometry, for short) if its diagram is of type C_n: where 0,1,2, ... n-2,n-1 are types (the words "point", "line", "plane" and "solid" are often used instead of "element of type 0" or "of type 1" or "of type 2" or "of type 3", respectively). See Figure 1.

A C_n-geometry has *thick lines* if each of its lines is incident with at least three points. C_n geometries with thick lines admit parameters as in Figure 2.

Just one example of non-building finite C_n-geometry with thick lines is presently known, namely the so called A_7-geometry, or 7-geometry (see [1], [15] or [6]). It has type C_3, parameters as in Figure 3 and it is *flat,* that is each of its points is incident with all of its planes. So, flat C_3-geometries have been taken into a particular consideration. For instance, the following proposition has been proved in [13].

PROPOSITION 1. *Let Γ be a finite C_n-geometry ($n \geq 4$) with thick lines and let us assume that every C_3-residue of Γ is either a building or flat. Then Γ is a building.*

So, if all finite C_3-geometries with thick lines were either buildings or flat, all finite C_n-geometries ($n \geq 4$) with thick lines would be buildings. Unfortunately, we are not able to prove the previous conjecture on C_3-geometries even if we assume to deal with geometries admitting *known parameters,* that

---

1980 *Mathematics Subject Classification* (1985 *Revision*). Primary 51B25; Secondary 51E25.
This paper is in final form and no version of it will be submitted for publication elsewhere.

A. PASINI

FIGURE 1

FIGURE 2

FIGURE 3

FIGURE 4

is parameters subject to some of the relations holding in known examples of finite generalized quadrangles (the reader is referred to §2 of this paper for a more precise definition of known parameters). Indeed, the case in Figure 4 gives trouble.

Anyway, a weaker statement holds in all finite C_n-geometries with thick lines and known parameters (see §2 of this paper).

PROPERTY 1. *Either the geometry* Γ *is a building or some of the irreducible representations of the Hecke algebra* $H(\Gamma)$ *of* Γ *has null multiplicity.*

In this paper we prove the following (see §5):

PROPOSITION 2. *Let* Γ *be a finite* C_n-*geometry* ($n \geq 4$ *with thick lines and let us assume that the property 1 above holds in all* C_3-*residues of* Γ. *Then* Γ *is a building.*

We get the following theorem as an easy consequence of Proposition 2 and

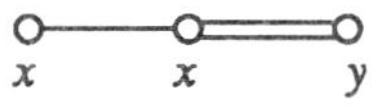

FIGURE 5

of the fact that the property 1 holds in all finite C_3-geometries with thick lines and known parameters.

THEOREM. *All finite C_n-geometries* $(n \geq 4)$ *with thick lines and known parameters are buildings.*

We have proved in [11] that also all finite C_n-geometries $(n \geq 4)$ with thick lines and flag-transitive automorphism groups are buildings. So, only the 'wild' case without known parameters and without flag-transitivity remains open when $n \geq 4$. Things are worse when $n = 3$, of course.

2. Further definitions and known results. 2.1. *The Ott-Liebler number.* Given a C_3-geometry Γ, let (a, u) be a point-plane flag of Γ and let $\alpha(a, u)$ be the number of planes v incident with a, collinear with u, distinct from u and such that the line incident with u and v does not pass through a. The number $\alpha(a, u)$ does not depend on the choice of the point-plane flag (a, u) of Γ (see [10]). It is called *the Ott-Liebler number* of Γ. We shall write α instead of $\alpha(a, u)$, for short.

We have $\alpha = 0$ if and only if Γ is a building (see [10]).

If Γ is finite and admits parameters as in Figure 5 then the multiplicities of the irreducible representations of the Hecke algebra H(Γ) of Γ can be expressed by formulas involving only the parameters x, y and the Ott-Liebler number α. The complete list of these multiplicities can be found in [12] (Table 3 of [12]). We do not rewrite it here. We just recall the definition of H(Γ). Let C be the set of chambers of Γ and let V be the vector space over the complex field having C as a basis. For each of the types $i = 0, 1, 2$ of Γ, let $s_i \in \text{End}(V)$ be the linear mapping defined on the basis C of V as follows:

$$s_i(c) = \sum_{d \underset{i}{\sim} c} d \qquad \begin{array}{l} \text{(where } c \in C \text{ and } d \text{ ranges over the set of chambers} \\ \text{of } \Gamma \text{ that are } i\text{-adjacent with } c\text{)}. \end{array}$$

The subalgebra of End (V) generated by $\{s_0, s_1, s_2\}$ is the Hecke algebra H(Γ) of Γ. The algebra H(Γ) is semisimple and has at most 10 irreducible components. That is, it has 10 irreducible components some of which may possibly have null multiplicity. All possible irreducible representations of H(Γ) can be computed explicitly, together with the multiplicities that they have as components of H(Γ) (expressed in terms of x, y and α, as we said before). Looking at the list of those multiplicities, we immediately see that some of them are null if and only if either $\alpha = x^2 y$ or $\alpha = y^3$.

We have $\alpha = x^2 y$ if and only if Γ is flat (see [10]). We have $x \leq y$ in that case (see [10]).

We have $a \leq x^2 y$ in any case (see [10]).

It is not clear at all what the geometric meaning of the relation $\alpha = y^3$ could be. Anyway, $\alpha = y^3$ only if $x \geq y$, because $\alpha \leq x^2 y$. Moreover, we have $\alpha \leq y^3$ in any case, because $y^3 - \alpha$ appears as a factor in the multiplicities of one of the irreducible representations of $H(\Gamma)$ (see Table 3 of [12]).

Let m be the minimum of x and y. It follows from the above that $\alpha \leq m^2 y$. Property 1 considered in the introduction of this paper can be restated as follows:

PROPERTY 2. *Either $\alpha = 0$ or $\alpha = m^2 y$. That is, either $\alpha = 0$ or α takes its maximal value.*

The generators s_0, s_1, s_2 of $H(\Gamma)$ satisfy the following relations (see [5] or [7]):

$$(*) \begin{cases} s_i^2 = x \cdot 1 + (x - 1)s_i & (i = 0, 1) \\ s_2^2 = y \cdot 1 + (y - 1)s_2 \\ s_0 s_1 s_0 = s_1 s_0 s_1 \\ s_1 s_2 s_1 s_2 = s_2 s_1 s_2 s_1 \\ s_0 s_2 = s_2 s_0 \end{cases}$$

This set of relations gives a presentation of $H(\Gamma)$ if and only if none of the irreducible representations of $H(\Gamma)$ vanish. So, it gives a presentation of $H(\Gamma)$ if and only if $\alpha \neq m^2 y$. In particular, it gives a presentation of $H(\Gamma)$ if Γ is a building (see [9]; see also [3]). Then Property 1 considered in the introduction of this paper (equivalently, Property 2 above) can be rephrased as follows:

PROPERTY 3. *The geometry Γ is a building if and only if the set of relations $(*)$ gives a presentation of $H(\Gamma)$.*

2.2. *Known parameters.* A finite C_n-geometry Γ is said to admit *known parameters* if it admits parameters as in Figure 6 where the pair (x,y) is subject to one of the following relations:

(1) $x = y > 1$.
(2) $y^2 = x^3$ and $x, y > 1$.
(3) $x^2 = y^3$ and $x, y > 1$.
(4) $y = x^2$ and $x > 1$.
(5) $x = y^2$ and $y > 1$.
(6) $y = x + 2$ and $x > 2$.
(7) $x = y + 2$ and $y > 2$.
(8) $1 = y < x$.
(9) $1 = x$.

Of course, the previous 9 cases reduce to 5, by duality, when $n = 2$. They cover all relations holding between parameters of known generalized quad-

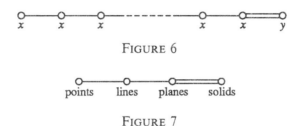

FIGURE 6

points   lines   planes   solids

FIGURE 7

rangles (see [14] and [4]). When $n = 3$ we have the following possibilities (see [10]):

1) $x = y > 1$. *The geometry* Γ *is either a building or flat* (Ott[7]). Just one flat example is known, namely the A_7-geometry, mentioned in the introduction of this paper. We observe that, when Γ is flat, the set of lines of Γ incident with two distinct points of Γ forms an ovoid in the residue of any of those two points. Then residues of points must admit a lot of ovoids when Γ is flat.

2) $y^2 = x^3$ *and* $x, y > 1$. *The geometry* Γ *is either a building or flat* (Rees and Scharlau [18]). No flat example is known. Indeed, in the flat case, residues of points cannot be isomorphic to any of the known generalized quadrangles (no generalized quadrangle of order (q^2, q^3) is known admitting ovoids).

3) $x^2 = y^3$ *and* $x, y > 1$. *No geometry exists admitting such parameters* (Rees and Scharlau [18]).

4) $y = x^2$ *and* $x > 1$. *The geometry* Γ *is a building* (Rees and Scharlau [18]).

5) $x = y^2$ *and* $y > 1$. *Either* Γ *is a building or we have* $\alpha = y^3$ (see [10]). No example is known where $\alpha = y^3$.

6) $y = x + 2$ *and* $x > 2$. *No geometry exists with such parameters* (see [10]).

7) $x = y + 2$ *and* $y > 2$. *No geometry exists with such parameters* (see [10]).

8) $1 = y < x$. *The geometry* Γ *is a building* (Rees [16]).

9) $x = 1$. A lot of different examples exist here (see [17] and [19]). The geometry Γ has thin lines in this case and we have $m = x = 1$. Then $0 \le \alpha \le y$. But we have $0 < \alpha < y$ in many examples. So the properties 1, 2 and 3 fail to hold in this class of geometries.

As we have observed above, cases (3), (6), and (7) never occur if $n = 3$ and Γ has thin lines in case (9). So, if $n = 3$ and Γ has thick lines, only cases (1), (2), (4), (5) and (8) survive. The property 1 of §1 of this paper (equivalently, Properties 2 and 3 of §2) holds in all of them.

3. A number analogous to the Ott-Liebler number in the C_4 case.
In this paragraph Γ will always by a C_4-geometry. See Figure 7.

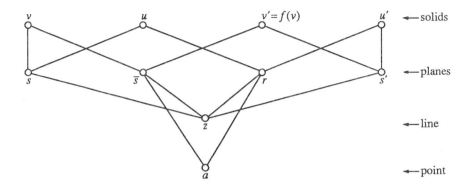

FIGURE 8

The geometry Γ may possibly be infinite, or without parameters, or with some thin lines.

For every point a of Γ, let α_a be the Ott-Liebler number of the residue Γ_a of a. We call it the *inner Ott-Liebler number of* Γ *at* a. Given a point-solid flag (a, u), let $A(a, u)$ be the number of solids v incident with a, distinct from u, coplanar with u and such that the plane incident with u and v does not pass through a (it is easily seen that such a plane is uniquely determined, if it exists). Given a non incident point-solid pair (b, w), let $B(b, w)$ be the number of solids incident with b and coplanar with w.

LEMMA 1. *Given a point* a, *let* u, u' *be any two solids incident with* a *and let* w *be any solid non incident with* a. *Then we have* $A(a, u) = A(a, u')$ *and* $B(a, w) = A(a, u) + 1$.

PROOF. First of all, let us assume that there is a plane r incident with both u and u'. Let v be a solid incident with a, different from u and such that there is a plane s incident with both v and u but not with a. Let z be the line incident with both r and s in the residue Γ_u of u. Let $\bar{s}$ be the plane incident with both z and a in Γ_v. The plane $\bar{s}$ is not incident with u', otherwise $s, \bar{s}, r$ and v, u, u' form a proper triangle in the upper part of the residue of z and this is impossible because that part is a generalized quadrangle. Then there is a (uniquely determined) plane-solid pair (s', v') in Γ_z such that $\bar{s} * v' * s' * u'$ (the symbol $*$ denotes the incidence relation, as in [20]). Of course, we have $v' \neq u'$ and $s' \neq \bar{s}$. Then s' is not incident with a, otherwise we get $a * z$ in $\Gamma_{v'}$ and we have the contradiction $a * s$. Let us set $f(v) = v'$. The picture in Figure 8 summarizes this construction.

So, we have defined a mapping f from the set V of all solids v incident with a, different from u and such that there is a plane s incident with both u and v but not with a, to the set V' of all solids v' incident with a, different from u' and such that there is a plane s' incident with both u' and v' but not with a. It is easily seen that the previous construction can be inverted (see Figure 8). So, we also get a mapping f' from V' to V

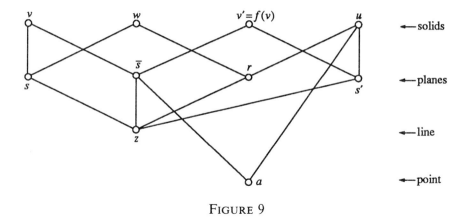

FIGURE 9

such that $f \cdot f'$ is the identity mapping on V' and $f' \cdot f$ is the identity mapping on V. Then we have $A(a, u) = A(a, u')$. The first claim of the lemma now follows from the residual connectedness of Γ. Let us prove the second claim now. Let a, u, w be as in our lemma.

By Lemma 1 of [13] and by the first claim of this lemma we can always assume to have chosen u so that there is a plane r incident with both u and w. Let v be any solid incident with a and such that there is a plane s incident with both v and w. First of all, we consider the case when $s \neq r$. Let z be the line incident with both s and r in Γ_w and let $\bar{s}$ be the plane incident with both z and a in Γ_v. The plane $\bar{s}$ is not incident with u, otherwise $r, s, \bar{s}$ and u, v, w form a proper triangle in the upper part of the residue of z and this is impossible because that part is a generalized quadrangle. So, there is a (uniquely determined) plane-solid pair (s', v') in Γ_z such that $\bar{s} * v' * s' * u$. Of course, we have $\bar{s} \neq s'$ and $u \neq v'$. Then s' is not incident with a, otherwise we get $a * z$ in $\Gamma_{v'}$ and we have the contradiction $a * w$. Let us set $f(v) = v'$.

When $s = r$ and $v \neq u$, then we set $f(v) = v$. See Figure 9. So, we have defined a mapping f from the set W of all solids v incident with a, different from u and such that there is a plane s incident with both v and w, to the set V of all solids v' different from u, incident with a and such that there is a plane s' incident with both u and v' but not with a. It is easily seen that f is a bijection of W onto V. Indeed, given $v' \in V$, let s' be the plane incident with both v' and u. Let z be the line incident with both r and s' in Γ_u and let $\bar{s}$ be the plane incident with both z and a in $\Gamma_{v'}$. Then a plane-solid flag (s, v) is uniquely determined in Γ_z connecting $\bar{s}$ to w. So, the relation $B(a, w) = A(a, u) + 1$ is proved.   Q.E.D.

By Lemma 1, the number $A(a, u)$ does not depend on the choice of the solid u incident with a. So, we write A_a instead of $A(a, u)$. The number A_a will be called *outer Ott-Liebler number of* Γ *at* a.

Given a line z, let a, b be distinct points on z and let u be a solid incident with z. Let X_a be the set of solids v incident with a but not with z and such that there is a plane w incident with both u and v but not with a (it is easily seen that w is uniquely determined, if it exists). Let Y_a be the set of solids v different from u, incident with z and such that there is a plane w incident with both u and v but not with a (as above, w is uniquely determined, if it exists). Let x_a, y_a be the numbers of solids in X_a and Y_a respectively. Let X_b, Y_b, x_b, y_b be defined in the same way (with respect to b instead of a).

LEMMA 2. *We have* $A_a = x_a + y_a$ *and* $\alpha_a = |Y_b - (Y_b \cap Y_a)|$.

PROOF. The first relation is trivial. Let us prove the second one. Let V be the set of solids v incident with z, different from u and such that there is a plane w incident with u, v and a but not with z. It is easily seen that $Y_b - (Y_b \cap Y_a) \subseteq V$. Conversely, let $v \in V$ and let w be as above. The plane w is not incident with b, otherwise we get $w * z$ in Γ_v. Then $v \in Y_b - (Y_b \cap Y_a)$.

So, we have $V = Y_b - (Y_b \cap Y_a)$. But we also have $\alpha_a = |V|$. We are done.   Q.E.D.

LEMMA 3. *We have* $x_a(\alpha_a + 1) + y_a\alpha_a = x_b(\alpha_b + 1) + y_b\alpha_b$.

PROOF. Let us set $Z_a = X_a \cup Y_a$ and $Z_b = X_b \cup Y_b$. Given a solid v in Z_a, let $V_a(v)$ be the set of all solids v_1 different from v, incident with z and such that there is a plane w_1 incident with a, v_1, and v but not with z. Given $v' \in Z_b$, let $V_b(v')$ be defined in the same way as $V_a(v)$, substituting a with b and v with v'. It is known that the number of solids in $V_a(v)$ equals α_a or $\alpha_a + 1$ according to whether v is incident with z or not (see [10]).

Let $v \in Z_a$ and let w be the plane incident with both u and v. Let $\bar{v} \in V_a(v)$ and let $\bar{w}$ be the plane incident with v and $\bar{v}$. Let r be the line incident with both w and $\bar{w}$ in Γ_v. The line r is not incident with b, otherwise we get the contradiction $\bar{w} * z$ in Γ_v. Then there is precisely one plane $\bar{w}'$ in Γ_v incident with both r and b. The plane $\bar{w}'$ is not incident with u, otherwise $w, \bar{w}, \bar{w}'$ and $v, u, \bar{v}$ form a proper triangle in the upper part of the residue of r and this is impossible because that part is a generalized quadrangle. Then we find a (uniquely determined) plane-solid flag (w', v') in Γ_r such that $\bar{w}' * v'$ and $w' * u$. The plane w' is not incident with b, otherwise we get $r*b$ in $\Gamma_{v'}$: this is impossible (see above). Then $v' \in Z_b$. The plane $\bar{w}'$ is not incident with z, otherwise we have two distinct planes $\bar{w}$ and $\bar{w}'$ in $\Gamma_{\bar{v}}$ both incident with both r and a, so we have $a * r$ in $\Gamma_{\bar{v}}$ and we get the contradiction $a * w$. Hence $\bar{v} \in V_b(v')$. We set $f(v, \bar{v}) = (v', \bar{v})$. See Figure 10.

Of course, we have $v * z$ (respectively $v' * z$) if and only if $v \in Y_a$ (if and only if $v' \in Y_b$).

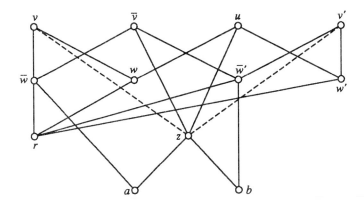

FIGURE 10

It is easily seen that the previous construction can be inverted (the picture above does not change if we interchange a and b, v and v', $\bar{w}$ and $\bar{w}'$ and w and w'). Therefore the mapping f defined above is a bijection from the set of all pairs $(v, \bar{v})$ where $v \in Z_a$ and $\bar{v} \in V_a(v)$ to the set of all pairs $(v', \bar{v})$ where $v' \in Z_b$ and $\bar{v} \in V_b(v')$. The lemma easily follows from this. Q.E.D.

Let X'_{ab} be the set of solids $v \in X_a$ such that the plane w incident with both u and v is incident also with b and let us set $X_{ab} = X_a - X'_{ab}$. Let us set $Y_{ab} = Y_a \cap Y_b$ and $Y'_{ab} = Y_a - Y_{ab}$. We keep the symbol $*$ to denote incidence, as above.

LEMMA 4. *We have*:

$$X_{ab} = \cup (X'_{ac} \mid c * z \text{ and } c \neq a, b)$$
$$Y_{ab} = \cup (Y'_{ac} \mid c * z \text{ and } c \neq a, b)$$
$$X'_{ac} \cap X'_{ad} = Y'_{ac} \cap Y'_{ad} = \varnothing \text{ for every choice of}$$

distinct points c, d *incident with z and different from a.*

PROOF. Let $v \in X_{ab}$ and let w be the plane incident with both u and v. As w is not incident with z, there is precisely one point c in Γ_u incident with both w and z. So, we have $v \in X'_{ac}$. Of course, $c \neq a, b$ because w is not incident to any of a or b. This proves the first relation. The equality $X'_{ac} \cap X'_{ad} = \varnothing$ follows from the uniqueness of c in the argument above. The second relation and the equality $Y'_{ac} \cap Y'_{ad} = \varnothing$ can be proved in a similar way.   Q.E.D.

Let us set $x_{ab} = |X_{ab}|$, $x'_{ab} = |X'_{ab}|$, $y_{ab} = |Y_{ab}|$, and $y'_{ab} = |Y'_{ab}|$. We have $x_a = x_{ab} + x'_{ab}$ and $y_a = y_{ab} + y'_{ab}$. Therefore:

(i)
$$x_a = \sum_{\substack{b * z \\ b \neq a}} x'_{ab} \quad \text{and} \quad y_a = \sum_{\substack{b * z \\ b \neq a}} y'_{ab}.$$

by Lemma 4. By Lemmas 2 and 3 we have:

(ii) $$A_a \alpha_a + x_a = A_b \alpha_b + x_b.$$

Let a, b be distinct collinear points and let z be a line incident with both of them, as above. Let L_{ab} be the number of lines incident with both a and b and distinct from z. Let u be a solid incident with z, as above.

LEMMA 5. *We have:*

(iii) $$x'_{ab} = (\alpha_b + 1)L_{ab} \quad and \quad y'_{ab} = \alpha_b.$$

(iv) $$x_{ab} = \sum_{\substack{c * z \\ c \neq a, b}} (\alpha_c + 1)L_{ac} \quad and \quad x_a = \sum_{\substack{c * z \\ c \neq a}} (\alpha_c + 1)L_{ac}.$$

(v) $$y_{ab} = y_{ba} = \sum_{\substack{c * z \\ c \neq a, b}} \alpha_c \quad and \quad y_a = \sum_{\substack{c * z \\ c \neq a}} \alpha_c.$$

(vi) $$A_a = \sum_{\substack{c * z \\ c \neq a}} [(L_{ac} + 1)\alpha_c + L_{ac}] = x_a + y_a.$$

PROOF. Let us prove (iii). The second relation of (iii) is just a restatement of the second part of Lemma 2. Let us prove the first relation.

Let $v \in X'_{ab}$ and let w be the plane incident with both u and v. Of course, we have $v * b$. Let r be the line in Γ_v through a and b. We have $r \neq z$ because v is not incident with r. Then, r is not incident with u and v is incident with r and coplanar with u in Γ_b. Conversely, given a line r incident with both a and b and different from z (so, r cannot be incident with u), let v be a solid incident with r and coplanar with u in Γ_b. The solid v cannot be incident with z, otherwise r and z are distinct lines meeting in two distinct points a, b in Γ_v. Let w be the plane in Γ_b incident with v and u. The plane w cannot be incident with a, otherwise we get $w * z$ in Γ_u and we have the contradiction $v * z$. Then $v \in X'_{ab}$. From this the first relation of (iii) follows.

The first equality of (iv) can be proved by a similar argument. All details are left to the reader. The second one easily follows from the first one and from (iii).

The first relation of (v) can be proved by an argument similar to the one used to prove the second part of Lemma 2. All details are left to the reader. The second relation of (v) follows from the first one and from (iii).

Finally, (vi) is a trivial consequence of (iv), (v) and of Lemma 2.   *Q.E.D.*

LEMMA 6. *Given any two points a and b we have:*

$$(A_a + 1)(\alpha_a + 1) = (A_b + 1)(\alpha_b + 1).$$

PROOF. As Γ is residually connected, it will be sufficie t to have proved this equality when a and b are collinear. In that case, he equality is an easy consequence of (i)–(vi).   *Q.E.D.*

Definition. Let us set $\alpha + 1 = (A_a + 1)(\alpha_a + 1)$. The number α defined in this way does not depend on the choice of the point a, by Lemma 6. We call it *Ott-Liebler number* of Γ.

PROPOSITION 3. *The geometry Γ is a building if and only if $\alpha = 0$.*

PROOF. The "only if" part is trivial. Let us prove the "if" part. Let $\alpha = 0$. Then $A_a = \alpha_a = 0$ for every point a. Then all C_3- residues of Γ are buildings (see [10]). Moreover, we have $L_{ab} = 0$ for every pair of collinear points a, b, by (vi). Then property (LL) of [20] holds in Γ. Hence, property $(LL)_{res}$ of [12] holds in Γ. Then Γ is a building by Lemma 1 of [12] and Proposition 9 of [20].   Q.E.D.

REMARK 1. THE HIGHER RANK CASE. Dr. G. Pica [22] has made me notice that the Ott-Liebler number can be defined for C_n-geometries of arbitrary rank n, by means of the following inductive definition.

If $n = 2$, then 0 is the Ott-Liebler number of Γ. Let $n > 2$ and let us assume to have defined the Ott-Liebler number for any $m = 2, 3, \ldots n - 1$. Lemma 1 and the first part of Lemma 2 can be proved in general (substituting solids and planes with elements of type $n - 1$ and $n - 2$, respectively). So, we can define the outer Ott-Liebler number A_a of Γ at a in any case. Given a point a, let α_a be the Ott-Liebler number of Γ_a (already defined by the inductive hypothesis). We set $\alpha + 1 = (A_a + 1)(\alpha_a + 1)$. We have to prove that $\alpha + 1$ does not depend on the choice of a. Given two collinear points a, b, let z be a line incident with both of them, let A_{az}, A_{bz} be the outer Ott-Liebler numbers of Γ_a and Γ_b, respectively, computed at z, and let α_{az} and α_{bz} be the Ott-Liebler numbers of the residues of the flags (a, z) and (b, z), respectively. Of course, we have $\alpha_{az} = \alpha_{bz}$. Furthermore, we can prove that $(A_a + 1)(A_{az} + 1) = (A_b + 1)(A_{bz} + 1)$, as in Lemma 6. Therefore, we have $(A_a + 1)(A_{az} + 1)(\alpha_{az} + 1) = (A_b + 1)(A_{bz} + 1)(\alpha_{bz} + 1)$. But $(A_{az} + 1)(\alpha_{az} + 1) = \alpha_a + 1$ and $(A_{bz} + 1)(\alpha_{bz} + 1) = (\alpha_b + 1)$, by the inductive hypothesis. We are done.

This more general definition of the Ott-Liebler number might be useful in a number of contexts. For instance, let $f : \Gamma \to \Gamma'$ be a 2-covering from a C_n-geometry Γ to another C_n-geometry Γ', let F be the group of deck transformations of f and let α, α' be the Ott-Liebler numbers of Γ and Γ' respectively, defined as above. Then we have $\alpha' + 1 = (\alpha + 1) \cdot |F|$.

But we shall not make any use of that general definition in this paper.

REMARK 2. RELATIONS BETWEEN THE OTT-LIEBLER NUMBER AND NUMBERS OF CLOSED GALLERIES OF GIVEN TYPE. It is well known that the Ott-Liebler number α of a C_3-geometry Γ equals the number $g((012)^3)$ of closed galleries of type $(012)^3$ of Γ based at a given chamber C of Γ (of course, the symbol $(012)^3$ is a shortening of 012012012). This fact can be easily seen looking at Figure 11. The relation $\alpha = g((012)^3)$ turns out to be rather useful in representation theory (see [7]).

Let Γ be a C_4-geometry, let a be a point of Γ and let x_a, y_a, α_a, A_a

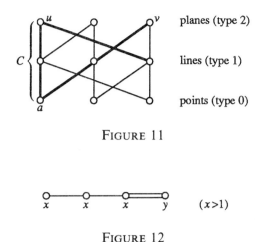

FIGURE 11

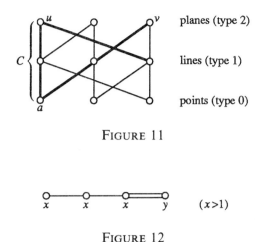

$$
\underset{x}{\circ}\!\!-\!\!-\!\!-\!\!-\underset{x}{\circ}\!\!-\!\!-\!\!-\!\!-\underset{x}{\circ}\!\!=\!\!=\!\!=\!\!=\underset{y}{\circ}\qquad (x>1)
$$

FIGURE 12

and α have the meaning stated in this paragraph (in particular, α is the Ott-Liebler number of Γ). Let C be a chamber of Γ containing a and, given a type τ for closed galleries, let $g(\tau)$ by the number of closed galleries of Γ of type τ based at C. Then, exploiting pictures like the one in Figure 11, we can see that $x_a(\alpha_a + 1) + y_a\alpha_a = g((0123)^4) + g(10(123)^301)$, $y_a = g((0123)^30)$ and $\alpha = g((0123))^4 + g(10(123)^301) + g((0123)^30) + g((123)^3)$.

Moreover, let w, z be the plane and the line of C, respectively and let t_C be the number of triplets of solids (v, v', v'') such that $v \neq v'$, $v * w * v'$, v'' is coplanar with both v and v' and, if w', w'' are the planes incident with v and v'' and with v' and v'' respectively, then $w' * z$, $w \neq w'$ and w'' is incident with a but not with z. Then we have $g((0123)^4) = x_a(\alpha_a + 1) + y_a(\alpha_a - t_C)$.

I do not know whether anything can be done with any of the relations above, or not.

4. Lemmas on finite C_4-geometries with thick lines.

In this section Γ will always be a finite C_4-geometry with thick lines. The geometry Γ admits parameters x, y as in Figure 12.

Given a point a of Γ, α_a and A_a will be the inner and outer Ott-Liebler numbers of Γ at a, respectively, as in the previous paragraph. Let S be the total number of solids of Γ and let S_a be the number of solids of Γ not incident with a. Given a point c of Γ different from a, let L'_{ac} be the number of lines incident with both a and c and let us set $L_{ac} = L'_{ac} - 1$, as in §3 (of course, $L'_{ac} = 0$ is allowed; in that case we have $L_{ac} = -1$). Finally, let S_{ac} be the number of solids incident with c but not with a.

Let α be the Ott-Liebler number of Γ. The symbol $*$ will denote incidence, as above.

LEMMA 7. *We have:*

$$S = \frac{(x^3y + 1)(x^2y + 1)(xy + 1)(y + 1)}{\alpha + 1},$$

$$S_a = \frac{(x^2y + 1)(xy + 1)(y + 1)(x^3y - A_a)}{\alpha + 1}, \quad and$$

$$S_{ac} = \frac{(x^2y + 1)(xy + 1)(y + 1)}{\alpha_c + 1} - (xy + 1)(y + 1)L'_{ac}.$$

PROOF. We know from [10] that the number of solids incident with a is

$$\frac{(x^2y + 1)(xy + 1)(y + 1)}{\alpha_a + 1}.$$

Given any solid u incident with a, there are x^3 planes w incident with u but not with a and y solids incident with w and different from u. Then we have

$$\frac{(x^2y + 1)(xy + 1)(y + 1)}{\alpha_a + 1} \cdot x^3 y = \frac{(x^2y + 1)(xy + 1)(y + 1)}{\alpha_a + 1}A_a + S_a(A_a + 1)$$

The first two relations of the lemma easily follow from this.

As for the third one, we remark that every solid incident with both a and c contains exactly one line incident with both a and c and every line incident with both a and c is incident with exactly $(xy + 1)(y + 1)$ solids. This is enough to establish that relation.   Q.E.D.

LEMMA 8. *We have* $L_{ac} \le (x^2y - \alpha_a)/(\alpha_a + 1)$.

PROOF. The equality is trivial if a and c are not collinear. Let us assume that they are collinear. Let z be a line through a and c and let u be a solid incident with z. In Γ_a there are α_a solids v incident with z, coplanar with u, distinct from u and such that the plane incident with u and v does not pass through z. Given any line r through a and c and different from z, there are $\alpha_a + 1$ solids v in Γ_a incident with r and coplanar with u. It is easily seen that no solid can appear more than one time in this computation (otherwise we find two distinct lines through a and c in its residue). So, we get $L_{ac}(\alpha_a + 1) + \alpha_a$ solids. Moreover, if v is one of those solids and w is the plane incident with u and v, then w does not pass through z. Then the number of those solids cannot exceed x^2y. We are done.   Q.E.D.

Let y_a have the meaning stated in §3.

LEMMA 9. *We have* $y_a \le xy^2$.

PROOF. Indeed, let z be a line through a and let u be a solid incident with z. Given two distinct points c, c' of z, different from a, let v, v' be solids incident with z, different from u and such that there are planes w, w' in Γ_c and $\Gamma_{c'}$ respectively such that $u * w * v$, $u * w' * v'$ and none of

w and w' passes through z. We have $v \neq v'$, otherwise $w = w'$ and we get the contradiction $w * z$ in Γ_v. Then there are exactly y_a solids as above (see (v) of Lemma 5). On the other hand, none of those solids is coplanar with u in Γ_z. Then their number cannot exceed xy^2.   Q.E.D.

Given a solid u, we set $\delta_{au} = 1$ or 0 according to whether $a * u$ or not. The letter c will always denote a point different from a, as above, even when it will appear as an index in a summation symbol.

The following lemma is due to Liebler.

LEMMA 10. (R. Liebler [5], (4.5)). *Let u, v be solids. We have*:

$$\frac{x^3 - 1}{\alpha_a + 1}(\delta_{au} - \delta_{av}) = \sum_{c*u} L'_{ac} - \sum_{c*v} L'_{ac}.$$

(The reader is referred to [5] for the proof.)

LEMMA 11. *Let u be a solid incident with a. We have*:

$$(\alpha_a + 1) \cdot \sum_c (L'_{ac})^2 = (x^2 y + 1) \cdot \sum_{c*u} L'_{ac}.$$

PROOF. The sum $\sum_{c*u} L'_{ac}$ does not depend on the choice of the solid $u * a$, by Lemma 10. There are

$$\frac{(x^2 y + 1)(xy + 1)(y + 1)}{\alpha_a + 1}$$

solids incident with a (see [**10**]). So, we get

$$\sum_{\substack{c*u \\ u*a}} L'_{ac} = \frac{(x^2 y + 1)(xy + 1)(y + 1)}{\alpha_a + 1} \cdot \sum_{c*u} L'_{ac}.$$

But, given a point $c \neq a$, there are exactly $(xy + 1)(y + 1)L'_{ac}$ solids incident with both a and c. Therefore,

$$\sum_{\substack{c*u \\ u*a}} L'_{ac} = (xy + 1)(y + 1) \cdot \sum_c (L'_{ac})^2.$$

The lemma follows easily from this.   Q.E.D.

LEMMA 12. *Let u be a solid incident with a. We have*:

$$\frac{(x^3 - 1)(x^3 y - A_a)}{\alpha + 1} + (\alpha_a + 1) \cdot \sum_c \frac{L'_{ac}}{A_c + 1} = \frac{x^3 y + 1}{A_a + 1} \cdot \sum_{c*u} L'_{ac}.$$

PROOF. If v is a solid not incident with a, then we have

$$\frac{x^3 - 1}{\alpha_a + 1} = \sum_{c*u} L'_{ac} - \sum_{c*v} L'_{ac},$$

by Lemma 10. Let v range over the set of solids not incident with a. We get

$$S_a \cdot \frac{x^3 - 1}{\alpha_a + 1} = S_a \cdot \sum_{c*u} L'_{ac} - \sum_c S_{ac} \cdot L'_{ac}.$$

The conclusion follows from this and from Lemma 7.   Q.E.D.

LEMMA 13. *We have*:

$$\frac{(x^3 - 1)(x^2y + 1)(x^3y - A_a)}{(\alpha + 1)(\alpha_a + 1)} + (x^2y + 1) \cdot \sum_c \frac{L'_{ac}}{\alpha_c + 1}$$

$$= \frac{(x^3y + 1)}{A_a + 1} \cdot \sum_c (L'_{ac})^2.$$

PROOF. Trivial, by Lemmas 11 and 12.

LEMMA 14. *We have*:

$$\sum_c L'_{ac} = \frac{(x^2 + x + 1)(x^2y + 1)x}{\alpha_a + 1}.$$

PROOF. Indeed, there are $(x^2 + x + 1)(x^2y + 1)/(\alpha_a + 1)$ lines through a (see [10]).   Q.E.D.

5. Proof of Proposition 2. Let Γ be as in Proposition 2. Let us assume that $n = 4$, to begin with. Let x, y be the parameters of Γ and let α_a, A_a have the meaning stated in §3. If $x \leq y$, then every C_3 residue of Γ is either a building or flat (indeed Property 1 holds in C_3 residues of Γ).Then Γ is a building by Proposition 1 (but we can get the same conclusion directly by Lemma 10, as Liebler does in [5]).

Let $x > y$. Then, for every point a of Γ, we have either $\alpha_a = 0$ or $\alpha_a = y^3$. If $\alpha_a = 0$ for every point a of Γ, then Γ is a building by Theorem 1 of [20] and by [2]. Let us assume that $\alpha = y^3$ for some point a of Γ. Then $y \neq 1$ (see §2).

Let z be a line through a. By Lemma 9, the number of points c on z different from a and such that $a_c = y^3$ cannot exceed x/y. Then there is some point b on z such that $\alpha_b = 0$, because $y > 1$. By Lemma 8 and (vi) of Lemma 5 we easily get that $x^3y \geq A_b$. Then we have

(1)
$$\frac{x^3y + 1}{y^3 + 1} \geq A_a + 1,$$

by Lemma 6.

On the other hand, substituting α_c with 0 and $(L'_{ac})^2$ with L'_{ac} in Lemma 13 and using Lemma 14, we get

$$\frac{(x^3 - 1)(x^2y + 1)(x^3y - A_a)}{(y^3 + 1)^2(A_a + 1)} \geq \frac{x^3y + 1}{A_a + 1} - x^2y - 1.$$

The following inequality easily follows from the above:

(2)
$$\frac{x^3y^4(x - 1)}{(x^2y + 1)(y^3 + 1) - x - 1} \leq A_a.$$

By (1) and (2) we have

$$\frac{x^3y^3(x - 1)}{(x^2y + 1)(y^3 + 1) - x - 1} \leq \frac{x^3 - y^2}{y^3 + 1},$$

then

$$\frac{x^3 y^3 (x-1)}{x^2 y + 1} < x^3 - y^2,$$

and, finally:

(3) $$x^4 y^3 - x^3 y^3 < x^5 y + x^3 - x^2 y^3 - y^3.$$

Either $x = y^2$ or $y^2 - y \geq x$ (see [14]). Let us assume that $y^2 - y \geq x$. Substituting y^2 with $y + x$ in the monomial $x^4 y^3$ of (3), we get the inequality

$$x^4 y^2 + x^2 y^3 + y^3 < x^3 y^3 + x^3,$$

which contradicts the assumption that $x > y$. Therefore $x = y^2$.

Substituting x with y^2 in (1), we can refine (1) as follows

(4) $$y^4 - y \geq A_a + 1.$$

Substituting x with y^2 in (2) and exploiting (4) we get

$$\frac{y^{12} - y^{10}}{y^8 + y^5 + y^3 - y^2 + 2} \geq y^4 - y - 1.$$

This inequality forces $y = 1$. But we have assumed $y > 1$. We have the contradiction.

Therefore Γ is a building.

Let us come to the general case now ($n \geq 4$). All C_4-residues of Γ are buildings, by the above. Then Γ is 2-covered by a building, by Theorem 1 of [20]. Then Γ is a building, by [2].

The proof is complete.

SUMMARIZING TABLE

Assumptions: finiteness, thick lines and	C_n, $n \geq 4$	C_3
(either) known parameters	buildings	see §2.2 of this paper
(or) flag-transitivity	buildings	See [11] or [12].
(or) nothing else	?	?

BIBLIOGRAPHY

1. M. Aschbacher, *Finite geometries of type C_3 with flag-transitive automorphism groups*, Geom. Ded. **16** (1984), 195–200.
2. A. Brouwer and A. Cohen, *Some remarks on Tits' geometries*, Indag. Math. **45** (1983), 393–402.
3. C. Curtis, N. Iwahori and R. Kilmoyer, *Hecke algebras and characters of parabolic type of finite groups with BN-pairs*, I.H.E.S. Publ. Math. **40** (1971), 81–116.
4. W. Kantor, *Generalized polygons, SCABs and GABs*, Buildings and the Geometry of Diagrams, L.N.1181, Springer, 1986, pp. 78–158.

5. R. Liebler, *A representation theoretic approach to finite geometries of spherical type*, preprint, April 1985.

6. A. Neumaier, *Some sporadic geometries related to* $PG(3, 2)$, Arch. Math. **42** (1984), 89–96.

7. U. Ott, *On finite geometries of type* B_3, J. Comb. Th., A, **39** (1985), 209–221.

8. _____, *Some remarks on representation theory in finite geometries*, Geometries and Groups, L.N.893, Springer, 1981, pp. 68–110.

9. _____, *Bericht Über Hecke algebren un Coxeter algebren endlicher geometrien*, Finite Geometries and Designs, Cameron, Hirschfeld and Hughes ed., Cambridge University Press 1981, pp. 260–271.

10. A. Pasini, *On geometries of type* C_3 *that are either buildings or flat*, Bull. Soc. Math. Belgique, B, **38** (1986), 75–99.

11. _____, *Geometries of type* C_n *and* F_4 *with flag-transitive automorphism groups*, Geom. Ded. **25** (1988), 317–337.

12. _____, *Geometric and algebraic methods in the classification of geometries belonging to Lie diagrams*, Annals of Discrete Mathematics **38** (1988), 315-356.

13. _____, *On Tits geometries of type* C_n, E. J. Comb. **8** (1987), 45-54.

14. S. Payne and J. Thas, *Finite Generalized Quadrangles*, Pitman, Boston, 1984.

15. S. Rees, C_3 *geometries arising from the Klein quadric*, Geom. Ded. **18** (1985), 67–85.

16. _____, *A classification of a class of* C_3 *geometries*, J. Comb. Th., A **44** (1987), 173–181.

17. _____, *Finite* C_3 *geometries in which all lines are thin*, Math. Zeit. **189** (1985), 263–271.

18. S. Rees and R. Scharlau, private communication, February 1985.

19. D. Surowski, *Decomposition and automorphisms of certain* C_n *geometries*, Algebras, Groups and Geometries **4**, 1987, pp. 421-437.

20. J. Tits, *A local approach to buildings*, The Geometric Vein, Springer, 1981, pp. 519–547.

21. _____, *Buildings of Spherical Type and Finite BN-pairs*, L.N. 386, Springer, 1974.

22. G. Pica, private communication.

FACULTY OF ENGINEERING, UNIVERSITY OF NAPLES, FACOLTA' DI INGEGNERIA, VIA CLAUDIO 21, 80125 NAPLES, ITALY

Current address: DEPARTMENT OF MATHEMATICS, UNIVERSITY OF SIENA, VIA DEL CAPITANO 15, 5310 SIENA, ITALY

Contemporary Mathematics
Volume 111, 1990

Cyclic Codes and Cyclic Configurations

VERA PLESS

ABSTRACT. We define cyclic configurations which include certain (v, k, λ) designs with a cyclic group. Using new results about cyclic codes, we give a simple proof for the non-existence of these configurations under certain conditions. We give number theoretic conditions on n in order that the lines of a cyclic projective plane can be vectors of minimum weight in a cyclic $(n, \frac{n+1}{2})$ code which is the sum of a self-orthogonal $(n, \frac{n-1}{2})$ code and the all-ones vector. If p is a prime $\equiv 3 \pmod 4$ and s is odd, we show that a cyclic projective plane of order p^s is not contained in a quadratic residue code over $GF(p)$ of length $n = p^{2s} + p^s + 1$.

I. Introduction. We define cyclic configurations which include certain (v, k, λ) designs with a cyclic group and also a vector and all its cyclic shifts in a cyclic self-orthogonal code. Using new results [**3, 4, 5, 6, 8, 9**] about cyclic codes, we give a simple proof for the non-existence of these configurations under certain conditions. This generalizes a result of Mann [**2**] for (v, k, λ) designs with a cyclic group.

We give number-theoretic conditions on n in order for self-orthogonal cyclic $(n, \frac{n-1}{2})$ codes over $GF(p)$ to exist. Analogous conditions are given for the existence of self-orthogonal or unitarily self-orthogonal cyclic $(n, \frac{n-1}{2})$ codes over $GF(p^2)$. These conditions are related to the non-existence conditions for cyclic configurations.

Finally, we give number theoretic conditions on n in order that the lines of a cyclic projective plane can be vectors of minimum weight in a cyclic $(n, \frac{n+1}{2})$ code (over either $GF(p)$ or $GF(p^2)$) which is the sum of a self-orthogonal $(n, \frac{n-1}{2})$ code and the all-ones vector $\underline{h} = (1, 1, \ldots, 1)$. In particular, if p is a prime $\equiv 3 \pmod 4$ and $n = p^{2s} + p^s + 1$, then a cyclic projective plane of order p^s is not contained in a quadratic residue code over $GF(p)$ of length n if s is odd.

1980 *Mathematics Subject Classification* (1985 *Revision*). Primary 94B25; Secondary 20B20.
This work was supported in part by NSA Grant MDA 904-85-H-0016.
This paper is in final form and no version of it will be submitted for publication elsewhere.

Some of these results generalize results in [**4, 5**] when $p = 2$.

II. Preliminaries. If $x = (x_0, \ldots, x_{n-1})$ and $y = (y_0, \ldots, y_{n-1})$ are vectors over $GF(p)$ ($GF(p^2)$) then the usual dot product $(x, y) = \sum_{i=0}^{n-1} x_i y_i$ evaluated in $GF(p)$ ($GF(p^2)$).

Let A be an $n \times n$ $0 - 1$ matrix whose rows r_i, $i = 1, \ldots, n$ have a fixed weight k. We regard the rows of A as vectors over $GF(p)$ and let $\lambda_{ij} = (r_i, r_j)$. We call A a *cyclic configuration for p* if the following conditions hold.

(1) $(n, p) = 1$

(2) Each row of A is a cyclic shift of the preceeding row, that is if $r_i = (a_0, a_1, \ldots, a_{n-1})$, then $r_{i+1} = (a_{n-1}, a_0, \ldots, a_{n-2})$. Also the first row of A is a cyclic shift of the last row of A.

(3) $p \mid (\lambda_{ij} - \lambda_{k\ell})$ for $1 \le i, j, k, \ell \le n$ where i, j, k, ℓ are not necessarily distinct. $\lambda_{ii} = k$.

If all λ_{ij} for $i \ne j$ equal some λ and we let $v = n$, then A is the incidence matrix of a (v, k, λ) design which has a cyclic group. Another example of a cyclic configuration for p is given by a vector and its $n - 1$ cyclic shifts in a length n cyclic code over $GF(p)$ which is self-orthogonal. Other examples may be possible.

We say that a cyclic configuration A is contained in a cyclic code C if the rows of A are vectors in C. If C is over $GF(p)$ or $GF(p^2)$ all scalar multiples of these rows would be in C also.

We assume the reader is familiar with error-correcting codes and our terminology is as in [7]. As duadic codes will play a major role here we will give their definition and some relevant properties.

From now on all codes are cyclic codes of odd length n over the field K which is either $GF(p)$ or $GF(p^2)$ for p a prime. As is usual we suppose $(n, p) = 1$. If C is a cyclic code with a generating idempotent e, we say $C = \langle e \rangle$.

A vector $v = (v_0, \ldots, v_{n-1})$ is called even-like if $\sum_{i=0}^{n} v_i = 0$ where this is computed in K. A vector which is not even-like is called odd-like. A code with only even-like vectors is called even-like, otherwise it is called odd-like.

FACT 1 [6]. The set of all even-like vectors is an $(n, n - 1)$ code $E = \langle 1 - 1/n\, h \rangle$. Clearly, $E^\perp = \langle 1/n\, h \rangle$.

FACT 2 [6]. A cyclic code C is odd-like if and ony if h is in C.

FACT 3 [6]. If $v = (v_0, \ldots, v_{n-1})$ and $\alpha = \sum_{i=0}^{n-1} v_i$, then $vh = \alpha h$.

If $(a, n) = 1$, we let μ_a denote the coordinate permutation $i \to ai$ (mod n). If $v = (v_0, v_1 \ldots, v_{n-1})$, then $\mu_a(v) = (v_0, v_{a^{-1}(1)}, \ldots, v_{a^{-1}(n-1)})$.

FACT 4 [6]. If $C_1 = \langle e_i \rangle$, $i = 1, 2$, then $\mu_a(C_1) = C_2$ if and only if $\mu_a(e_1) = e_2$.

FACT 5. If $C = \langle e \rangle$ is a cyclic code over $GF(p)$ ($GF(p^2)$)

then $\mu_p(e) = e \, (\mu_{p^2}(e) = e)$.

If $C_1 = \langle e_1 \rangle$ and $C_2 = \langle e_2 \rangle$ are even-like codes, then C_1 and C_2 are even-like duadic codes if

(1) there exists a μ_a with $\mu_a(C_1) = C_2$ and
(2) $e_1 + e_2 = 1 - 1/n \, h$

If C_i, $i = 1, 2$ are even-like duadic codes, then $C_1' = \langle 1 - e_2 \rangle$ and $C_2' = \langle 1 - e_1 \rangle$ are odd-like duadic codes.

FACT 6 [6]. If C_i' are even-like duadic codes of length n, then dim $C_i = \frac{n-1}{2}$ and $C_1 \cap C_2 = 0$.

The odd-like duadic codes have dimension $\frac{n+1}{2}$.

Further, $C_i' = C_i \perp \langle 1/n \, h \rangle$.

FACT 7 [6]. If $C = \langle e \rangle$, $C^\perp = \langle 1 - \mu_{-1}(e) \rangle$.

FACT 8 [6]. If $C_1 = \langle e \rangle$ is a cyclic, self-orthogonal $(n, \frac{n-1}{2})$ code, then C_1 is an even-like duadic code and $C_2 = \langle \mu_{-1}(e) \rangle$. Hence $C_1' = \langle 1 - \mu_{-1}(e) \rangle = C_1^\perp$.

If C_1 is an even-like duadic code and $C_2 = \langle \mu_{-1}(e) \rangle$, then C_1 is a self-orthogonal $(n, \frac{n-1}{2})$ cyclic code.

If $K = GF(p^2)$, then in addition to the usual dot product a unitary dot product is available. If $x = (x_0, \ldots, x_{n-1})$ and $y = (y_0, \ldots, y_{n-1})$, then $(x, y)^u = \sum_{i=0}^{n-1} x_i y_i^p$ evaluated over $GF(p^2)$. The orthogonal of C with respect to this dot product is denoted by $C^{u(\perp)}$. If $C \subset C^{u(\perp)}$, we call C unitarily self-orthogonal.

PROPOSITION 1. If $C = \langle e \rangle$, $C^{u(\perp)} = \langle 1 - \mu_{-p}(e) \rangle$.

PROPOSITION 2. If $C_1 = \langle e \rangle$ is a cyclic, unitarily self-orthogonal code over $GF(p^2)$, then C_1 is an even-like duadic code and $C_2 = \langle \mu_{-p}(e) \rangle$. Hence $C_1' = \langle 1 - \mu_{-p}(e) \rangle = C_1^{u(\perp)}$.

If C_1 is an even-like duadic code and $C_2 = \langle \mu_{-p}(e) \rangle$, then C_1 is a cyclic, unitarily self-orthogonal $(n, \frac{n-1}{2})$ code.

PROOF. Suppose first that $C_1 = \langle e \rangle$ is a unitarily self-orthogonal $(n, \frac{n-1}{2})$ code. Then $C_1 \subset C_1^{u(\perp)} = \langle 1 - \mu_{-p}(e) \rangle$. As $(n, p) = 1$, h is not in C_1. By Fact 2, C_1 is even-like. By Fact 1, h is in $C_1^{u(\perp)}$. As dim $C_1 = \frac{n-1}{2}$, $C_1^{u(\perp)} = C_1 \perp \langle 1/n \, h \rangle$. Thus the idempotent of $C_1^{u(\perp)}$, $1 - \mu_{-p}(e) = e + 1/n \, h - e(1/n \, h) = e + 1/n \, h$ as e is even-like. Thus $C_1 = \langle e \rangle$ and $C_2 = \langle \mu_{-p}(e) \rangle$ are even-like duadic codes.

Suppose now that C_1 is an even-like duadic code and $C_2 = \langle \mu_{-p}(e) \rangle$. By Fact 6, $C_1 \subset C_1' = \langle 1 - \mu_{-p}(e) \rangle = C^{u(\perp)}$. Hence C_1 is a cyclic, unitarily self-orthogonal $(n, \frac{n-1}{2})$ code.

A cyclic code which has the form $C \perp \langle 1/n \, h \rangle$ where C is an $(n, \frac{n-1}{2})$ cyclic, self-orthogonal (unitarily self-orthogonal) code is called a code of type

$S(S_u)$. Such a code has dimension $\frac{n+1}{2}$. We relate these to cyclic configurations for p.

Let D be the code generated by the set $\{r_i - r_j \mid r_i \text{ and } r_j \text{ are rows of a cyclic configuration for } p\}$. Then as

$$(r_i - r_j, r_i - r_j) = 2k - 2\lambda_{ij} \equiv 0 \quad (\text{mod } p),$$
$$(r_i - r_j, r_i - r_k) = k - \lambda_{ij} - \lambda_{ik} + \lambda_{jk} \equiv 0 \quad (\text{mod } p), \quad and$$
$$(r_i - r_j, r_k - r_\ell) = \lambda_{ik} - \lambda_{jk} + \lambda_{j\ell} - \lambda_{i\ell} \equiv 0 \quad (\text{mod } p),$$

D is a self-orthogonal cyclic code over $GF(p)$ or $GF(p^2)$ and also a unitarily self-orthogonal code over $GF(p^2)$.

PROPOSITION 3. *Let D' be the code over $GF(p)$ or $GF(p^2)$ generated by a cyclic configuration for p. Then D' is contained in a code of type S or S_u when such a code exists.*

PROOF. D is a self-orthogonal cyclic code and so must be contained in a self-orthogonal cyclic code of maximum dimension as the orthogonal group acts transitively on self-orthogonal codes of a fixed dimension. An $(n, \frac{n-1}{2})$ self-orthogonal code has maximum dimension and D is contained in one when they exist. A similar statement holds for unitarily self-orthogonal codes.

If we sum the rows of A, we obtain kh. If $p \mid k$, then D' itself is self-orthogonal. If p does not divide k, then $D' = D \perp \langle 1/n\, h \rangle$. In either case D' is contained in a code of type S or S_u whenever they exist.

III. Existence and containment.

THEOREM 1. *If $-1 = p^i$ (mod n), there do not exist cyclic self-orthogonal codes over $GF(p)$.*

PROOF. If $-1 = p^i$ (mod n), $\mu_{-1} = \mu_{p^i}$. Suppose $C = \langle e \rangle$ is self-orthogonal. Then $C \subset C^\perp = \langle 1 - \mu_{-1}(e) \rangle$. Hence $e(1 - \mu_{-1}(e)) = e$ so that $e\mu_{-1}(e) = 0$. But $e\mu_{-1}(e) = e\mu_{p^i}(e) = e^2 = e = 0$.

COROLLARY. *If $-1 = p^i$ (mod n), there do not exist cyclic configurations for p.*

If $-1 = p^i$ (mod v), p does not divide v and $p \mid (k - \lambda)$, Mann [2] has shown that there are no (v, k, λ) designs with a cyclic group. This corollary gives a simple coding theoretic proof of this result and it also extends it to cyclic configurations for p.

We denote the order of a (mod n) by $\underline{ord_n a}$. Let $n = \Pi p_i^{a_i}$, where the p_i are primes and let $t_i = ord_n p_i$. Call t_i $\underline{\text{singly-even}}$ if $2|t_i$ but 4 does not divide t_i. Call t_i $\underline{\text{doubly-even}}$ if $4|t_i$. If t_i is doubly-even but 2^3 does not divide t_i, we say $4\|t_i$.

THEOREM 2. *Cyclic self-orthogonal* $(n, \frac{n-1}{2})$ *codes over* $GF(p)$ *exist if and only if each* t_i *is odd.*

Cyclic self-orthogonal $(n, \frac{n-1}{2})$ *codes over* $GF(p^2)$ *exist if and only if each* t_i *is either odd or singly-even.*

PROOF. By Fact 8 a cyclic, self-orthogonal $(n, \frac{n-1}{2})$ code is an even-like duadic code $C_1 = \langle e \rangle$ with $C_2 = \langle \mu_{-1}(e) \rangle$. It follows from [6, Theorem 2] that this occurs over $GF(p)$ if and only if $ord_n p$ is odd. The latter is equivalent to the condition that each t_i is odd. This proves the first statement.

The proof of the second statement is similar to the first except that even-like duadic codes $C_1 = \langle e \rangle$ and $C_2 = \langle \mu_{-1}(e) \rangle$ exist over $GF(p^2)$ if and only if $ord_n p$ is either odd or singly-even. The condition for $ord_n p$ to be odd is given above. It can be shown for odd n that $ord_n p$ is singly-even if and only if each t_i is either odd or singly even and at least one t_i is singly-even.

COROLLARY 1. *If each* t_i *is odd, then any cyclic configuration is contained in a type* S *code over* $GF(p)$. *If each* t_i *is either odd or singly-even, then any cyclic configuration is contained in a type* S *code over* $GF(p^2)$.

PROOF. This follows directly from Proposition 3 and the theorem.

COROLLARY 2. *If each* t_i *is singly-even, then no cyclic configurations exist.*

PROOF. It was demonstrated in [5, Theorem 15] that when each t_i is singly-even and C is a self-orthogonal $(n, \frac{n-1}{2})$ code over $GF(4)$ that $\mu_{-1}(C) = \mu_2(C)$. The same proof holds for a general p and shows that $\mu_{-1}(C) = \mu_p(C)$. If D' is a cyclic configuration for p, then Corollary 1 shows that D is contained in some self-orthogonal code C over $GF(p^2)$ of dimension $\frac{n-1}{2}$. Hence $\mu_{-1}(D) = \mu_p(D)$. As D is generated by vectors whose components are in $GF(p)$, $\mu_p(D) = D$. As C is an even-like duadic code (Fact 8) and $C_2 = \langle \mu_{-1}(C) \rangle$, $\mu_{-1}(C) \cap C = 0$ (Fact 6). It follows that $D = 0$ so that $D' = 0$.

Note that each t_i being singly-even implies that $-1 = p^i \pmod{n}$ for some i so that we could have gotten non-existence directly by Theorem 1.

COROLLARY 3. *If* $n = p^{2s} + p^s + 1$, *then a cyclic projective plane of order* p^s *is contained in a code* C *of type* S *over* $GF(p)$ ($GF(p^2)$) *if* s *is odd (if* s *is odd or singly-even). When containment occurs the minimum weight vectors of* C *are precisely the multiples of the lines of the plane.*

PROOF. As $(p^{3s} - 1) = (p - 1)(p^{2s} + p^s + 1)$, $ord_n p$ divides $3s$. Hence if s is odd, $ord_n p$ is odd and if s is singly-even, then $ord_n p$ is either odd or singly-even. By the proof of the theorem a code of type S over $GF(p)$ exists when $ord_n p$ is odd and a code of type S over $GF(p^2)$ exists when

$ord_n p$ is either odd or singly-even. Containment follows from Proposition 3.

The statement about the minimum weight vectors follows by standard arguments, see [5, Theorem 18] for example.

It is of some interest to know when a projective plane or (v, k, λ) design is contained in a quadratic residue code. A quadratic residue code over $GF(p)$ is a duadic code C of prime length n with the additional property that $\mu_r(C) = C$ for r any quadratic residue. It is known that quadratic residue codes over $GF(p)$ exist of prime length n whenever p is a quadratic residue mod n. If $p = 2$, this occurs when n is a prime $\equiv \pm 1 \pmod 8$. If $p = 3$, the condition is that n is a prime $\equiv \pm 1 \pmod{12}$. When quadratic residue codes exist, there are two even-like ones and two odd-like ones. If $C_1 = \langle e \rangle$ is a quadratic residue code, then $C_2 = \langle \mu_a(e) \rangle$ is the other one (of the same even or odd likeness) where a can be any non-residue.

THEOREM 3. *A cyclic configuration for p is not contained in any quadratic residue code over $GF(p)$ of length $n \equiv 1 \pmod 4$.*

If $n = p^{2s} + p^s + 1$, then a cyclic projective plane of order p^s is not contained in a quadratic residue code over $GF(p)$ of length n if s is odd and $p \equiv 3 \pmod 4$.

PROOF. Any cyclic code over $GF(p)$ which contains a cyclic configuration for p must contain the code D' generated by the configuration and also the associated self-orthogonal code D. If -1 is a quadratic residue $\pmod n$ and $C_1 = \langle e \rangle$ is an even-like quadratic residue code, then the odd-like quadratic residue code $C_2 = \langle 1 - e \rangle = C_1^\perp$. Thus $C_1 \cap C_1^\perp = 0$ so that D must equal 0. Hence a cyclic configuration cannot be contained in a quadratic residue code when -1 is a quadratic residue $\pmod n$. The first statement is demonstrated by noting that -1 is a quadratic residue $\pmod n$ if and only if $n \equiv 1 \pmod 4$.

To prove the second statement we notice that $p^{2s} + p^s + 1 \equiv 1 \pmod 4$ if and only if $p^{2s} + p^s = p^s(p^s + 1) \equiv 0 \pmod 4$. If s is odd and $p \equiv 3 \pmod 4$, then $p^s + 1 \equiv 0 \pmod 4$.

COROLLARY [1, 4]. *A cyclic projective plane of order 2^s is contained in a binary quadratic residue code only when $s = 1$.*

PROOF. This is so since $2^{2s} + 2^s + 1 \equiv 1 \pmod 4$ when $s \geq 2$.

When $s = 1$ the projective plane of order 2 is contained in the quadratic residue $(7,4,3)$ code. The projective plane of order 8 is contained in a $(71,36,9)$ binary code of type S which is not a quadratic residue code [3]. The projective plane of order 4 is contained in a $(21,11,5)$ code of type S over $GF(4)$ [5]. The projective plane of order 3 is contained in a $(13,7,4)$ ternary code of type S which is not a quadratic residue code.

Again we let $n = \prod p_i^{a_i}$ where the p_i are primes and $ord_n p_i = t_i$.

THEOREM 4. *Unitarily self-orthogonal* $(n, \frac{n-1}{2})$ *codes over* $GF(p^2)$ *exist if and only if each* t_i *is either odd or doubly-even.*

If each t_i *is odd and* $C = \langle e \rangle$ *is a unitarily self-orthogonal code of length* n, *then* e *is a vector over* $GF(p)$.

PROOF. This theorem was demonstrated in [5] for $p = 2$. The same proof works for any prime p.

COROLLARY 1. *If each* t_i *is either odd or doubly-even, any cyclic configuration is contained in a code of type* S_u *over* $GF(p^2)$. *If each* t_i *is odd this code is also of type* S.

COROLLARY 2. *If* $4\|$ *each* t_i, *no cyclic configuration can exist.*

PROOF. As for the case $p = 2$ [5] it can be seen that $\mu_{-p}(C) = \mu_p(C)$ if C is a unitarily self-orthogonal $(n, \frac{n-1}{2})$ code and $4\|$ each t_i. If D' is the code of a cyclic configuration and D is its associated self-orthogonal code, then D must be contained in a unitarily self-orthogonal $(n, \frac{n-1}{2})$ code C. Hence $\mu_{-p}(D) = \mu_p(D) = D$ as the generating vectors of D have coefficients in $GF(p)$. As before C is a duadic code and $\mu_{-p}(C) \cap C = 0$ so that $D = 0$. Note that in this situation also, the conditions on t_i imply that $-1 = p^i \pmod{n}$ for some i and so this corollary could have been derived from Theorem 1.

REFERENCES

1. R. Calderbank and D. B. Wales, *Multiplying vectors in binary quadratic residue codes*, SIAM J. Algebraic Discrete Methods, **3** (1982), 43–55.
2. H. B. Mann, *Balanced incomplete block designs and abelian difference sets*, III, J. Math. **8** (1964), 252–261.
3. V. Pless, J. M. Masley, and J. S. Leon, *On weights in duadic codes*, J. Combin. Theory Ser. A, **44** (1987), 6–21.
4. _____, *Cyclic projective planes and binary, extended cyclic self-dual codes*, J. Comb. Theory A, **43** (1986), 331-333.
5. _____, *Q-codes*, J. Comb. Theory A, **43** (1986), 258–276.
6. _____, *Duadic codes revisited*, Proceedings of the 1987 Southeastern Conference on Combinatorics and Computing, Boca Raton, Florida.
7. _____, *Introduction to the Theory of Error-Correcting Codes*, revised edition, John Wiley and Sons, New York, 1989.
8. J. J. Rushanon, *Generalized Q-Codes*, Ph.D. Thesis, Caltech (1986).
9. M. H. M. Smid, *On Duadic codes*, master's thesis, Eindhoven University of Technology (1986).

DEPARTMENT OF MATHEMATICS, UNIVERSITY OF ILLINOIS AT CHICAGO, CHICAGO, ILLINOIS 60680

Contemporary Mathematics
Volume **111**, 1990

Designs and Approximation

J. J. SEIDEL

1. Summary. Each of the notions of a well-defined $t - (v, k, \lambda)$ design, $t - (v, K, \lambda)$ design, spherical t-design, cubature formula of strength t for the unit sphere S in Euclidean space $\mathcal{R}^d$, amounts to approximation of the set of all blocks (resp. vectors) by a representative subcollection. This is explained in the Sections 2, 3 and 4. Similar statements hold for the recently defined Euclidean design and measure of strength t, cf. [7], [4]. These notions are introduced in Sections 5 and 7. Fisher-type inequalities for Euclidean designs are considered in Section 6, and Section 8 indicates how the present notions apply to extremal lattices in Euclidean space $\mathcal{R}^d$.

2. $t - (v, k, \lambda)$ design. Let V denote a finite set of order v. Its elements are denoted by $x_1, x_2, \ldots, x_v$, to be interpreted as variables which can take only the real values 0 and 1. The k-subsets of V are described by the vectors of the *discrete sphere*

$$S_k := \{x \in \mathcal{R}^v : x_1^2 + x_2^2 + \cdots + x_v^2 = k, \ x_i \in \{0, 1\}\}.$$

(Sometimes it is more convenient to view S_k as a linear flat: then drop the squares). The t-subsets of V, for $t < k$, are described by the *square free monomials*

$$x_{\sigma(1)} x_{\sigma(2)} \cdots x_{\sigma(t)}, \qquad \sigma \in \mathrm{Sym}(v).$$

Clearly, any t-subset is contained in any k-subset whenever the monomial of the t-subset takes the value 1 for the vector of the k-subset. Since a $t - (v, k, \lambda)$ design is a collection of k-subsets of V such that every t-subset of V is contained in a constant number λ of the k-subsets, we may translate as follows.

A $t - (v, k, \lambda)$ design is a subset X of the discrete sphere S_k such that

$$\sum_{x \in X} x_{\sigma(1)} x_{\sigma(2)} \cdots x_{\sigma(t)} = \lambda, \qquad \forall \sigma \in \mathrm{Sym}(v).$$

1980 *Mathematics Subject Classification* (1985 *Revision*). Primary 05B30; Secondary 62K10.
This paper is in final form and no version of it will be submitted for publication elsewhere.

For the complete t-design, consisting of all k-subsets of V, we have

$$\sum_{x \in S_k} x_{\sigma(1)} x_{\sigma(2)} \cdots x_{\sigma(t)} = \binom{v-t}{k-t}, \quad \forall \sigma \in \mathrm{Sym}(v).$$

Counting in two ways the incident pairs of t-subsets and k-subsets gives the relations

$$\lambda \binom{v}{t} = |X| \binom{k}{t}, \qquad \binom{v-t}{k-t}\binom{v}{t} = \binom{v}{k}\binom{k}{t}.$$

By elimination, the four relations above yield

$$|X|^{-1} \sum_{x \in X} x_{\sigma(1)} \cdots x_{\sigma(t)} = \binom{v}{k}^{-1} \sum_{x \in S_k} x_{\sigma(1)} \cdots x_{\sigma(t)},$$

for all $\sigma \in \mathrm{Sym}\,(v)$. Taking $\mathcal{R}$-linear combinations of the various monomials we finally arrive at

$$(2) \qquad |X|^{-1} \sum_{x \in X} f(x) = \binom{v}{k}^{-1} \sum_{x \in S_k} f(x), \quad \forall f \in \mathrm{Hom}_t(S_k).$$

Since trivially each t-design is an i-design, for all $i \leq t$, the relation (2) also holds for all $f \in \mathrm{Pol}_t(S_k)$. Here $\mathrm{Pol}_t(S_k)$ denotes the linear space over $\mathcal{R}$ of all polynomials with square free monomials in v variables of degree $\leq t$, restricted to S_k, and $\mathrm{Hom}_t(S_k)$ denotes the linear subspace of these polynomials of degree equal to t. Notice that $\mathrm{Hom}_t(S_k) = \mathrm{Pol}_t(S_k)$ because

$$\mathrm{Pol}_t(S_k) = \sum_{i=0}^{t} \mathrm{Hom}_i(S_k), \qquad \mathrm{Hom}_i(S_k) \subset \mathrm{Hom}_{i+1}(S_k),$$

for $i < t$, since $(k-i)x_1 \ldots x_i = x_1 \ldots x_i (x_{i+1} + \cdots + x_v)$.

From the above we infer that a $t - (v, k, \lambda)$ design is an *approximation* of the set S_k of all k-subsets by the subcollection X. Indeed, relation (2) expresses the equality of the averages over S_k and over X, for all homogeneous polynomials of degree t (with square free monomials).

3. $t - (v, K, \lambda)$ **design.** Dropping the constancy of the block sizes we allow sizes from $K = \{k_1, \ldots, k_p\}$. So we consider p discrete spheres, and put

$$S_K := S_{k_1} \cup \cdots \cup S_{k_p}.$$

In considering subsets $Y \subset S_K$ we assume that $Y_i := Y \cap S_{k_i} \neq \emptyset$, for $i = 1, \ldots, p$. A *t-wise balanced design* $t - (v, K, \lambda)$ is a collection Y of subsets of V of sizes $\in K$, such that every t-subset of V is contained in λ of them. By use of notation of 2. this is expressed by

$$\sum_{y \in Y} y_{\sigma(1)} \cdots y_{\sigma(t)} = \lambda, \quad \forall \sigma \in \mathrm{Sym}(v).$$

Double counting now yields

$$\lambda \binom{v}{t} = \sum_{i=1}^{p} |Y_i| \binom{k_i}{t}.$$

Using the complete designs consisting of all k_i-subsets, $i = 1, \ldots, p$, and

$$\sum_{y \in S_{k_i}} y_{\sigma(1)} \cdots y_{\sigma(t)} = \binom{v-t}{k_i-t} = \binom{v}{t}^{-1} \binom{v}{k_i} \binom{k_i}{t},$$

we obtain by elimination, for all $\sigma \in \mathrm{Sym}\,(v)$,

$$\sum_{y \in Y} y_{\sigma(1)} \cdots y_{\sigma(t)} = \sum_{i=1}^{p} |Y_i| \binom{v}{k_i}^{-1} \sum_{y \in S_{k_i}} y_{\sigma(1)} \cdots y_{\sigma(t)}.$$

Introducing *weights* $w_y := |Y|^{-1} \binom{v}{k_i}^{-1} |Y_i|$ for $|y| = k_i$ we get

(3) $$|Y|^{-1} \sum_{y \in Y} f(y) = \sum_{y \in S_K} w_y f(y), \quad \forall f \in \mathrm{Hom}_t (S_K).$$

Again we conclude that a $t - (v, K, \lambda)$ design is an *approximation* of the set S_K of all blocks of size $\in K$ by the subcollection Y. Indeed, relation (3) expresses the equality of the average over Y and the weighted average over S_K, for all homogeneous polynomials of degree t. It is by no means true that a t-wise balanced design is also i-wise balanced, for $i < t$. In fact, Kramer [6] states an example of a t-wise balanced design which is not i-wise balanced for any $0 < i < t$.

As an example we mention the following well-known [8] pairwise balanced design with $|Y| = v$. Its $v \times v$ point-block incidence matrix M is obtained from the $v \times v$ point-block incidence matrix N of a symmetric $2 - (v, k, \lambda)$ design as follows:

$$N = \begin{bmatrix} j' & o' \\ A & B \end{bmatrix}, \qquad M = \begin{bmatrix} o' & j' \\ J - A & B \end{bmatrix}.$$

$$\begin{matrix} \downarrow & \downarrow & & \downarrow & \downarrow \\ k & k & & v-k & k+1 \end{matrix}$$

It is easily seen that M is the incidence matrix of a $2 - (v, \{v - k, k + 1\}, k - \lambda)$ design. The famous λ-conjecture [8] states that each $2 - (v, K, \lambda)$ design is obtained in this way. In addition, the example agrees with the Fisher inequality $|Y| \geq v$ for pairwise balanced designs [8].

4. Spherical t-design. We now turn to real Euclidean geometry $\mathcal{R}^d$, which is provided with a positive definite inner product $(,)$. Let $S = \{x \in \mathcal{R}^d : (x, x) = 1\}$ denote the unit sphere in $\mathcal{R}^d$, and let σ denote the standard normalized Borel-measure on S.

A finite n-subset X of S is called a *spherical design of index t*, whenever

(4) $$|X|^{-1} \sum_{x \in X} f(x) = \int_S f(x) \, d\,\sigma(x),$$

for all $f \in \mathrm{Hom}_t(\mathcal{R}^d)$, cf. [3] and [9]. Here

$$\mathrm{Hom}_t(\mathcal{R}^d) = < x_1^{t_1} x_2^{t_2} \ldots x_d^{t_d} : t_1 + \cdots + t_d = t >_{\mathcal{R}}$$

denotes the linear space of the real homogeneous polynomials of degree t in d-variables, which has dimension $\binom{d+t-1}{d-1}$. Thus a design of index t again is an *approximation*, now of the set S of all unit-vectors by a finite subcollection X, since the averages over X and over S are equal, for all homogeneous polynomials of degree t. In other words, the t-th moments of X and of S agree.

Obviously, a spherical design of index t also has index $t - 2$ since the restrictions to S satisfy $\mathrm{Hom}_{t-2}(S) \subset \mathrm{Hom}_t(S)$. We will call $X \subset S$ a spherical design of *strength* t, or a *spherical t-design* [3], if the defining property (4) holds for all polynomials of degree $\leq t$, that is, for all

$$f \in \mathrm{Pol}_t(S) = \mathrm{Hom}_t(S) + \mathrm{Hom}_{t-1}(S).$$

EXAMPLE. The regular pentagon in $\mathcal{R}^2$ has indices 1, 2, 3, 4, 6, 8, hence is a spherical 4-design, not 5-design, cf. [9].

Examples for (d, n, t) of spherical t-designs consisting of n points on the unit sphere in $\mathcal{R}^d$ are the following:

$$(d, n, t) = (3, 12, 5), (8, 240, 7), (24, 2\binom{28}{5}, 11),$$

obtained from the icosahedron, the E8- and the Leech lattice, respectively, cf [3].

By introducing weights we obtain the following generalization. A *cubature formula* for $X \subset S$, $w : X \to \mathcal{R}^+ : x \mapsto w_x$, is a formula of the type

$$(4^*) \qquad \sum_{x \in X} w_x f(x) = (\sum_{x \in X} w_x) \int_S f(x) d\ \sigma(x),$$

which holds true for all $f \in \mathrm{Pol}_t(S)$, cf. [5].

This again is an *approximation* of S by X, since the average over S equals the weighted average over X for all polynomials of degree $\leq t$. Cubature formulae are useful for the evaluation of integrals over S.

5. Euclidean design. The notion of spherical design now will be generalized to several spheres, following [7] and [4]. We use polar coordinates $y = r\, x$, for $x \in S$, $\|y\| = r$, $y \in r\, S \subset \mathcal{R}^d$. For a finite weighted subset (Y, w) in $\mathcal{R}^d$ with $0 \notin Y \subset \mathcal{R}^d$ and $w : Y \to \mathcal{R}^+ : y \mapsto w_y$, its radial support R, its spherical support $R\, S$, and its mass $m(r)$ on the sphere $r\, S$ of radius r, are defined by

$$R := \{\|y\| : y \in Y\}, \ R\, S := \bigcup_{r \in R} r\, S, \ m(r) := w(Y \cap r\, S) = \sum_{y \in Y \cap r\, S} w_y.$$

We call (Y, w) a *Euclidean design* of *index* t [strength t] whenever, for all $f \in \mathrm{Hom}_t(RS)[\in \mathrm{Pol}_t(RS)]$,

$$\sum_{y \in Y} w_y f(y) = \sum_{r \in R} m(r) \int_S f(rx)d\,\sigma(x)$$

holds, that is, whenever

(5) $$\sum_{y \in Y} w_y f(y) = \int_{RS} f(y)\,d\,\mu(y)$$

holds, where we put $d\,\mu(y) := m(r)d\,\sigma(x)$. This formula expresses an *approximation* of the set of spheres RS by the finite set Y, and represents the analogue of a t-wise balanced design in the discrete case.

EXAMPLE. A Euclidean design of index 2 is equivalent to a eutactic star. Indeed,

$$\sum_{y \in Y} w_y\, y_i\, y_j = \int_{RS} y_i\, y_j\, d\,\mu(y)$$

equals 0 for $i \neq j$ and equals a constant c for $i = j \in \{1, 2, \ldots, d\}$. Hence the n vectors

$$c^{-\frac{1}{2}} y \sqrt{w_y} : y \in Y \subset \mathcal{R}^d$$

have a coordinate matrix H satisfying $H^t H = I_d$. This implies that these vectors are the orthogonal projections onto $\mathcal{R}^d$ of an orthonormal basis in $\mathcal{R}^n \supset \mathcal{R}^d$, that is, that they form a eutactic star, cf. [9] and [7].

6. Fisher-type inequalities. For the cardinality of a Euclidean design (Y, w) of strength t [index t] the following Fisher-type inequalities have been obtained, cf. [4], Theorem 5.4.

THEOREM.

$$|Y| \geq \binom{d+e-1}{d-1}, \ if\ (Y, w)\ has\ index\ 2e.$$

$$|Y| \geq \sum_{i=0}^{2p-1} \binom{d+e-i-1}{d-1}, \ if\ (Y, w)\ has\ strength\ 2e\ and\ has\ p\ radii.$$

$$|Y| \geq 2\sum_{i=0}^{p-1} \binom{d+e-2i-1}{d-1}, \ if\ (Y, w)\ is\ antipodal,\ has\ strength\ 2e+1$$

and has p radii.

For spherical t-designs (the case $p = 1$) these formulae read

$$|Y| \geq \binom{d+e-1}{d-1} + \binom{d+e-2}{d-1}, \qquad |Y| \geq 2\binom{d+e-1}{d-1},$$

for strength $2e$ and $2e+1$, respectively. Several examples are known where the last two bounds are tight [3]. For $p \geq \frac{1}{2}(e+1)$ the theorem yields a formula, first proved in [7], which fits nicely with the upper bound for an s-distance set in $\mathcal{R}^d$, cf. [2],

$$|Y| \geq \binom{d+e}{e}, \qquad |Y| \leq \binom{d+s}{s}.$$

We sketch the proof of the second formula of the theorem. The definition of a Euclidean design of strength $2e$ amounts to the isometry of the linear spaces $\mathrm{Pol}_e(Y)$ and $\mathrm{Pol}_e(RS)$. This implies

$$|Y| \geq \dim \mathrm{Pol}_e(R\,S).$$

The right hand side is determined from the restriction map $\mathrm{Pol}_e(\mathcal{R}^d) \to \mathrm{Pol}_e(R\,S)$, whose *kernel* is the space

$$\left(\prod_{r \in R}((x,\,x) - r^2) \right) \mathrm{Pol}_{e-2p}(\mathcal{R}^d).$$

We refer to [4] for details.

REMARK. The paper [4] also contains a discussion of the difficulties which arise in the analogous case of t-wise balanced block designs $t - (v,\,K,\,\lambda)$, $|K| = p$, $t = 2e$. The analogous Fisher inequalities

$$|Y| \geq \binom{v}{e} \quad \text{and} \quad |Y| \geq \sum_{i=0}^{p-1} \binom{v}{e-i}$$

for index $2e$ and strength $2e$, respectively, may be viewed as 'plausible conjectures'.

7. Measure of strength t. We wish to compare the measure ξ in $\mathcal{R}^d$ with the product measure

$$\bar{\xi} := \sigma \times \xi(r\,B),$$

where σ is the standard Borel measure on the unit sphere S in $\mathcal{R}^d$, $\xi(r\,B)$ is the mass of the ball of radius r, and

$$d\,\bar{\xi}(r\,x) = d\,\xi(r\,B).d\,\sigma(x).$$

We say that the measure ξ has strength t whenever

(7) $$\int_{\mathcal{R}^d} f(y)\,d\,\xi(y) = \int_{\mathcal{R}^d} f(y)\,d\,\bar{\xi}(y),$$

for all $f \in \mathrm{Pol}_t(\mathcal{R}^d)$. This definition was given in [4], and illustrates again the idea of *approximation*. The present definition is equivalent to the earlier definition in [7], stated in terms of monomials as follows

(7^*) $$\int_{\mathcal{R}^d} \underline{y}^t\,d\,\xi(y) = \mu_t \int_S \underline{x}^t\,d\,\sigma(x).$$

From this it easily follows that

$$\mu_t = \int_{\mathcal{R}^d} \|y\|^t\,d\,\xi(y),$$

which shows the equivalence with (7).

In terms of the orthogonal group $O(d)$ a measure ξ in $\mathcal{R}^d$ of strength t may be defined as follows, as was observed by Bagchi [1]:

$$(7^{**}) \qquad\qquad \int f \, d\xi = \int f \, d\xi \circ \phi,$$

for all $\phi \in O(d)$, all $f \in \mathrm{Hom}_j(\mathcal{R}^d)$, all $j \leq t$. Indeed,

$$\int_{\mathcal{R}^d} f \, d\xi = \int_{\mathcal{R}^d} f \, d\xi \circ \phi = \int_{O(d)} \left[\int_{\mathcal{R}^d} f \, d\xi \circ \phi \right] d\phi$$

$$= \int_{\mathcal{R}^d} \left[\int_{O(d)} f(\phi^{-1} y) d\phi \right] d\xi(y) = \int_{\mathcal{R}^d} \|y\|^j d\xi(y) \int_{O(d)} f(\phi^{-1}(\frac{y}{\|y\|})) d\phi$$

$$= \int_{\mathcal{R}^d} \|y\|^j \, d\, \xi(y) \int_S f(x) d\, \sigma(x),$$

and this is equivalent to (7^*).

Examples of infinite weighted sets in $\mathcal{R}^d$ which give rise to measures of strength t are provided by the integral lattices of the next section.

8. Euclidean lattices of strength t. Let Λ denote a lattice in $\mathcal{R}^d$ containing the origin, and let R denote the set of the radii occurring in Λ. Let $w : R \to R : x \mapsto w_x$ denote a weight function which is constant on the layers Λ_r of the lattice:

$$\Lambda_r := \{ y \in \Lambda : \|y\| = r \}, \qquad \Lambda = \bigcup_{r \in R} \Lambda_r,$$

and let $\sum_{r \in R} r^l w(r) < \infty$ for integer l with $0 \leq 2l \leq t$. Here, $w(r)$ stands for w_x for $x \in \Lambda_r$. The weighted lattice (Λ, w) is a measure of discrete support, and the strength of (Λ, w) is defined to be the strength of ξ. Introducing the notation

$$f(\Lambda_r) := \sum_{x \in \Lambda_r} f(x),$$

for any polynomial f and layer Λ_r, we apply the definition of Section 7 in terms of harmonic polynomials as follows. (Λ, w) is a *lattice of strength t* whenever

$$(8) \qquad\qquad \sum_{r \in R} r^{2l} w(r) h_j(\Lambda_r) = 0,$$

for integer j and l with $0 < j \leq t - 2l$, for $h_j \in \mathrm{Harm}_j(\mathcal{R}^d)$.

From this definition the following conclusion may be drawn. If each layer Λ_r is a spherical t-design, then the lattice Λ has strength t.

This raises two questions. Firstly, which lattices have this property that every layer is a spherical t-design? Secondly, in this situation, is it possible to

select a suitable weight function w such that (Λ, w) has strength larger than
t? To the second question no answer is known. However, the first question
has been settled for a special class of lattices by the following theorem which
is due to Venkov [10], cf. [7].

A lattice is called *extremal* whenever it is integral, even, unimodular and
has largest possible minimal norm, that is, whenever for all $x, y \in \Lambda$:

$$(x, y) \in Z, (x, x) \in 2Z, \det \Lambda = 1, \min_{0 \neq x \in \Lambda} (x, x) = 2\left[\frac{d}{24}\right] + 2 = 2\delta + 2.$$

Here we took for the dimension d

$$d = 24(\delta + 1) - 8\varepsilon, \qquad \varepsilon \in \{3, 2, 1\}.$$

THEOREM. *For an extremal lattice in* R^d *every layer has strength* $4\varepsilon - 1$.

Examples of extremal lattices are the E_8 lattice in R^8, the lattices E_{16}
and $E_8 + E_8$ in R^{16}, and the Leech lattice in R^{24}. Application of the
theorem yields that every layer has strength 7 for E_8, strength 3 for E_{16} and
for $E_8 + E_8$, and strength 11 for the Leech lattice. It would be interesting
to answer the second question for the strength of the lattices of dimension 8,
16, 24 just mentioned.

REFERENCES

1. B. Bagchi, Private communication.
2. A. Blokhuis, Few-distance sets, CWI tract 7, Math. Centre 1980.
3. P. Delsarte, J.M. Goethals, J.J. Seidel, Spherical codes and designs, Geom. Dedic. **6** (1977), 363–388.
4. P. Delsarte, J.J. Seidel, Fisher-type inequalities for Euclidean t-designs, Linear Algebra Appl. **114/115** (1989), 213–230.
5. J.M. Goethals, J.J. Seidel, *Cubature formulae, polytopes and spherical designs*, The Geometric Vein, the Coxeter Festschrift, (C. Davis, B. Grünbaum, F.A. Sherk, Eds.), Springer 1982, pp. 203–218.
6. E.S. Kramer, Some results on t-wise balanced designs, Ars Combin. **15**, (1983), 179–192.
7. A. Neumaier, J.J. Seidel, Discrete measures for spherical designs, eutactic stars and lattices, Proc. Nederl. Akad. Wetensch. Ag1($\equiv$ Indagationes Math. **50**) (1988), 321–334.
8. H.J. Ryser, New types of combinatorial designs, Actes Congrès Intern. Math., tome 3, 1970, 235–239.
9. J.J. Seidel, Eutactic stars, Colloq. Math. Soc. János Bolyai **18**(1976), 983–999.
10. B.B. Venkov, On even unimodular extremal lattices, Proc. Steklov Inst. Math. **165**(1984), 43–48, (trl. AMS 1985, 47-52).

DEPARTMENT OF MATHEMATICS, TECHNICAL UNIVERSITY EINDHOVEN, 5600 MB EINDHOVEN, THE NETHERLANDS.

Contemporary Mathematics
Volume 111, 1990

Flocks, Maximal Exterior Sets, and Inversive Planes

J. A. THAS

ABSTRACT. A flock F of the hyperbolic quadric $Q^+(3, q)$ of $PG(3, q)$ is a set of $q + 1$ disjoint irreducible conics of $Q^+(3, q)$. A maximal exterior set (MES) X of the hyperbolic quadric $Q^+(2n - 1, q)$, $n \geq 3$, of $PG(2n - 1, q)$ is a set of $(q^n - 1)/(q - 1)$ points of $PG(2n - 1, q) - Q^+(2n - 1, q)$, such that each line joining two distinct elements of X has no point in common with $Q^+(2n - 1, q)$. For q even all flocks of $Q^+(3, q)$ and all MES of $Q^+(2n - 1, q)$ have been determined. Our main results on flocks and MES here are: (a) For $q = 3, 7$ and each $q \equiv 1 (mod\ 4)$ the quadric $Q^+(3, q)$ has only two flocks (up to a projectivity) (b) If not all conics of the flock F of $Q^+(3, q)$, $q \equiv 3 (mod\ 4)$, belong to a same equivalence class (it is well-known that all irreducible conics of $Q^+(3, q)$ can be divided into two equivalence classes), then F is linear, i.e., the planes of the conics of F all contain a common line (c) For $q = 3, 7$ and each $q \equiv 1 (mod\ 4)$, the quadric $Q^+(2n - 1, q)$, $n \geq 3$, has no MES (d) The hyperbolic quadric $Q^+(2n - 1, q)$, $n \geq 3$, has no MES for each value of q for which $Q^+(3, q)$ admits only two flocks (up to a projectivity). As an important application we solved a classical problem on inversive planes: If for at least one point p of the inversive plane I of order q, $q \equiv 1 (mod\ 4)$, the internal plane I_p is Desarguesian, then I is Miquelian. In an appendix recent results on flocks, MES and inversive planes are mentioned.

1. Introduction

1.1. Definitions. An *exterior set* with respect to the hyperbolic quadric $Q^+(2n - 1, q)$ of $PG(2n - 1, q)$, $n \geq 2$, is a set X of points such that each line joining two distinct elements of X has no point in common with $Q^+(2n - 1, q)$. It is easy to show that for an exterior set X of $Q^+(2n - 1, q)$ we have $|X| \leq (q^n - 1)/(q - 1)$ [5]. If $|X| = (q^n - 1)/(q - 1)$, then X is called a *maximal exterior set* (MES).

Let $n = 2$ and let X be a MES of $Q^+(3, q)$. Then $|X| = q + 1$. If

---

1980 *Mathematics Subject Classification* (1985 *Revision*). Primary 51E20, 51E25, 51E15, 05B25, 05B05.

This paper is in final form and no version of it will be submitted for publication elsewhere.

$X = \{x_1, x_2, \ldots, x_{q+1}\}$, then let π_i be the polar plane of x_i with respect to $Q^+(3, q)$ and let $\pi_i \cap Q^+(3, q) = C_i$. Since $x_i x_j \cap Q^+(3, q) = \varnothing$ for all $i \neq j$, we have $C_i \cap C_j = \varnothing$ for all $i \neq j$. So $\{C_1, C_2, \ldots, C_{q+1}\}$ is a partition of $Q^+(3, q)$ by irreducible conics. A partition of $Q^+(3, q)$ by irreducible conics is called a *flock* of $Q^+(3, q)$. Hence each MES of $Q^+(3, q)$ defines a flock of $Q^+(3, q)$. Conversely, each flock of $Q^+(3, q)$ defines a MES of $Q^+(3, q)$.

Let L be a line of $PG(3, q)$ having no points in common with $Q^+(3, q)$. Then L is a MES of $Q^+(3, q)$. Here the planes of the corresponding conics $C_1, C_2, \ldots, C_{q+1}$ all contain the polar line M of L with respect to $Q^+(3, q)$. In such a case the MES and the corresponding flock are called *linear*.

1.2. Flocks. Let F be a flock of the hyperbolic quadric $Q^+(3, q)$, q even. Then it was shown by J. A. Thas [10, 23] that F is necessarily linear.

Next, let $Q^+(3, q)$ be the hyperbolic quadric of $PG(3, q)$, with q odd. In the set of all irreducible conics of $Q^+(3, q)$ we define the following equivalence relation [10, 23]: two conics C_1 and C_2 are equivalent if and only if there is an irreducible conic C on $Q^+(3, q)$ which is tangent to both C_1 and C_2. There are two equivalence classes, denoted by I and II. Let L be a line having no point in common with $Q^+(3, q)$, and let L' be the polar line of L with respect to $Q^+(3, q)$. The set of all conics of class I (resp. II) containing L is denoted by V (resp. V'). For $q \equiv 1 \pmod 4$ the set of all conics of class II (resp. I) containing L' is denoted by W (resp. W'); for $q \equiv -1 \pmod 4$ the set of all conics of class I (resp. II) containing L' is denoted by W (resp. W'). Then it was shown by J. A. Thas [10, 23] that $V \cup W$ (resp. $V' \cup W'$) is a (non-linear) flock of $Q^+(3, q)$. By most authors these flocks are called *Thas flocks*.

Let $F = \{C_1, C_2, \ldots, C_{q+1}\}$ be a flock of $Q^+(3, q)$, q odd, and let π_i be the plane containing C_i. If all planes $\pi_1, \pi_2, \ldots, \pi_{q+1}$ contain a common point, then, relying on a theorem of B. Sègre and G. Korchmáros [21], J. C. Fisher and J. A. Thas [10] show that F is linear. If at least $(q+1)/2$ of the planes $\pi_1, \pi_2, \ldots, \pi_{q+1}$ contain a common line, then P.-H. Zieschang proves that F is a linear flock or a Thas flock [28].

In [1] L. Bader shows that for $q = 11, 23, 59$ the quadric $Q^+(3, q)$ has a flock which is neither linear nor a Thas flock. These flocks were independently discovered by N.L. Johnson [14], and for $q = 11, 23$ also by R.D. Baker and G.L. Ebert [2].

By the theorem of P.-H. Zieschang it is immediately clear that each flock of $Q^+(3, 3)$ is a linear flock or a Thas flock. Relying on D. J. Oakden's classification [18] of all line spreads of $PG(3, 5)$ containing a regulus, N. L. Johnson [15] proved that also for $q = 5$ each flock of $Q^+(3, 5)$ is linear or a Thas flock.

Finally, we notice that independently M. Walker [26] and J. A. Thas [10, 23] proved that with each flock F of $Q^+(3, q)$ there corresponds a translation plane π or order q^2 with dimension at most two over its kernel. Moreover, the plane π is Desarguesian if and only if F is linear. Recently, N. L. Johnson wrote an extensive paper on these translation planes arising from flocks of hyperbolic quadrics [14].

1.3. Maximal exterior sets. In [5] F. De Clerck and J. A. Thas show that the existence of a MES of $Q^+(2n - 1, q)$ implies the existence of a MES of $Q^+(2m - 1, q)$, $2 \leq m \leq n$.

Relying on a theorem of W. M. Kantor [16], F. De Clerck and J. A. Thas [5] proved that $Q^+(2n - 1, 2)$ with $n \geq 4$ has no MES and, relying on the fact that each flock of $Q^+(3, 2^h)$ is linear, J. A. Thas [5, 20] proved that $Q^+(5, q)$, q even and $q \neq 2$, has no MES. Hence $Q^+(2n - 1, q)$, $n \geq 3, q$ even and $q \neq 2$, has no MES. G. M. Conwell proved that $Q^+(5, 2)$ admits a unique (up to a projectivity) MES [17, 32]. Together with the results in 1.1 we conclude that for q even the only MES are the linear MES of $Q^+(3, q)$ and the unique MES of $Q^+(5, 2)$.

Now, let q be odd. S. Rees [20] proved that $Q^+(5, 3)$, and so $Q^+(2n - 1, 3)$ with $n \geq 3$, has no MES, and P.-H. Zieschang [29] obtained the same result for $q = 5$.

Finally, S. Rees [20] noticed that the existence of a MES of $Q^+(5, q)$ implies the existence of a non-building C_3-geometry.

2. Flocks

In this section we always suppose that q is odd.

2.1. THEOREM 1. Let $F = \{C_1, C_2, \ldots, C_{q+1}\}$ be a flock of the hyperbolic quadric $Q^+(3, q)$. Let p_1, p_2, p_3, p_4 be four distinct points of $Q^+(3, q)$ such that p_1p_2, p_1p_3, p_2p_4 and p_3p_4 are distinct lines of $Q^+(3, q)$.

(a) *The conics of F containing the respective points p_1, p_2, p_3, p_4 are all distinct.* Let notations be chosen in such a way that p_i is on the conic $C_i \in F$, $i = 1, 2, 3, 4$. The tangent of C_i at p_i is denoted by L_i. Then $L_1 \cap L_4 = \varnothing$ if and only if $L_2 \cap L_3 = \varnothing$.

If $L_1 \cap L_4 = \varnothing$ and $L_2 \cap L_3 = \varnothing$, then L_1, L_2, L_3, L_4 are elements of a regulus R, i.e., these lines are generators of a same family of some hyperbolic quadric. The lines p_1p_4 and p_2p_3 are elements of the complementary regulus R' of R.

If $L_1 \cap L_4 = \{u\}$ and $L_2 \cap L_3 = \{v\}$, then uv is the intersection of the planes L_1L_4 and L_2L_3. Notice that $u \in p_2p_3$ and $v \in p_1p_4$.

(b) *Just two of the conics of F containing the respective points p_1, p_2, p_3, p_4 coincide.* Let $p_1 \in C_1$, $p_4 \in C_4$ and $p_2, p_3 \in C_2$. Then the tangent L_1 of C_1 at p_1 and the tangent L_4 of C_4 at p_4 have a point u in common. This point u is on the line p_2p_3.

PROOF. Let θ be the polarity defined by the quadric $Q^+(3, q)$, and let $F' = \{x_1, x_2, \ldots, x_{q+1}\}$ with $x_i = \pi_i^\theta$ and π_i the plane containing C_i, $i = 1, 2, \ldots, q+1$. Coordinates are chosen in such a way that $Q^+(3, q)$ has equation $X_0 X_3 = X_1 X_2$ and $p_1(1, 0, 0, 0), p_2(0, 1, 0, 0), p_3(0, 0, 1, 0)$, $p_4(0, 0, 0, 1)$. Let $x_i(a_0^i, a_1^i, a_2^i, a_3^i)$, $i = 1, 2, \ldots, q + 1$.

Since F' is a MES of $Q^+(3, q)$ each plane containing $p_1 p_2$ has at most one point in common with F'. Since $|F'| = q + 1$, each plane containing $p_1 p_2$ has exactly one point in common with F'. Analogously, each plane containing any of $p_1 p_3, p_3 p_4, p_2 p_4$ intersects F' in exactly one point.

(a) First, suppose that the conics of F containing the respective points p_1, p_2, p_3, p_4 are all distinct. Notations are chosen in such a way that p_i is on the conic C_i, $i = 1, 2, 3, 4$. The tangent of C_i at p_i is denoted by L_i. We have $p_1 p_2 p_3 \cap F' = \{x_1\}$, $p_1 p_2 p_4 \cap F' = \{x_2\}$, $p_1 p_3 p_4 \cap F' = \{x_3\}$, and $p_2 p_3 p_4 \cap F' = \{x_4\}$.

$$\begin{aligned}
\text{Put } x_1 &= (a_0^1, a_1^1, a_2^1, a_3^1) = (\alpha_1, \beta_1, \gamma_1, 0), &&\text{with } \alpha_1 \beta_1 \gamma_1 \neq 0, \\
x_2 &= (a_0^2, a_1^2, a_2^2, a_3^2) = (\alpha_2, \beta_2, 0, \delta_2), &&\text{with } \alpha_2 \beta_2 \delta_2 \neq 0, \\
x_3 &= (a_0^3, a_1^3, a_2^3, a_3^3) = (\alpha_3, 0, \gamma_3, \delta_3), &&\text{with } \alpha_3 \gamma_3 \delta_3 \neq 0, \\
x_4 &= (a_0^4, a_1^4, a_2^4, a_3^4) = (0, \beta_4, \gamma_4, \delta_4), &&\text{with } \beta_4 \gamma_4 \delta_4 \neq 0.
\end{aligned}$$

The plane $p_1 p_2 x_i$, with $i \neq 1, 2$, has equation $X_3 = (a_3^i / a_2^i) X_2$,

the plane $p_1 p_3 x_i$, with $i \neq 1, 3$, has equation $X_1 = (a_1^i / a_3^i) X_3$,

the plane $p_2 p_4 x_i$, with $i \neq 2, 4$, has equation $X_2 = (a_2^i / a_0^i) X_0$,

the plane $p_3 p_4 x_i$, with $i \neq 3, 4$, has equation $X_0 = (a_0^i / a_1^i) X_1$.

Since $\{p_1 p_2 x_1, p_1 p_2 x_2, \ldots, p_1 p_2 x_{q+1}\}$ is the set of all planes through $p_1 p_2$ and since $p_1 p_2 x_1 = p_1 p_2 p_3$ and $p_1 p_2 x_2 = p_1 p_2 p_4$, we have

$$\{a_3^3 / a_2^3, a_3^4 / a_2^4, \ldots, a_3^{q+1} / a_2^{q+1}\} = GF(q) \backslash \{0\}.$$

Hence

$$\prod_{i \neq 1, 2} a_3^i / a_2^i = -1.$$

Analogously

$$\prod_{i \neq 1, 3} a_1^i / a_3^i = -1, \qquad \prod_{i \neq 2, 4} a_2^i / a_0^i = -1, \qquad \prod_{i \neq 3, 4} a_0^i / a_1^i = -1.$$

Consequently

$$\left(\prod_{i=5}^{q+1} \frac{a_3^i}{a_2^i} \cdot \frac{a_1^i}{a_3^i} \cdot \frac{a_2^i}{a_0^i} \cdot \frac{a_0^i}{a_1^i} \right) \cdot \frac{\delta_3}{\gamma_3} \cdot \frac{\delta_4}{\gamma_4} \cdot \frac{\beta_2}{\delta_2} \cdot \frac{\beta_4}{\delta_4} \cdot \frac{\gamma_1}{\alpha_1} \cdot \frac{\gamma_3}{\alpha_3} \cdot \frac{\alpha_1}{\beta_1} \cdot \frac{\alpha_2}{\beta_2} = 1.$$

Hence

(1) $$\alpha_2 \beta_4 \gamma_1 \delta_3 = \alpha_3 \beta_1 \gamma_4 \delta_2.$$

Let us choose $\alpha_1 = \beta_2 = \gamma_3 = \delta_4 = 1$. The Plücker coordinates [13] of the line $p_1 x_1$ are determined by the matrix

$$\begin{pmatrix} 1 & \beta_1 & \gamma_1 & 0 \\ 1 & 0 & 0 & 0 \end{pmatrix}.$$

Hence the coordinates of $p_1 x_1$ are: $p_{12} = -\beta_1$, $p_{13} = -\gamma_1$, $p_{14} = 0$, $p_{23} = 0$, $p_{24} = 0$, $p_{34} = 0$. If u_1 is the image of $p_1 x_1$ on the Klein quadric $Q^+(5, q)$ [13], then

$$u_1 (p_{12}, p_{13}, p_{14}, p_{23}, p_{24}, p_{34}) = u_1 (-\beta_1, -\gamma_1, 0, 0, 0, 0).$$

Analogously, if u_i is the image of $p_i x_i$ on $Q^+(5, q)$, $i = 2, 3, 4$, then $u_2(\alpha_2, 0, 0, 0, -\delta_2, 0)$, $u_3(0, \alpha_3, 0, 0, 0, -\delta_3)$, $u_4(0, 0, 0, 0, \beta_4, \gamma_4)$. We have

$$\alpha_2 \alpha_3 \beta_4 (-\beta_1, -\gamma_1, 0, 0, 0, 0) + \alpha_3 \beta_1 \beta_4 (\alpha_2, 0, 0, 0, -\delta_2, 0) +$$
$$\alpha_2 \beta_4 \gamma_1 (0, \alpha_3, 0, 0, 0, -\delta_3) + \alpha_3 \beta_1 \delta_2 (0, 0, 0, 0, \beta_4, \gamma_4) =$$
$$(0, 0, 0, 0, 0, -\alpha_2 \beta_4 \gamma_1 \delta_3 + \alpha_3 \beta_1 \delta_2 \gamma_4) \overset{(1)}{=} (0, 0, 0, 0, 0, 0).$$

Hence the points u_1, u_2, u_3, u_4 are contained in a common plane π of $PG(5, q)$. It is clear that $p_1 x_1$ and $p_2 x_2$ have no common point, so u_1 and u_2 are not collinear on $Q^+(5, q)$. Analogously not one of the lines $u_1 u_3$, $u_2 u_4$, $u_3 u_4$ is contained in $Q^+(5, q)$.

Suppose that $\beta_4 \gamma_1 = \beta_1 \gamma_4$. By (1) this is equivalent to $\alpha_2 \delta_3 = \alpha_3 \delta_2$. Geometrically this means that $p_1 x_1 \cap p_4 x_4 \neq \varnothing$ and $p_2 x_2 \cap p_3 x_3 \neq \varnothing$. Applying the polarity θ with respect to $Q^+(3, q)$ we see that the tangents L_1 and L_4 have a common point u, and that the tangents L_2 and L_3 have a common point v. Clearly $u \in p_1 p_2 p_3 \cap p_2 p_3 p_4 = p_2 p_3$ and $v \in p_1 p_3 p_4 \cap p_1 p_2 p_4 = p_1 p_4$. Hence uv is the intersection of the planes $L_1 L_4$ and $L_2 L_3$.

Next, suppose that $\beta_4 \gamma_1 \neq \beta_1 \gamma_4$ which is equivalent to $\alpha_2 \delta_3 \neq \alpha_3 \delta_2$. Then $p_1 x_1 \cap p_4 x_4 = \varnothing$ and $p_2 x_2 \cap p_3 x_3 = \varnothing$. In this case $\pi \cap Q^+(5, q)$ is an irreducible conic C. Consequently $p_1 x_1$, $p_2 x_2$, $p_3 x_3$, $p_4 x_4$ are elements of a regulus $\bar{R}$. Since $p_1 p_4$ (resp. $p_2 p_3$) has a point in common with each of $p_1 x_1, \ldots, p_4 x_4$ it belongs to the complementary regulus $\bar{R}'$. Applying the polarity θ, we see that the tangents L_1, L_2, L_3, L_4 belong to the regulus $R = \bar{R}^\theta$, and that $p_2 p_3$ and $p_1 p_4$ belong to the complementary regulus $R' = \bar{R}'^\theta$.

(b) Next, suppose that just two of the conics of F containing the respective points p_1, p_2, p_3, p_4 coincide. Let $p_1 \in C_1$, $p_4 \in C_4$ and $p_2, p_3 \in C_2$. Then we have $p_1 p_2 p_3 \cap F' = \{x_1\}$, $p_1 p_2 p_4 \cap F' = \{x_2\}$, $p_1 p_3 p_4 \cap F' =$

$\{x_2\}$, $p_2 p_3 p_4 \cap F' = \{x_4\}$.

Put $x_1 = (a_0^1, a_1^1, a_2^1, a_3^1) = (\alpha_1, \beta_1, \gamma_1, 0)$, with $\alpha_1 \beta_1 \gamma_1 \neq 0$,

$\quad\quad x_4 = (a_0^4, a_1^4, a_2^4, a_3^4) = (0, \beta_4, \gamma_4, \delta_4)$, with $\beta_4 \gamma_4 \delta_4 \neq 0$,

$\quad\quad x_2 = (a_0^2, a_1^2, a_2^2, a_3^2) = (\alpha_2, 0, 0, \delta_2)$, with $\alpha_2 \delta_2 \neq 0$.

The plane $p_1 p_2 x_i$, with $i \neq 1, 2$, has equation $X_3 = (a_3^i / a_2^i) X_2$,

the plane $p_1 p_3 x_i$, with $i \neq 1, 2$, has equation $X_1 = (a_1^i / a_3^i) X_3$,

the plane $p_2 p_4 x_i$, with $i \neq 2, 4$, has equation $X_2 = (a_2^i / a_0^i) X_0$,

the plane $p_3 p_4 x_i$, with $i \neq 2, 4$, has equation $X_0 = (a_0^i / a_1^i) X_1$.

Since $\{p_1 p_2 x_1, p_1 p_2 x_2, \ldots, p_1 p_2 x_{q+1}\}$ is the set of all planes through $p_1 p_2$ and since $p_1 p_2 x_1 = p_1 p_2 p_3$ and $p_1 p_2 x_2 = p_1 p_2 p_4$, we have

$$\{a_3^3 / a_2^3, a_3^4 / a_2^4, \ldots, a_3^{q+1} / a_2^{q+1}\} = GF(q) \backslash \{0\}.$$

Hence

$$\prod_{i \neq 1,2} a_3^i / a_2^i = -1.$$

Analogously,

$$\prod_{i \neq 1,2} a_1^i / a_3^i = -1, \quad \prod_{i \neq 2,4} a_2^i / a_0^i = -1, \quad \prod_{i \neq 2,4} a_0^i / a_1^i = -1.$$

Consequently

$$\left(\prod_{i \neq 1,2,4} \frac{a_3^i}{a_2^i} \cdot \frac{a_1^i}{a_3^i} \cdot \frac{a_2^i}{a_0^i} \cdot \frac{a_0^i}{a_1^i} \right) \cdot \frac{\delta_4}{\gamma_4} \cdot \frac{\beta_4}{\delta_4} \cdot \frac{\gamma_1}{\alpha_1} \cdot \frac{\alpha_1}{\beta_1} = 1.$$

Hence $\beta_4 \gamma_1 = \beta_1 \gamma_4$. Geometrically this means that $p_1 x_1 \cap p_4 x_4 \neq \varnothing$. Applying the polarity θ we see that the tangents L_1 and L_4 have a common point u. Clearly $u \in p_1 p_2 p_3 \cap p_2 p_3 p_4 = p_2 p_3$.

Let $F = \{C_1, C_2, \ldots, C_{q+1}\}$ be a flock of the hyperbolic quadric $Q^+(3, q)$. Let θ be the polarity defined by the quadric $Q^+(3, q)$, and let $F' = \{x_1, x_2, \ldots, x_{q+1}\}$ with $x_i = \pi_i^\theta$ and π_i the plane containing C_i, $i = 1, 2, \ldots, q + 1$. The involution of $PG(3, q)$ with axis π_i and center x_i will be denoted by $\theta(i)$, $i = 1, 2, \ldots, q + 1$. This involution $\theta(i)$ fixes $Q^+(3, q)$ setwise and C_i pointwise.

THEOREM 2. *There holds $F^{\theta(i)} = F$ and $F'^{\theta(i)} = F'$, for all i.*

PROOF. Consider C_i and C_j, with $i \neq j$. If C_i and C_j are orthogonal, i.e., if π_j contains x_i, then $C_j^{\theta(i)} = C_j$. Now suppose that C_i and C_j are not orthogonal. Let $p \in C_j$. Then $p^{\theta(i)}$ is the unique point of $Q^+(3, q) - \{p\}$, such that the lines of $Q^+(3, q)$ through p and the lines of $Q^+(3, q)$

through $p^{\theta(i)}$ contain the same points of C_i. Since C_i and C_j are not orthogonal, we have $p^{\theta(i)} \notin C_j$. The conic of F through $p^{\theta(i)}$ will be denoted by C_p. By Theorem 1 the tangent of C_j at p and the tangent of C_p at $p^{\theta(i)}$ have a point u in common. Moreover $u \in L = \pi_i \cap \pi_j$. Since $|C_j| = q + 1$ and $\left|F - \{C_i, C_j\}\right| = q - 1$, there are points p and p' on C_j, $p \neq p'$, for which $C_p = C_{p'}$. Let u' be the common point of L and the tangent of C_j at p' (possibly $u = u'$). This point u' also belongs to the tangent of C_p at $p'^{\theta(i)}$. The plane $\pi_j^{\theta(i)}$ contains $p^{\theta(i)}$, $p'^{\theta(i)}$, u, u', so $\pi_j^{\theta(i)}$ is the plane of the conic C_p. Hence $C_j^{\theta(i)} = C_p$. It follows that $F^{\theta(i)} = F$. Consequently also $F'^{\theta(i)} = F'$.

COROLLARY 1. *Let L be a line of $PG(3, q)$ and let $x_i \in L \cap F'$. Then*
$$(L \cap F')^{\theta(i)} = L \cap F'.$$

PROOF. Since $x_i \in L$ we have $L^{\theta(i)} = L$. So $(L \cap F')^{\theta(i)} = L^{\theta(i)} \cap F'^{\theta(i)} = L \cap F'$.

In 1.2 we mentioned that the set of all irreducible conics of $Q^+(3, q)$ can be divided into two equivalence classes, denoted by I and II. Let $\bar{I}$ (resp. $\overline{II}$) be the set of all planes containing the conics of class I (resp. II). Then the set of the poles with respect to $Q^+(3, q)$ of the elements of $\bar{I}$ (resp. $\overline{II}$) will be denoted by I' (resp. II'). Let $F' \cap I' = F_1'$ and $F' \cap II' = F_2'$.

COROLLARY 2. *If $x_i \in F'$, then $F_1'^{\theta(i)} = F_1'$ and $F_2'^{\theta(i)} = F_2'$. If the line L contains $x_i \in F'$, then $(L \cap F_1')^{\theta(i)} = L \cap F_1'$ and $(L \cap F_2')^{\theta(i)} = L \cap F_2'$.*

PROOF. Let $C_j \in F$ with $x_j \in F_1'$, and let C_i and C_j be not orthogonal. Let $p \in C_j$. By Theorem 1 the tangent of C_j at p and the tangent of $C_j^{\theta(i)}$ at $p^{\theta(i)}$ intersect in a point u of the plane π_i. The intersection of the plane $pp^{\theta(i)}u$ and $Q^+(3, q)$ is an irreducible conic which is tangent to both C_j and $C_j^{\theta(i)}$. Hence C_j and $C_j^{\theta(i)}$ belong to a same equivalence class. Consequently $x_j^{\theta(i)} \in F_1'$, and so $F_1'^{\theta(i)} = F_1'$. Analogously $F_2'^{\theta(i)} = F_2'$.

Suppose now that the line L contains $x_i \in F'$. Then $(L \cap F_1')^{\theta(i)} = L^{\theta(i)} \cap F_1'^{\theta(i)} = L \cap F_1'$. Analogously $(L \cap F_2')^{\theta(i)} = L \cap F_2'$.

If $q \equiv 1 \pmod 4$, then orthogonal circles intersect if and only if they are in the same class [10]; if $q \equiv -1 \pmod 4$, then orthogonal circles intersect if and only if they are in different classes [10]. This is used to prove the following corollary of Theorem 2.

COROLLARY 3. *Let $q \equiv 1 \pmod 4$. If for the line L we have $\left|L \cap F_t'\right| \neq 0$, then $\left|L \cap F_t'\right|$ is odd, $t \in \{1, 2\}$. If $L \cap F_1' \neq \varnothing$ and $L \cap F_2' \neq \varnothing$, then for each point $x_i \in L \cap F'$, the point of L orthogonal to x_i belongs to F';*

moreover $\left|L \cap F_1'\right| = \left|L \cap F_2'\right|$.

Let $q \equiv -1 \pmod 4$. *If* $\left|L \cap F_t'\right|$, $t \in \{1, 2\}$, *is even, then for each point* $x_i \in L \cap F_t'$, *the point of* L *orthogonal to* x_i *belongs to* F_t'; *if* $\left|L \cap F_t'\right|$, $t \in \{1, 2\}$, *is odd, then for each point* $x_i \in L \cap F_t'$, *the point of* L *orthogonal to* x_i *does not belong to* F_t'. *If* $L \cap F_1' \neq \varnothing$ *and* $L \cap F_2' \neq \varnothing$, *then* $\left|L \cap F_1'\right|$ *and* $\left|L \cap F_2'\right|$ *are even.*

PROOF. Let $q \equiv 1 \pmod 4$. Suppose that $\left|L \cap F_t'\right| \neq 0$, $t \in \{1, 2\}$. Let $x_i \in L \cap F_t'$. Since any two distinct circles of F are disjoint, the point of L orthogonal to x_i does not belong to F_t'. Hence x_i is the only point of $L \cap F_t'$ which is fixed by the involution $\theta(i)$. Since $(L \cap F_t')^{\theta(i)} = L \cap F_t'$, it now follows immediately that $\left|L \cap F_t'\right|$ is odd. Suppose now that $L \cap F_1' \neq \varnothing$ and $L \cap F_2' \neq \varnothing$. Then $\left|L \cap F_1'\right|$ and $\left|L \cap F_2'\right|$ are odd. Let $x_i \in L \cap F_1'$. Since $(L \cap F_2')^{\theta(i)} = L \cap F_2'$ and $\left|L \cap F_2'\right|$ is odd, it now follows that the point of L orthogonal to x_i belongs to $L \cap F_2'$. Consequently $\left|L \cap F_1'\right| = \left|L \cap F_2'\right|$.

Next, let $q \equiv -1 \pmod 4$. Suppose that $\left|L \cap F_t'\right|$, $t \in \{1, 2\}$, is odd. Let $x_i \in L \cap F_t'$. Since $(L \cap F_t')^{\theta(i)} = L \cap F_t'$ and $\left|L \cap F_t'\right|$ is odd, it follows that x_i is the only point of $L \cap F_t'$ fixed by the involution $\theta(i)$. Hence the point of L orthogonal to x_i does not belong to F_t'. If $\left|L \cap F_t'\right|$, $t \in \{1, 2\}$, is even, then analogously we show that for each point $x_i \in L \cap F_t'$, the point of L orthogonal to x_i belongs to F_t'. Finally, assume that $L \cap F_1' \neq \varnothing$ and $L \cap F_2' \neq \varnothing$. Let $x_i \in L \cap F_1'$. Since $q \equiv -1 \pmod 4$ and any two distinct circles of F are disjoint, the point of L orthogonal to x_i does not belong to F_2'. So no point of $L \cap F_2'$ is fixed by the involution $\theta(i)$. Since $(L \cap F_2')^{\theta(i)} = L \cap F_2'$, it follows that $\left|L \cap F_2'\right|$ is even. Analogously, $\left|L \cap F_1'\right|$ is even.

2.2. THEOREM 3. *Let* F *be a flock of the hyperbolic quadric* $Q^+(3, q)$. *Then either* $|F \cap I| = |F \cap II| = (q + 1)/2$, *or all the elements of* F *belong to the same class. In the first case, each irreducible conic of* $Q^+(3, q)$ *not in* F *is tangent to* 0 *or* 2 *elements of* F. *In the second case* $q \equiv -1 \pmod 4$.

PROOF. Let $F = \{C_1, C_2, \ldots, C_{q+1}\}$ and let C_i be contained in the plane π_i, $i = 1, 2, \ldots, q + 1$. Now we fix a conic C_i. Let C_j^*, $C_j^* \neq C_i$, be an irreducible conic of $Q^+(3, q)$ which is tangent to C_i and let θ_j be the number of elements of $F - \{C_i\}$ which is tangent to C_j^*. The point common to C_i and C_j^* is denoted by p. Since $C_j^* - \{p\}$ is the union of the disjoint sets $C_1 \cap C_j^*, \ldots, C_{i-1} \cap C_j^*, C_{i+1} \cap C_j^*, \ldots, C_{q+1} \cap C_j^*$, and since $\left|C_j^* - \{p\}\right| = q$ is odd, there is a conic $C_{i'} \in F$, $i' \neq i$, for which $\left|C_{i'} \cap C_j^*\right| = 1$. Hence $\theta_j \geq 1$. Clearly C_i and $C_{i'}$ belong to the same class, say class I. Conversely, let us consider a conic $C_{i'} \in F$, $i' \neq i$, which belongs to class I. Then there are exactly $2(q+1)$ irreducible conics C_j^* on $Q^+(3, q)$ which are tangent to both C_i and $C_{i'}$ [23]. Now we count in different ways the number of ordered pairs $(C_j^*, C_{i'})$, with C_j^* an irreducible conic on $Q^+(3, q)$ for which $\left|C_j^* \cap C_i\right| = 1$ and with $\left|C_j^* \cap C_{i'}\right| = 1$, $C_{i'} \in F - \{C_i\}$. We obtain

(2)
$$\sum_j \theta_j = 2(\alpha - 1)(q + 1), \quad \text{with } \alpha = |F \cap I|.$$

Since $\theta_j \geq 1$ and the number of indices j is equal to $(q+1)(q-1)$, we also have

(3)
$$\sum_j \theta_j \geq (q+1)(q-1).$$

From (2) and (3) it follows that $\alpha \geq (q+1)/2$.

Suppose that $|F \cap I| \neq 0$ and $|F \cap II| \neq 0$. Then by the previous paragraph $|F \cap I| \geq (q+1)/2$ and $|F \cap II| \geq (q+1)/2$. Hence $|F \cap I| = |F \cap II| = (q+1)/2$. With the notations of the preceding paragraph we then have $\sum_j \theta_j = (q+1)(q-1)$ and so $\theta_j = 1$ for all j. Hence each irreducible conic of $Q^+(3, q)$ which is not in F, is tangent to 0 or 2 elements of F.

Next, let all conics of F be in the same class, say class I, and suppose that $q \equiv 1 \pmod 4$. Let $C_i \in F$. If C_i and $C_j \in F$ are orthogonal, then $\left|C_i \cap C_j\right| \neq 0$ since $q \equiv 1 \pmod 4$ and $C_i, C_j \in I$. Hence no element of F is orthogonal to C_i. Since $F^{\theta(i)} = F$ and C_i is the only fixed element in F, the involution $\theta(i)$ divides $F - \{C_i\}$ into pairs. Consequently $\left|F - \{C_i\}\right| = q$ is even, a contradiction.

2.3. In this section we give a short proof of a theorem due to P.-H. Zieschang [28].

Let $F = \{C_1, C_2, \ldots, C_{q+1}\}$ be a flock of $Q^+(3, q)$ and let π_i be the plane containing C_i, $i = 1, 2, \ldots, q+1$.

THEOREM 4. *If at least $(q+1)/2$ of the planes $\pi_1, \pi_2, \ldots, \pi_{q+1}$ contain a common line L, then F is linear or a Thas flock.*

PROOF. Suppose that the planes $\pi_1, \pi_2, \ldots, \pi_{(q+1)/2}$ contain the line L. Let L' be the polar line of L with respect to $Q^+(3, q)$. Assume that C_i, $i \notin \{1, 2, \ldots, (q+1)/2\}$ does not contain L nor L'. Then π_i is orthogonal to at most one of $\pi_1, \pi_2, \ldots, \pi_{(q+1)/2}$. So we may assume that π_i is not orthogonal to $\pi_2, \pi_3, \ldots, \pi_{(q+1)/2}$. Now we consider the conics $C_2^{\theta(i)}, C_3^{\theta(i)}, \ldots, C_{(q+1)/2}^{\theta(i)} \in F$. If $C_j = C_{j'}^{\theta(i)}$, $j \neq j'$, $j \in \{1, 2, \ldots, (q +$

$1)/2\}$ and $j' \in \{2, 3, \ldots, (q + 1)/2\}$, then π_j, $\pi_{j'}$ and π_i have a line in common, so π_i contains L, a contradiction. If $C_j = C_j^{\theta(i)}$, $j \in \{2, 3, \ldots, (q+1)/2\}$, then π_i is orthogonal to π_j, again a contradiction. So there arise $q + 1$ distinct conics $C_1, C_2, \ldots, C_{(q+1)/2}, C_2^{\theta(i)}, C_3^{\theta(i)}, \ldots, C_{(q+1)/2}^{\theta(i)}, C_i$, i.e., all conics of F. Let $\{p\} = L \cap \pi_i$. Then p belongs to $\pi_1, \pi_2, \ldots, \pi_{(q+1)/2}, \pi_2^{\theta(i)}, \pi_3^{\theta(i)}, \ldots, \pi_{(q+1)/2}^{\theta(i)}, \pi_i$. Hence each of the planes $\pi_1, \pi_2, \ldots, \pi_{q+1}$ contains p. Now by a theorem of J. C. Fisher and J. A. Thas [10] the flock F is linear, hence π_i contains L, a contradiction. Consequently each plane π_i, $i \notin \{1, 2, \ldots, (q+1)/2\}$, contains either L or L'.

First, suppose that all conics of F belong to the same equivalence class, say class I. The line L is contained in exactly $(q+1)/2$ planes intersecting $Q^+(3, q)$ in conics of class I. Clearly these planes are the planes $\pi_1, \pi_2, \ldots, \pi_{(q+1)/2}$. Hence $\pi_{(q+3)/2}, \ldots, \pi_{q+1}$ are the $(q+1)/2$ planes through L' which intersect $Q^+(3, q)$ in conics of class I. Consequently F is a Thas flock.

Next, suppose that $|F \cap I| = |F \cap II| = (q+1)/2$. Assume that C_j and $C_{j'}$, with $j, j' \in \{1, 2, \ldots, (q+1)/2\}$, belong to different classes. If π_i contains L', $i \notin \{1, 2, \ldots, (q+1)/2\}$, then, since π_i is orthogonal to both π_j and $\pi_{j'}$, the conic C_i intersects one of C_j, $C_{j'}$, a contradiction. So each π_i, $i \notin \{1, 2, \ldots, (q+1)/2\}$, contains L, i.e., the flock F is linear. Finally, assume that $C_1, C_2, \ldots, C_{(q+1)/2}$ all belong to the same class, say class I. If π_i, with $i \notin \{1, 2, \ldots, (q+1)/2\}$, contains L, then C_i is of class II, and by the previous argument F is linear. To conclude, assume that $\pi_{(q+3)/2}, \ldots, \pi_{q+1}$ all contain L'. Since $C_{(q+3)/2} \cap C_1 = \varnothing, \ldots, C_{q+1} \cap C_1 = \varnothing$ and C_1, C_i with $i \notin \{1, 2, \ldots, (q+1)/2\}$ are orthogonal, the conics $C_{(q+3)/2}, \ldots, C_{q+1}$ all belong to a same equivalence class. Now it is clear that in this final case F is a Thas flock.

2.4. THEOREM 5. *If F is a flock of the hyperbolic quadric $Q^+(3, q)$, with $q \equiv 1 \pmod 4$, then F is linear or a Thas flock.*

PROOF. Let $F = \{C_1, C_2, \ldots, C_{q+1}\}$, let π_i be the plane of C_i, and let x_i be the pole of π_i with respect to $Q^+(3, q)$, $i = 1, 2, \ldots, q + 1$. Further, let $F' = \{x_1, x_2, \ldots, x_{q+1}\}$. By Theorem 3 we have $|F'_1| = |F'_2| = (q+1)/2$. In Corollary 2 we have shown that $F_1'^{\theta(i)} = F_1'$ and $F_2'^{\theta(i)} = F_2'$ for all $x_i \in F'$. Moreover for each line L containing $x_i \in F'$, we have $(L \cap F_1')^{\theta(i)} = L \cap F_1'$ and $(L \cap F_2')^{\theta(i)} = L \cap F_2'$. In Corollary 3 we have shown that $|L \cap F_j'|$ is always zero or odd, $j = 1, 2$.

Let $x_i, x_j \in F_1'$, with $i \neq j$. Now we consider all points $x_i^{\theta(t)}$ with $x_t \in (x_i x_j \cap F_1') - \{x_i\}$. Since $q \equiv 1 \pmod 4$ the points x_i and x_t are not orthogonal, so $x_i^{\theta(t)} \neq x_i$. Assume that $x_i^{\theta(t)} = x_i^{\theta(t')}$, with $t \neq t'$.

Let $x_i x_j \cap Q^+(3, q) = \{r, s\}$ (r and s belong to $PG(3, q^2) - PG(3, q)$), let $(r\ s\ x_t\ x_t^*) = -1$ and $(r\ s\ x_{t'}\ x_{t'}^*) = -1$. Then $(x_t\ x_t^*\ x_i\ x_i^{\theta(t)}) = -1$ and $(x_{t'}\ x_{t'}^*\ x_i\ x_i^{\theta(t)}) = -1$. Hence $\{x_t, x_t^*\} = \{x_{t'}, x_{t'}^*\}$ is the unique pair harmonic conjugate to the pairs $\{r, s\}$ and $\{x_i, x_i^{\theta(t)}\}$. Hence $x_t = x_{t'}^*$ and $x_{t'} = x_t^*$, so x_t and $x_{t'}$ are orthogonal, contradicting $q \equiv 1 \pmod 4$. Consequently $x_i^{\theta(t)} \neq x_i^{\theta(t')}$ for x_t and $x_{t'}$ different points of $(x_i x_j \cap F_1') - \{x_i\}$. So there is exactly one point $x_t \in (x_i x_j \cap F_1') - \{x_i\}$ for which $x_i^{\theta(t)} = x_j$. The involution $\theta(t)$ maps a line containing x_i and a point of F_2' onto a line containing x_j and a point of F_2'. A line joining a point of F_1' to a point of F_2' will be called a line of type 12. So we have proved that the number of lines of type 12 containing a given point of F_1' is independent of the choice of that point. That number will be denoted by τ_1. Analogously, the number of lines of type 12 containing a given point of F_2' will be denoted by τ_2.

Let L be a line of type 12. By Corollary 3, for each point $x_i \in L \cap F'$ the point of L orthogonal to x_i belongs to F', and $|L \cap F_1'| = |L \cap F_2'|$.

Let L_k be a general line of type 12, and let $\alpha_k = |L_k \cap F_1'| = |L_k \cap F_2'|$. Now we count in different ways the number of ordered pairs (x_i, L_k), with $x_i \in F_1'$, L_k of type 12, and $x_i \in L_k$. We obtain

$$(4) \qquad (q + 1)\tau_1/2 = \sum_k \alpha_k.$$

Next, we count in different ways the number of ordered pairs (x_j, L_k), with $x_j \in F_2'$, L_k of type 12, and $x_j \in L_k$. We obtain

$$(5) \qquad (q + 1)\tau_2/2 = \sum_k \alpha_k.$$

From (4) and (5) it follows that $\tau_1 = \tau_2 = \tau$.

Consider the plane π_i, with $x_i \in F_1'$. Since $q \equiv 1 \pmod 4$ all points of F' orthogonal to x_i belong to F_2'. Hence $\pi_i \cap F_1' = \varnothing$. For each line L of type 12 through x_i, the intersection of L and π_i belongs to F'. Conversely, each point of $\pi_i \cap F'$ is joined to x_i by a line of type 12. Hence $|\pi_i \cap F'| = |\pi_i \cap F_2'| = \tau$. Analogously, for the plane π_j, with $x_j \in F_2'$, we have $|\pi_j \cap F'| = |\pi_j \cap F_1'| = \tau$.

Let L be a line of type 12 and let $x_i \in F'$, say $x_i \in F_1'$. Further, let $x_j \in L \cap F_1'$. If we join x_j to the τ points of $\pi_i \cap F' = \pi_i \cap F_2'$, then we obtain τ lines of type 12 through x_j. Hence one of these lines is L, so L contains a point of $\pi_i \cap F'$. Consequently L intersects each of the planes $\pi_1, \pi_2, \ldots, \pi_{q+1}$ in a point of F'.

Next, let $x_i \in F_1'$ and $x_j \in F_2'$. Then $\pi_i \cap F' = \pi_i \cap F_2'$ has no point in common with $\pi_j \cap F' = \pi_j \cap F_1'$. By the previous paragraph all lines of type 12 are obtained by joining the τ points of $\pi_i \cap F'$ to the τ points of $\pi_j \cap F'$. Hence there are exactly τ^2 lines of type 12.

Further, let $x_i, x_k \in F_1'$ with $x_i \neq x_k$, and let $\left| \pi_i \cap \pi_k \cap F' \right| = \alpha$. Since each of the τ^2 lines of type 12 intersects both of $\pi_i \cap F'$ and $\pi_k \cap F'$, and since each of the α points of $\pi_i \cap \pi_k \cap F'$ is contained in τ lines of type 12, we obtain

$$\tau^2 \leq \alpha\tau + (\tau - \alpha)^2, \quad \text{i.e., } \alpha(\tau - \alpha) \leq 0.$$

Clearly $\alpha \leq \tau$ and $\alpha \leq 0$. Hence either $\alpha = 0$ or $\alpha = \tau$.

First, suppose that for $x_i, x_k \in F_1'$ with $x_i \neq x_k$, we have $\alpha = \tau$. Let $\pi_i \cap \pi_k \cap F' = \pi_i \cap \pi_k \cap F_2' = \{x_{j_1}, x_{j_2}, \ldots, x_{j_\tau}\}$. Further, consider an arbitrary point $x_s \in F_1'$. Assume that $x_{j_l} x_s$, with $l \in \{1, 2, \ldots, \tau\}$, contains a point x_r of $F_2' - \{x_{j_l}\}$. If x_i and x_k both belong to the plane Mx_r, with $M = \pi_i \cap \pi_k$, then $x_i x_k$ and its polar line M with respect to $Q^+(3, q)$ have a non-empty intersection, contradicting $x_i x_k \cap Q^+(3, q) = \varnothing$. So we may assume that $x_i \notin Mx_r$. Since $x_i x_r$ is of type 12 it must have a point in common with $\pi_i \cap F' = M \cap F'$, hence $x_i \in Mx_r$, a contradiction. It follows that $x_{j_l} x_s \cap F_2' = \{x_{j_l}\}$, $l = 1, 2, \ldots, \tau$. Hence $\left| x_{j_l} x_s \cap F_1' \right| = \left| x_{j_l} x_s \cap F_2' \right| = 1$. Each of the τ^2 lines of type 12 contains one of $x_{j_1}, x_{j_2}, \ldots, x_{j_\tau}$, each of these lines contains exactly one point of F_1', and each point of F_1' is contained in exactly τ lines of type 12. Consequently $\left| F_1' \right| = \tau^2 / \tau = \tau$. Since $\left| F_1' \right| = (q+1)/2$, it follows that $\tau = (q+1)/2$. Hence $F_2' = \{x_{j_1}, x_{j_2}, \ldots, x_{j_\tau}\}$, i.e., $M \cap F' = F_2'$. Now by Theorem 4 F is a Thas flock.

Clearly we have the same conclusion if $\left| \pi_{i'} \cap \pi_{k'} \cap F' \right| = \tau$ for some $x_{i'}, x_{k'} \in F_2'$ with $x_{i'} \neq x_{k'}$.

Finally, assume that for each two distinct points x_i, x_j of the same class we have $\pi_i \cap \pi_j \cap F' = \varnothing$. Then $\pi_r \cap \pi_s \cap F' = \varnothing$ for all $r, s \in \{1, 2, \ldots, q+1\}$ with $r \neq s$. Let L be a line of type 12 and let $x_i \in L \cap F'$. Let $x_{i'}$ be the point of L which is orthogonal to x_i. Then $x_{i'} \in \pi_i \cap F'$ and $x_i \in \pi_{i'} \cap F'$. So x_i belongs to exactly one of the planes $\pi_1, \pi_2, \ldots, \pi_{q+1}$. Since L intersects each of the planes $\pi_1, \pi_2, \ldots, \pi_{q+1}$ in one point of F', it follows that $\left| L \cap F' \right|$ is equal to the number of planes $\pi_1, \pi_2, \ldots, \pi_{q+1}$. Hence $\left| L \cap F' \right| = q + 1$, so $L = F'$, which means that F is linear.

2.5. THEOREM 6. *If the flock F of the hyperbolic quadric $Q^+(3, q)$, with $q \equiv -1 \pmod 4$, contains conics of both equivalence classes, then F is linear.*

PROOF. Let $F = \{C_1, C_2, \ldots, C_{q+1}\}$, let π_i be the plane of C_i, and let x_i be the pole of π_i with respect to $Q^+(3, q)$, $i = 1, 2, \ldots, q+1$. Further, let $F' = \{x_1, x_2, \ldots, x_{q+1}\}$. By Theorem 3 we have $|F'_1| = |F'_2| = (q+1)/2$. Let L be a line of type 12, i.e., let L be a line for which $|L \cap F'_1| \neq 0$ and $|L \cap F'_2| \neq 0$. By Corollary 3, since $q \equiv -1 \pmod 4$ we have that $|L \cap F'_1|$ and $|L \cap F'_2|$ are even, and for each point $x_i \in L \cap F'_t$, $t = 1, 2$, the point of L orthogonal to x_i belongs to F'_t.

Let L be a line of type 12, let $|L \cap F'_t| = \mu_t$ with $t = 1, 2$, and let $x_i \in L \cap F'_1$. Further, let $x_j, x_k \in L \cap F'$, $j \neq k$. Put $L \cap Q^+(3, q) = \{r, s\}$ (r and s belong to $PG(3, q^2) - PG(3, q)$). Let $(r\, s\, x_j\, x_j^*) = -1$ and $(r\, s\, x_k\, x_k^*) = -1$. If x_i is one of x_j, x_j^*, x_k, x_k^*, then it is easy to see that $x_i^{\theta(j)} = x_i^{\theta(k)}$ if and only if x_j and x_k are orthogonal. Now let x_i be different from x_j, x_j^*, x_k, x_k^*. We have $(x_j\, x_j^*\, x_i\, x_i^{\theta(j)}) = -1$ and $(x_k\, x_k^*\, x_i\, x_i^{\theta(k)}) = -1$. If $x_i^{\theta(j)} = x_i^{\theta(k)}$, then $\{x_j, x_j^*\} = \{x_k, x_k^*\}$ is the unique pair harmonic conjugate to the pairs $\{r, s\}$ and $\{x_i, x_i^{\theta(j)}\}$. Hence $x_j = x_k^*$ and $x_j^* = x_k$, i.e., x_j and x_k are orthogonal. Conversely, suppose that x_j and x_k are orthogonal. Then $x_j = x_k^*$ and $x_j^* = x_k$, so $x_i^{\theta(j)} = x_i^{\theta(k)}$. Hence in all cases $x_i^{\theta(j)} = x_i^{\theta(k)}$ if and only if x_j and x_k are orthogonal. Consequently $\mu_1 = |L \cap F'_1| \geq |\{x_i^{\theta(j)} \mid\mid x_j \in L \cap F'\}| = (\mu_1 + \mu_2)/2$, from which $\mu_1 \geq \mu_2$. Analogously $\mu_2 \geq \mu_1$. So we conclude that $\mu_1 = \mu_2$, i.e., $|L \cap F'_1| = |L \cap F'_2|$.

Let $x_i \in F'_1$ and $x_j \in F'_2$. The number of lines of type 12 containing x_i (resp. x_j) is denoted by τ_1 (resp. τ_2). Further, let $\pi_i \cap F' = \rho_1$ and $\pi_j \cap F' = \rho_2$. Since $q \equiv -1 \pmod 4$ and since any two distinct conics of F are disjoint, we have $\pi_i \cap F' = \pi_i \cap F'_1$ and $\pi_j \cap F' = \pi_j \cap F'_2$. By joining x_i to the ρ_2 points of $\pi_j \cap F'_2$, we obtain ρ_2 lines of type 12 containing x_i. Hence $\tau_1 \geq \rho_2$. Moreover on each of these ρ_2 lines the point orthogonal to x_i belongs to F'. Hence each of these ρ_2 lines has a point in common with $\pi_i \cap F'_1$. It follows that $\rho_1 \geq \rho_2$. Analogously $\rho_2 \geq \rho_1$. Consequently $\rho_1 = \rho_2$. Since each line of type 12 on x_i contains a point of $\pi_i \cap F'_1$ we have also $\rho_1 \geq \tau_1$. Hence $\rho_2 \leq \tau_1 \leq \rho_1$, implying $\tau_1 = \rho_1 = \rho_2$. Analogously $\tau_2 = \rho_1 = \rho_2$. So $\rho_1 = \rho_2 = \tau_1 = \tau_2 = \tau$. There immediately follows that each point of F' is on τ lines of type 12, and that each plane π_k, $k = 1, 2, \ldots, q+1$, contains τ points of F'.

Let L be a line of type 12 and consider the plane π_k, $k \in \{1, 2, \ldots, q+1\}$. Suppose e.g. that $x_k \in F'_1$. Then $\pi_k \cap F' = \pi_k \cap F'_1$. Let $x_j \in L \cap F'_2$.

Joining x_j to the τ points of $\pi_k \cap F'$ there arise the τ lines of type 12 through x_j. It follows that the intersection of L and π_k is a point of F'. Hence L intersects each of the planes $\pi_1, \pi_2, \ldots, \pi_k$ in a point of F'.

Such as in the proof of Theorem 5 we now see that there are exactly τ^2 lines of type 12, and that planes π_i and π_k, with x_i and x_k distinct points of the same class, intersect in 0 or τ points.

First, assume that there exist distinct planes π_i and π_k, with x_i and x_k of the same class, which have τ points in common. Let e.g. $x_i, x_k \in F_1'$. Consider now a point $x_j \in F_2'$. If x_i and x_k both belong to the plane Mx_j, with $M = \pi_i \cap \pi_k$, then $x_i x_k$ and its polar line M with respect to $Q^+(3, q)$ have a non-empty intersection, contradicting $x_i x_k \cap Q^+(3, q) = \varnothing$. So we may assume that $x_i \notin Mx_j$. Then $x_i x_j$ is of type 12, so $x_i x_j$ intersects π_i in a point of F', so $x_i x_j$ contains a point of M, so $x_i \in Mx_j$, a contradiction. Consequently, this case cannot occur.

Hence for each two distinct points x_i, x_k of the same class we have $\pi_i \cap \pi_k \cap F' = \varnothing$. Now in the same way as in the proof of Theorem 5 we show that F is linear.

THEOREM 7. *If F is a flock of the hyperbolic quadric $Q^+(3, q)$, with $q = 3$ or 7, then F is linear or a Thas flock.*

PROOF. First, let $q = 3$. For any flock F of $Q^+(3, 3)$ at least $(q+1)/2 = 2$ of the planes $\pi_1, \pi_2, \pi_3, \pi_4$ contain a common line. Now by Theorem 4 the flock F is linear or a Thas flock.

Next, let $q = 7$. Let $F = \{C_1, C_2, \ldots, C_8\}$ be a flock of $Q^+(3, 7)$, let π_i be the plane of C_i, and let x_i be the pole of π_i with respect to $Q^+(3, 7)$. As always, let $F' = \{x_1, x_2, \ldots, x_8\}$. If $F_1' \neq \varnothing$ and $F_2' \neq \varnothing$, then by Theorem 6 the flock F is linear. So we may assume that one of F_1', F_2' is empty, say $F_2' = \varnothing$. By way of contradiction assume now that F is not a Thas flock. It is easy to see that no every two points x_i, x_j are orthogonal, say x_1 and x_2 are not orthogonal. Notations can be chosen in such a way that $x_1^{\theta(2)} = x_3$. Clearly $x_3 \in x_1 x_2 \cap F'$. Let $x_1 x_2 = L'$ and $\pi_1 \cap \pi_2 = L$. By Theorem 4 we have $L' \cap F' = \{x_1, x_2, x_3\}$ and $|L \cap F'| \leq 3$. So we may assume that $x_4 \notin L \cup L'$. The point x_4 is orthogonal to at most one of x_1, x_2, x_3. Assume e.g. that x_4 is not orthogonal to x_1 and x_2. Let $x_4^{\theta(1)} = x_5$ and $x_4^{\theta(2)} = x_6$. The points $x_1, x_2, \ldots, x_6$ are contained in a common plane π. By a theorem of J.C. Fisher and J.A. Thas [10] the points x_7 and x_8 are not both contained in π. Assume e.g. that $x_8 \notin \pi$. First, let us suppose that $\pi_8 \neq \pi$. Then x_8 is orthogonal to at most three points of $\pi \cap F'$. In such a case there arise at least three new points of the form $x_8^{\theta(j)}$ with $x_j \in \{x_1, x_2, \ldots, x_6\}$. Hence $|F'| \geq 6 + 1 + 3 = 10$, a contradiction. Consequently we necessarily have $\pi = \pi_8$. If $x_7 \notin \pi$, then analogously $\pi = \pi_7$, contradicting $\pi_7 \neq \pi_8$. So $x_7 \in \pi$. The point x_7 is orthogonal to at most two of the points x_1, x_2, x_4, say x_7 is not orthogonal to x_4. Then necessarily $x_4^{\theta(7)} = x_7^{\theta(4)} = x_3$. Hence x_3 and x_7 are not orthogonal. It follows that x_7 is orthogonal to at most one of x_1, x_2, say x_7 is not orthogonal to x_1. Then necessarily $x_1^{\theta(7)} = x_7^{\theta(1)} = x_6$. Since x_1, x_7, x_6 are collinear, also $x_1^{\theta(4)}, x_7^{\theta(4)}, x_6^{\theta(4)}$ are collinear. We have $x_1^{\theta(4)} = x_4^{\theta(1)} = x_5$, $x_7^{\theta(4)} = x_4^{\theta(7)} = x_3$ and $x_6^{\theta(4)} = x_4^{\theta(6)} = x_2$ (remind that, if for a line M the number $\left| M \cap F_1' \right|$ is odd, then no two points of $M \cap F_1'$ are orthogonal). So x_2, x_3, x_5 are collinear, the final contradiction.

REMARK 1. The next case $q = 11$ is the first case where there is a flock not projectively equivalent to a linear or a Thas flock. Recently we proved the following theorem [33]: *If F is a flock of $Q^+(3, 11)$, then F is either a linear flock, or a Thas flock, or the flock of Bader-Baker-Ebert-Johnson.*

3. Maximal exterior sets

In this section we always suppose that q is odd.

3.1. Introduction. Let X be a maximal exterior set (MES) of the hyperbolic quadric $Q^+(5, q)$. Two points x and y of $PG(5, q) - Q^+(5, q)$ are called equivalent if there exists a point $z \in PG(5, q) - Q^+(5, q)$ for which xz and yz are tangent lines of $Q^+(5, q)$. This relation is an equivalence relation [24] and there are two equivalence classes [24] denoted by I' and II' respectively.

Let $Q^+(3, q)$ be an hyperbolic quadric embedded in $Q^+(5, q)$, and let $Q^+(3, q) \subset PG(3, q)$. From [24] it easily follows that $\{PG(3, q) \cap I', PG(3, q) \cap II'\}$ is the partition of $PG(3, q) - Q^+(3, q)$ introduced in section 2.1 (and in this section denoted by $\{I', II''\}$).

3.2. THEOREM 8. *If X is a MES of $Q^+(5, q)$, then either $X \subset I'$ or $X \subset II'$.*

PROOF. Let X be a MES of $Q^+(5, q)$, and let $X \cap I' = X_1$, $X \cap II' = X_2$. Choose a plane π on $Q^+(5, q)$. Each solid $PG(3, q)$ containing π has at most one point in common with X. Since π is contained in exactly $q^2 + q + 1 = |X|$ solids, each solid through π has exactly one point in common with X. Let $x \in \pi$ and let π_x be the tangent hyperplane of $Q^+(5, q)$ at x. From the preceding it immediately follows that $|\pi_x \cap X| = q + 1$. If $x_i, x_j \in X$, with $x_i \neq x_j$, then the polar solid of $x_i x_j$ with respect to $Q^+(5, q)$ contains exactly one point x of π, i.e., the points x_i and x_j belong to exactly one of the sets $\pi_x \cap X$. So X provided with the $q^2 + q + 1$ sets $\pi_x \cap X$ is a $2 - (q^2 + q + 1, q + 1, 1)$ design, i.e., a projective plane $\mathcal{P}$ of order q.

With the notations of the previous paragraph, let $\pi_x \cap X = \{x_1, x_2, \ldots, x_{q+1}\}$ and let $PG(3, q)$ be a solid of π_x which does not contain x. Let

$xx_i \cap PG(3, q) = x_i'$, $i = 1, 2, \ldots, q+1$, and let $PG(3, q) \cap Q^+(5, q) = Q^+(3, q)$. Further, let π_i' be the polar plane of x_i' with respect to $Q^+(3, q)$, and let $\pi_i' \cap Q^+(3, q) = C_i$, $i = 1, 2, \ldots, q+1$. Then $F' = \{x_1', x_2', \ldots, x_{q+1}'\}$ is a MES of $Q^+(3, q)$ and so $F = \{C_1, C_2, \ldots, C_{q+1}\}$ is a flock of $Q^+(3, q)$. By Theorem 3 we have either $|F' \cap I'| = |F' \cap II''| = (q+1)/2$, or $F' \subset I'$, or $F' \subset II'$. Since any two points of $PG(5, q) - Q^+(5, q)$ on a common tangent line of $Q^+(5, q)$ belong to the same class, there arises either $|\pi_x \cap X_1| = |\pi_x \cap X_2| = (q+1)/2$,or $\pi_x \cap X_1 = \varnothing$, or $\pi_x \cap X_2 = \varnothing$. Hence each line of the projective plane P contains 0, $(q+1)/2$, or $q+1$ points of X_i, $i = 1, 2$.

Suppose that $X_1 \neq \varnothing$ and $X_2 \neq \varnothing$. Let $x_i \in X_1$ and $x_j \in X_2$. The number of lines of P through x_i (resp. x_j) containing $(q+1)/2$ points of X_2 (resp. X_1) will be denoted by α_1 (resp. α_2). Then

$$\alpha_1(q+1)/2 = |X_2|, \qquad \alpha_2(q+1)/2 = |X_1|.$$

Hence

$$(\alpha_1 + \alpha_2)(q+1) = 2(|X_1| + |X_2|) = 2(q^2 + q + 1).$$

Consequently $q + 1$ divides $2(q^2 + q + 1)$, a contradiction. It follows that either $X_1 = \varnothing$ or $X_2 = \varnothing$. We conclude that $X \subset I'$ or $X \subset II'$.

THEOREM 9. *The hyperbolic quadric* $Q^+(5, q)$, *with* $q \equiv 1 \pmod 4$, *has no MES.*

PROOF. Let X be a MES of $Q^+(5, q)$. By Theorem 8, either $X \subset I'$ or $X \subset II'$. With the notations of the proof of Theorem 8, it follows that for any point $x \in \pi$ all elements of the flock F belong to a same equivalence class of conics on $Q^+(3, q)$. Now by Theorem 3 $q \equiv -1 \pmod 4$.

COROLLARY 4. *The hyperbolic quadric* $Q^+(2n - 1, q)$, *with* $n \geq 3$ *and* $q \equiv 1 \pmod 4$, *has no MES.*

PROOF. Immediate from Theorem 9 and a theorem of F. De Clerck and J. A. Thas [5].

3.3. THEOREM 10. *The hyperbolic quadric* $Q^+(5, q)$ *has no MES for each value of* q *for which* $Q^+(3, q)$ *admits only the linear and the Thas flocks.*

PROOF. Assume that $Q^+(3, q)$ admits only the linear and the Thas flock.

By way of contradiction, let $X = \{x_1, x_2, \ldots, x_{q^2+q+1}\}$ be a MES of $Q^+(5, q)$. By Theorem 8, we have either $X \subset I'$ or $X \subset II'$, say $X \subset I'$. Let $|x_i x_j \cap X| = \alpha(i, j)$, with $i \neq j$ and $i, j \in \{1, 2, \ldots, q^2 + q + 1\}$.

First, suppose that $\alpha(i, j) > 2$ for at least one pair $\{i, j\}$. Let $PG(3, q)$ be the polar space of $x_i x_j$ with respect to $Q^+(5, q)$. Since $x_i x_j \cap Q^+(5, q) = \varnothing$, the intersection $PG(3, q) \cap Q^+(5, q)$ is an elliptic quadric $Q^-(3, q)$. Let $x \in Q^-(3, q)$, let π_x be the tangent hyperplane of $Q^+(5, q)$ at x, let

π be a solid of π_x which does not contain x, and let $\pi \cap Q^+(5, q) = Q^+(3, q)$. Further, let $\pi_x \cap X = \{x_{i_1}, x_{i_2}, \ldots, x_{i_{q+1}}\}$ and $xx_{i_k} \cap \pi = x'_{i_k}$, $k = 1, 2, \ldots, q+1$. Then $F' = \{x'_{i_1}, x'_{i_2}, \ldots, x'_{i_{q+1}}\}$ is a MES of $Q^+(3, q)$. Clearly $x_i, x_j \in \pi_x \cap X$, say $x_i = x_{i_1}$ and $x_j = x_{i_2}$. We have $|x_{i_1} x_{i_2} \cap X| = \alpha(i, j) > 2$, hence $|x'_{i_1} x'_{i_2} \cap F'| \geq \alpha(i, j) > 2$. Since all elements of F' belong to class I', the flock F defined by F' is a Thas flock and so $|x'_{i_1} x'_{i_2} \cap F'| = (q+1)/2$. Consequently $|xx_i x_j \cap X| = (q+1)/2$. Now we count in two ways the number θ of ordered pairs (x, x_k), with $x \in Q^-(3, q)$, $x_k \in X$, and x_k in the tangent hyperplane π_x of $Q^+(5, q)$ at x. For a given point $x \in Q^-(3, q)$, the number of points of X contained in π_x is equal to $q+1$. Hence

(6) $$\theta = (q^2 + 1)(q + 1).$$

Each point of $x_i x_j \cap X$ is contained in $q^2 + 1$ spaces π_x, $x \in Q^-(3, q)$; each point $x_k \in X$ contained in a set $(xx_i x_j \cap X) - x_i x_j$, $x \in Q^-(3, q)$, belongs to exactly one space π_x; any of the remaining points of X belongs to exactly $q+1$ tangent hyperplanes π_x, $x \in Q^-(3, q)$. Hence

$$\theta = \alpha(i, j)(q^2 + 1) + \left(\frac{q+1}{2} - \alpha(i, j)\right)(q^2 + 1) +$$

(7) $$(q^2 + q + 1 - \alpha(i, j) - \left(\frac{q+1}{2} - \alpha(i, j)\right)(q^2 + 1))(q + 1).$$

From (6) and (7)

$$\alpha(i, j) = (q^2 - 1)/2q,$$

a contradiction since $\alpha(i, j)$ is an integer.

Consequently $\alpha(i, j) = 2$ for each pair $\{i, j\}$. For a given pair $\{i, j\}$, let $PG(3, q)$ be the polar space of $x_i x_j$ with respect to $Q^+(5, q)$, and let $PG(3, q) \cap Q^+(5, q) = Q^-(3, q)$. Such as in the previous paragraph, we see that $|xx_i x_j \cap X| = 2$ or $(q+1)/2$ for all $x \in Q^-(3, q)$. Let δ be the number of points $x \in Q^-(3, q)$ for which $|xx_i x_j \cap X| = (q+1)/2$. We count in two ways the number θ of ordered pairs (x, x_k), with $x \in Q^-(3, q)$, $x_k \in X$, and x_k in the tangent hyperplane π_x of $Q^+(5, q)$ at x. For a given point $x \in Q^-(3, q)$, the number of points of X contained in π_x is equal to $q+1$. Hence

(8) $$\theta = (q^2 + 1)(q + 1).$$

The points x_i and x_j are contained in $q^2 + 1$ spaces π_x, $x \in Q^-(3, q)$; each point $x_k \in X$ contained in a set $(xx_i x_j \cap X) - \{x_i, x_j\}$, $x \in Q^-(3, q)$, belongs to exactly one space π_x; any of the remaining points of X belongs to exactly $q + 1$ tangent hyperplanes π_x, $x \in Q^-(3, q)$. Hence

(9) $\theta = 2(q^2 + 1) + \delta(q - 3)/2 + (q^2 + q + 1 - 2 - \delta(q - 3)/2)(q + 1)$.

From (8) and (9)
$$\delta = (6q - 2)/(q - 3).$$

Since δ is an integer, we have $\delta \in \{5, 7, 11, 19\}$. Now we count in two ways the number ζ of ordered pairs (x_i, x), with $x_i \in X - \{x_1\}$, x a point of $Q^+(5, q)$ in the polar space of $x_1 x_i$, $|xx_1 x_i \cap X| = (q+1)/2$, and $q \neq 3$. We remark that for such a pair (x_i, x), the point x is contained in the polar hyperplane of x_1 with respect to $Q^+(5, q)$. For a given point $x_i \in X - \{x_1\}$, the number of pairs (x_i, x) is equal to δ. Hence

(10) $\zeta = (q^2 + q)\,\delta$.

For a given point x of $Q^+(5, q)$ in the polar hyperplane of x_1 with respect to $Q^+(5, q)$, the number of pairs (x_i, x) is equal to $(q - 1)/2$. Hence

(11) $\zeta = (q^2 + 1)(q + 1)(q - 1)/2$.

From (10) and (11)
$$2\delta = (q^2 + 1)(q - 1)/q,$$

a contradiction since δ is an integer.

COROLLARY 5. *The hyperbolic quadric* $Q^+(2n - 1, q)$, *with* $n \geq 3$, *has no* MES *for each value of* q *for which* $Q^+(3, q)$ *admits only the linear and the Thas flocks.*

PROOF. Immediate from Theorem 10 and a theorem of F. De Clerck and J. A. Thas [5].

THEOREM 11. *The hyperbolic quadric* $Q^+(2n - 1, q)$, *with* $n \geq 3$ *and* $q \in \{3, 7\}$, *has no* MES.

PROOF. This is an immediate corollary of Theorem 7 and Corollary 5.

4. Inversive planes

In this section we always suppose that q is odd.

4.1. Introduction. A $3-(q^2+1, q+1, 1)$ design is called an *inversive plane* of order q [7, 11]. The blocks of an inversive plane are called *circles*. Up to isomorphism, there is a unique inversive plane of order q, with $q \in \{3, 5, 7\}$ [3, 8, 9, 27]. For $q = 7$ this result was obtained by R. H. F. Denniston with the aid of a computer.

Let I be an inversive plane of order q. For any point p of I, the points of I different from p together with the circles containing p (minus p) form a $2 - (q^2, q, 1)$ design, i.e., an *affine plane* of order q. That affine plane is denoted by I_p, and is called the *internal plane* of I at p.

If O is an elliptic quadric of $PG(3, q)$, then the points of O together with the intersections $\pi \cap O$, with π a non-tangent plane of O, form an inversive plane of order q. Such an inversive plane is called a *classical* or *Miquelian* inversive plane. For odd order q no other inversive planes are known. For a Miquelian inversive plane I each internal plane I_p is Desarguesian.

The following question is an old, but fundamental open problem: if each internal plane I_p of the inversive plane I is Desarguesian, prove then that I is Miquelian. Here we give an answer on that question under the even weaker assumption that I_p is Desarguesian for at least one point p. We remark that the corresponding problem for Laguerre and Minkowski planes has been solved by Y. Chen and G. Kaerlein in 1973 [4, 19].

Let O be an elliptic quadric of $PG(3, q)$, let $p \in O$, and let π be a plane of $PG(3, q)$ not containing p. The intersection of π and the tangent plane π_p of O at p is denoted by L. By projection ζ of $O - \{p\}$ from p onto π, the points of $O - \{p\}$ are mapped onto the q^2 points of $\pi - L$, the circles of O through p (minus p) are mapped onto the $q^2 + q$ lines of π different from L, and the circles of O not through p are mapped by ζ onto the $q^3 - q^2$ irreducible conics of π containing two points $r, s \in L$ which are conjugate with respect to the quadratic extension $GF(q^2)$ of $GF(q)$. In this way we obtain the well-known plane model of the classical inversive plane.

Let H be a hyperbolic quadric of $PG(3, q)$, let $p \in H$, and let π be a plane of $PG(3, q)$ not containing p. The intersection of π and the tangent plane π_p of H at p is denoted by L. The lines R and S on H through p intersect L in the resp. points $r, s \in L$. By projection ζ of $H - \{p\}$ from p onto π, the points of $H - (R \cup S)$ are mapped onto the q^2 points of $\pi - L$, all points of $R - \{p\}$ (resp. $S - \{p\}$) are mapped onto r (resp. s), all lines of H not through p are mapped onto the $2q$ lines of π containing just one of r, s, all irreducible conics of H through p (minus p) are mapped onto the $q^2 - q$ lines of π containing neither r nor s and the irreducible conics of H not through p are mapped by ζ onto the $q^3 - q^2$ irreducible conics of π through r and s. In fact, we described here the well-known plane model of the classical Minkowski plane.

Let K be a quadratic cone with vertex v of $PG(3, q)$, let $p \in K - \{v\}$, and let π be a plane not containing p. The intersection of π and the tangent plane π_p of K at p is denoted by L. The line R on K through p intersects L in the point $r \in L$. By projection ζ of $K - \{p\}$ from p onto π, the points of $K - R$ are mapped onto the q^2 points of $\pi - L$, all points of $R - \{p\}$ are mapped onto r, all lines of K not through p are mapped onto the q lines of π through r, but distinct from L, all irreducible conics of K through p (minus p) are mapped onto the q^2 lines of π not containing r, and the irreducible conics of K not through p are mapped by ζ onto the $q^3 - q^2$ irreducible conics of π which are tangent to L at r. In fact, we described here the well-known plane model of the classical Laguerre plane.

4.2. LEMMA 1. *Let H be a hyperbolic quadric of $PG(3, q)$, let $p \in H$ and let R and S be the lines on H through p. If the irreducible conics $C_1, C_2, \ldots, C_{q+1}$ on H define a partition of $H - (R \cup S)$ and if exactly one of these conics contains p, then $\{C_1, C_2, \ldots, C_{q+1}\}$ is a flock of H.*

PROOF. Suppose that $F = \{C_1, C_2, \ldots, C_{q+1}\}$ is not a flock of H. Then at least one of the points of $(R \cup S) - \{p\}$, say p', is on at least two elements of F, say C_1 and C_2. Let $p' \in S$ and call R' the second line on H through p'. Through each point of $R' - \{p'\}$ there passes a unique element of $F - \{C_1, C_2\}$, and conversely, each element of $F - \{C_1, C_2\}$ contains at most one point of $R' - \{p'\}$. Hence $q = |R' - \{p'\}| \leq |F - \{C_1, C_2\}| = q - 1$, a contradiction. We conclude that F is a flock of H.

Let K be a quadratic cone with vertex v of $PG(3, q)$. A partition of $K - \{v\}$ into q irreducible conics is called a *flock* of K. If the q planes of the conics of the flock contain a common line, then the flock is called *linear*.

LEMMA 2. *Let K be a quadratic cone of $PG(3, q)$, and let R be a line on K. If the irreducible conics $C_1, C_2, \ldots, C_q$ on K define a partition of $K - R$ and if $C_1, C_2, \ldots, C_{(q+1)/2}$ are elements of a flock F of K, then $F = \{C_1, C_2, \ldots, C_q\}$.*

PROOF. Let $C \in F - \{C_1, C_2, \ldots, C_{(q+1)/2}\}$ and suppose that $C \notin \{C_1, C_2, \ldots, C_q\}$. The common point of C and R is denoted by p. Each point of $C - \{p\}$ is contained in just one of the conics $C_{(q+3)/2}, C_{(q+5)/2}, \ldots, C_q$, and, conversely, each of these $(q - 1)/2$ conics contains at most two points of $C - \{p\}$. Hence $q = |C - \{p\}| \leq 2.(q - 1)/2 = q - 1$, a contradiction. We conclude that each conic C of $F - \{C_1, C_2, \ldots, C_{(q+1)/2}\}$ is element of $\{C_1, C_2, \ldots, C_q\}$, i.e., $F = \{C_1, C_2, \ldots, C_q\}$.

LEMMA 3. *Let a, a', b, b', c, c' be distinct points of $PG(1, q^2) - PG(1, q)$, where a, a', resp. b, b', resp. c, c', are conjugate with respect to the quadratic extension $GF(q^2)$ of $GF(q)$. If $(a a' b b') = -1$, $(b b' c c') = -1$ and $(c c' a a') = -1$, then $q \equiv -1 \pmod 4$.*

PROOF. Let $GF(q^2) = \{t_1 + t_2\omega \,\|\, t_1, t_2 \in GF(q)\}$, with $\omega^2 = \nu$ and ν a non-square in $GF(q)$. Then $PG(1, q^2) - PG(1, q) = \{p(t_1 + t_2\omega) \,\|\, t_1 \in GF(q), t_2 \in GF(q) - \{0\}\}$.

Let $a(a_1 + a_2\omega), a'(a_1 - a_2\omega), b(b_1 + b_2\omega), b'(b_1 - b_2\omega), c(c_1 + c_2\omega), c'(c_1 - c_2\omega)$, with $a_2 b_2 c_2 \neq 0$. Then $(a a' b b') = -1$ is equivalent to

$$\frac{b_1 + b_2\omega - a_1 - a_2\omega}{b_1 + b_2\omega - a_1 + a_2\omega} : \frac{b_1 - b_2\omega - a_1 - a_2\omega}{b_1 - b_2\omega - a_1 + a_2\omega} = -1, \text{ i.e.,}$$

(11)
$$(b_1 - a_1)^2 = \nu (b_2^2 + a_2^2).$$

Analogously, $(b b' c c') = -1$ is equivalent to

(12) $$(c_1 - b_1)^2 = \nu\,(c_2^2 + b_2^2),$$

and $(cc'aa') = -1$ is equivalent to

(13) $$(a_1 - c_1)^2 = \nu(a_2^2 + c_2^2).$$

Calculating (11)+(12)−(13), we obtain

$$\nu\,b_2^2 = (b_1 - a_1)(b_1 - c_1).$$

Since ν is a non-square in $GF(q)$, also $(b_1 - a_1)(b_1 - c_1)$ is a non-square in $GF(q)$. Analogously, $(c_1 - b_1)(c_1 - a_1)$ and $(a_1 - c_1)(a_1 - b_1)$ are non-squares in $GF(q)$. Hence the product $-(b_1 - a_1)^2(c_1 - b_1)^2(a_1 - c_1)^2$ of these three non-squares is a non-square of $GF(q)$. It follows that -1 is a non-square in $GF(q)$, and so $q \equiv -1 \pmod 4$.

4.3. THEOREM 12. *Let I be an inversive plane of odd order q, with $q \equiv 1$ (mod 4). If for at least one point p of I the internal plane I_p is Desarguesian, then I is Miquelian.*

PROOF. Since there is a unique inversive plane of order 5, we may assume that $q \geq 9$. The proof will be subdivided into six parts.
Part 1. Let I be an inversive plane of order q, with $q \equiv 1$ (mod 4). Suppose that there is at least one point p of I for which the internal plane I_p is the Desarguesian affine plane $AG(2, q)$. The line at infinity of $I_p = AG(2, q)$ is denoted by L, and the projective plane defined by $AG(2, q)$ is $PG(2, q)$. The circles of I through p (minus p) are the lines of $AG(2, q)$. The $q^3 - q^2$ circles not containing p are $(q+1)-$ arcs [12] of $PG(2, q)$. By a celebrated theorem of B. Segre [12], any $(q + 1)-$ arc of $PG(2, q)$, with q odd, is an irreducible conic. Hence the $q^3 - q^2$ circles not containing p are irreducible conics of $PG(2, q)$. Each of these conics intersects L in distinct points which are conjugate with respect to the quadratic extension $GF(q^2)$ of $GF(q)$.

Let $r, s \in I - \{p\}$ and consider the $q + 1$ circles of I through r and s. In the plane $AG(2, q)$ these circles are the line rs and q irreducible conics $C_1, C_2, \ldots, C_q$ through r and s. Now we consider the plane model of the hyperbolic quadric H of $PG(3, q)$, defined by the points r and s of $PG(2, q)$. By Lemma 1 the line L together with the conics $C_1, C_2, \ldots, C_q$ define a flock F of H. By Theorem 5 the flock F is linear or a Thas flock.

First suppose that F is linear. In such a case we have $C_1 \cap L = C_2 \cap L = \cdots = C_q \cap L = \{\bar{r}, \bar{s}\}$. The points $\bar{r}$ and $\bar{s}$ will be called the *carriers* of the flock F or the *carriers* defined by r and s.

Next, assume that F is a Thas flock. In such a case $(q - 1)/2$ of the conics $C_1, C_2, \ldots, C_q$ contain two common points $\bar{r}$ and $\bar{s}$ of L, say $C_1 \cap L = C_2 \cap L = \cdots = C_{(q-1)/2} \cap L = \{\bar{r}, \bar{s}\}$. Let $r\bar{r} \cap s\bar{s} = \{\bar{r}\}$ and $r\bar{s} \cap \bar{r}s = \{\bar{s}\}$. Then the conics $C_{(q+1)/2}, C_{(q+3)/2}, \ldots, C_q$ contain the points

$\bar{r}$ and $\bar{s}$. Clearly $\bar{r}$ and $\bar{s}$ are conjugate with respect to the quadratic extension $GF(q^2)$ of $GF(q)$. Let $C_i \cap L = \{\bar{r}_i, \bar{s}_i\}$, $i = (q+1)/2, (q+3)/2, \ldots, q$. The bundle consisting of all conics of $PG(2, q^2) \supset PG(2, q)$ through $r, s, \bar{r}, \bar{s}$ intersects L in an involution with fixed points $\bar{r}$ and $\bar{s}$. It follows that $(\bar{r}\bar{s}\bar{r}_i\bar{s}_i) = -1$, with $i = (q+1)/2, (q+3)/2, \ldots, q$. Hence $\{\bar{r}_{(q+1)/2}, \bar{s}_{(q+1)/2}\}, \ldots, \{\bar{r}_q, \bar{s}_q\}$ are nothing else than the $(q+1)/2$ pairs of conjugate points on L (with respect to the quadratic extension of $GF(q)$) which are harmonic conjugate to $\bar{r}$ and $\bar{s}$. Also in this case the points $\bar{r}$ and $\bar{s}$ will be called the *carriers* of the flock F or the *carriers* defined by r and s.

We now fix a point $o \in AG(2, q)$. If $a \in AG(2, q) - \{o\}$, then the carriers defined by o and a will be denoted by a' and a''.

Part 2. *Assume that a and b are distinct points of $AG(2, q) - \{o\}$, with $\{o, a\}$ and $\{o, b\}$ defining linear flocks. Assume moreover that $\{a', a''\} \neq \{b', b''\}$.*

If o, a, b **are not collinear, then the circle** oab **contains the carriers** a', a'', **but also the carriers** b', b'', **a contradiction. Hence** o, a, b **are collinear in** $PG(2, q)$. **Let** c **be a point of the plane** $AG(2, q)$ **which is not on the line** ab. **Then the circle** oac **contains the carriers** a', a'', **and the circle** obc **contains the carriers** b', b''. **It follows that** $\{o, c\}$ **defines a Thas flock.**

(1) $(a'a''b'b'') \neq -1$.

Let c be a point of the plane $AG(2, q)$ which is not on the line ab. The circle oac contains the carriers a', a'', and the circle obc contains the carriers b', b''. If a', a'' are the carriers defined by $\{o, c\}$, then $(a'a''b'b'') = -1$, a contradiction. Analogously, the carriers defined by $\{o, c\}$ are not b', b''. Let c', c'' be the carriers defined by $\{o, c\}$. Then $\{c', c''\}$ is the unique pair for which $(a'a''c'c'') = -1$ and $(b'b''c'c'') = -1$.

Let $c = c_1, c_2, \ldots, c_{q-1}$ be the $q-1$ affine points of $M - \{o\}$, with $M = oc$. The carriers defined by $\{o, c_i\}$ are c' and c'', $i = 1, 2, \ldots, q-1$. For each pair $\{d', d''\}$ of conjugate points on L with $(c'c''d'd'') = -1$, there is a unique circle through o, c_i, d', d'', with $i = 1, 2, \ldots, q-1$. Let C_i be the circle through o, c_i, d', d'', and let C_j be the circle through o, c_j, d', d'', with $i \neq j$. If d is a point of $AG(2, q) - ab$ on C_i and C_j, then d' and d'' are the carriers defined by $\{o, d\}$, a contradiction. Hence C_i and C_j are tangent at o, or the common points of C_i and C_j are o, d', d'' and a point of $AG(2, q)$ on $ab - \{o\}$. If for some pair $\{i, j\}$, with $i, j \in \{1, 2, \ldots, q-1\}$, the circles C_i and C_j contain a common point d of $AG(2, q)$ on $ab - \{o\}$, then it is easy to see that $C_1, C_2, \ldots, C_{q-1}$ all contain d. Hence $C_1, C_2, \ldots, C_{q-1}$ are mutually tangent at o, or $C_1, C_2, \ldots, C_{q-1}$ all contain a common point d of $AG(2, q)$ on $ab - \{o\}$. It is clear that if $\{d', d''\} = \{a', a''\}$ (resp.

$\{d', d''\} = \{b', b''\})$ then $C_1, C_2, \ldots, C_{q-1}$ all contain a (resp. b).

If $C_1, C_2, \ldots, C_{q-1}$ all contain a common point d of $AG(2, q)$ on $ab - \{o\}$, then the flock defined by $\{o, d\}$ is linear with carriers d' and d''.

Now suppose that $C_1, C_2, \ldots, C_{q-1}$ are mutually tangent at o. By way of contradiction assume that the common tangent of $C_1, C_2, \ldots, C_{q-1}$ at o is not the line ab. Then the circles $C_1, C_2, \ldots, C_{q-1}$ intersect the affine line ab in o and $q-1$ distinct affine points of $ab - \{o, a, b\}$, a contradiction. It follows that ab is the common tangent of $C_1, C_2, \ldots, C_{q-1}$ at o. Hence the set $\{C_1, C_2, \ldots, C_{q-1}\}$ is uniquely defined by the line ab, so for at most one pair $\{d', d''\}$ the circles $C_1, C_2, \ldots, C_{q-1}$ are mutually tangent at o.

Consequently the $(q + 1)/2$ pairs $\{d', d''\}$ define either $(q + 1)/2$ or $(q-1)/2$ points of $AG(2, q)$ on $ab - \{o\}$. The set of these points is denoted by D. Let c_i be an affine point of $oc - \{o\}$ and let $oc'' \cap c_i c' = \{\bar{c}_i\}$ and $oc' \cap c_i c'' = \{\bar{\bar{c}}_i\}$. Let the conic through $o, c_i, \bar{c}_i, \bar{\bar{c}}_i, d', d''$ be denoted by C_i. Then $C_1, C_2, \ldots, C_{q-1}$ either are tangent to oa at o, or contain a point $d \in D$. It follows that in $PG(2, q)$ the points c', c'', o together with the line oc define uniquely the line oa. In fact, in $PG(2, q)$, any line through o, but distinct from oa, together with c', c'', o define uniquely the line oa. Let M and M' be distinct lines through o, with $M \neq oa \neq M'$. Hence, if ζ is any projectivity of $PG(2, q)$ for which $\{c', c''\}^\zeta = \{c', c''\}$, $o^\zeta = o$, $M^\zeta = M'$, then necessarily $(oa)^\zeta = oa$. This clearly yields a contradiction.

(2) $(a'a''b'b'') = -1$.

Let c be a point of the plane $AG(2, q)$ which is not on the line ab. Then $\{o, c\}$ defines a Thas flock with carriers c' and c''. By way of contradiction assume that $\{a', a''\} \neq \{c', c''\} \neq \{b', b''\}$. The circle oac contains a', a'' and the circle obc contains b', b'', so $(a'a''c'c'') = -1$ and $(b'b''c'c'') = -1$. Hence $(a'a''b'b'') = (b'b''c'c'') = (c'c''a'a'') = -1$, contradicting Lemma 3. It follows that $\{c', c''\} = \{a', a''\}$ or $\{c', c''\} = \{b', b''\}$.

The affine points of the line $M = oc$ are denoted by $o, c = c_1, c_2, \ldots, c_{q-1}$. By way of contradiction assume that a', a'' are the carriers defined by at least $(q + 1)/2$ pairs $\{o, c_i\}$, say by $\{o, c_1\}, \{o, c_2\}, \ldots, \{o, c_{(q+1)/2}\}$. For each pair $\{d', d''\}$ of conjugate points on L with $(a'a''d'd'') = -1$, there is a unique circle through o, c_i, d', d'', $i = 1, 2, \ldots, (q + 1)/2$. Let $\{d', d''\} \neq \{b', b''\}$. The circle through o, c_i, d', d'' is denoted by C_i, $i = 1, 2, \ldots, (q + 1)/2$. Such as in (1) we show that $C_1, C_2, \ldots, C_{(q+1)/2}$ are mutually tangent at o, or $C_1, C_2, \ldots, C_{(q+1)/2}$ all contain a common point d of $AG(2, q)$ on $ab - \{o\}$.

Suppose that $C_1, C_2, \ldots, C_{(q+1)/2}$ all contain a common point d of $AG(2, q)$ on $ab - \{o\}$. In such a case $\{o, d\}$ defines a linear flock with carriers d' and d''. By Lemma 3 we have $(b'b''d'd'') \neq -1$, but in (1) this was proved to be impossible.

Hence $C_1, C_2, \ldots, C_{(q+1)/2}$ are mutually tangent at o. The common tangent will be denoted by N. Let $C'_{(q+3)/2}, C'_{(q+5)/2}, \ldots, C'_{q-1}$ be the remaining circles which are tangent to N at o. Now we consider the plane model of the quadratic cone K of $PG(3, q)$, defined by the line N and the point $o \in N$ of $PG(2, q)$. The line L together with the conics $C_1, \ldots, C_{(q+1)/2}, C'_{(q+3)/2}, \ldots, C'_{q-1}$ define a partition of $PG(2, q) - N$. The conics $C_1, C_2, \ldots, C_{(q+1)/2}$ all contain d' and d''. Now by Lemma 2 the line L together with the conics $C_1, \ldots, C_{(q+1)/2}, C'_{(q+3)/2}, \ldots, C'_{q-1}$ define a linear flock F of K. If $N \neq ab$, then one of the $q-1$ conics contains a, d', d'', a contradiction. Hence $N = ab$. Now, let $\{\bar{d}', \bar{d}''\}$ be a pair of conjugate points on L with $(a'a''\bar{d}'\bar{d}'') = -1$, $\{d', d''\} \neq \{\bar{d}', \bar{d}''\}$ and $\{b', b''\} \neq \{\bar{d}', \bar{d}''\}$. Again we find $q-1$ circles $\bar{C}_1, \ldots, \bar{C}_{(q+1)/2}, \bar{C}'_{(q+3)/2}, \bar{C}'_{(q+5)/2}, \ldots, \bar{C}'_{q-1}$ tangent to ab at o. Hence $\{C_1, \ldots, C_{(q+1)/2}, C'_{(q+3)/2}, \ldots, C'_{q-1}\} = \{\bar{C}_1, \ldots, \bar{C}_{(q+1)/2}, \bar{C}'_{(q+3)/2}, \ldots, \bar{C}'_{q-1}\}$. It follows that $C_1, \ldots, C'_{q-1}$ all contain d', d'' and $\bar{d}', \bar{d}''$, a contradiction.

Consequently exactly $(q-1)/2$ pairs $\{o, c_i\}$ have a', a'' as carriers, and exactly $(q-1)/2$ pairs $\{o, c_j\}$ have b', b'' as carriers. Let d', d'' be conjugate points on L with $(a'a''d'd'') = -1$ and $\{d', d''\} \neq \{b', b''\}$. Suppose that a', a'' are the carriers defined by $\{o, c_1\}, \{o, c_2\}, \ldots, \{o, c_{(q-1)/2}\}$. Let C_i be the circle through o, c_i, d', d'', with $i = 1, 2, \ldots, (q-1)/2$. Again $C_1, C_2, \ldots, C_{(q-1)/2}$ are mutually tangent at o, or $C_1, C_2, \ldots, C_{(q-1)/2}$ all contain a common point d of $AG(2, q)$ on $ab - \{o, a, b\}$.

First, suppose that $C_1, C_2, \ldots, C_{(q-1)/2}$ are mutually tangent at o. The common tangent will be denoted by N. Let $C'_{(q+1)/2}, C'_{(q+3)/2}, \ldots, C'_{q-1}$ be the remaining circles which are tangent to N at o. Now we consider the plane model of the quadratic cone K of $PG(3, q)$, defined by the line N and the point $o \in N$ of $PG(2, q)$. The circles $C_1, C_2, \ldots, C_{(q-1)/2}$ all contain the points d', d'' of L, so, by Lemma 2, the line L together with the conics $C_1, \ldots, C_{(q-1)/2}, C'_{(q+1)/2}, \ldots, C'_{q-1}$ define a linear flock F of K. If $N \neq ab$, then one of the $q-1$ conics contains a, d', d'', a contradiction. Hence $N = ab$. One of the conics $C_1, \ldots, C'_{q-1}$, say $C'_{(q+1)/2}$, contains $c_{(q+1)/2}$. The carriers defined by $\{o, c_{(q+1)/2}\}$ are b', b'', and $C'_{(q+1)/2}$ contains d', d''. Hence $(b'b''d'd'') = -1$, contradicting Lemma 3.

Consequently $C_1, C_2, \ldots, C_{(q-1)/2}$ all contain a common point d of $AG(2, q)$ on $ab - \{o, a, b\}$. Since there are $(q-1)/2$ pairs $\{d', d''\}$, there arise $(q-1)/2$ distinct points of $AG(2, q)$ on $ab - \{o, a, b\}$. Interchanging roles of a and b, we again consider $(q-1)2$ pairs $\{e', e''\}$ on L. By Lemma 3 any pair $\{d', d''\}$ is distinct from any pair $\{e', e''\}$.

Hence we find $(q-1)/2$ new points of $AG(2, q)$ on $ab - \{o, a, b\}$. Hence $ab - \{o, a, b\}$ contains at least $q - 1$ affine points, a contradiction.

Conclusion of this Part 2. All pairs $\{o, a\}$ defining a linear flock yield the same carriers a', a'' on L.

Part 3. *Assume that a and b are distinct points of $AG(2, q) - \{o\}$, with $o \notin ab$, with $\{o, a\}$ defining a linear flock, and with $\{o, b\}$ defining a Thas flock. Assume moreover that $\{a', a''\} \neq \{b', b''\}$. The circle oab intersects L in a' and a''. Since b', b'' are the carriers defined by $\{o, b\}$, we have $(a'a''b'b'') = -1$.*

Let c be a point of $AG(2, q)$, with $c \notin oa$ and $c \notin ob$, and suppose that $\{c', c''\} \neq \{a', a''\}$ and $\{c', c''\} \neq \{b', b''\}$. By Part 2 the pair $\{o, c\}$ defines a Thas flock. Considering the circle oac, we see that $(a'a''c'c'') = -1$. By Lemma 3 we have $(b'b''c'c'') \neq -1$. Hence obc does neither contain c', c'', nor b', b''. If $obc \cap L = \{d', d''\}$, then $(c'c''d'd'') = -1$ and $(b'b''d'd'') = -1$, so $\{a', a''\} = \{d', d''\}$. There is a unique circle through o, b, a', a'', namely the circle oab. Since obc contains o, b, a', a'', we have $c \in oab$. Consider now a line T through o, with $T \notin \{oa, ob, oc\}$. For each affine point t of T not on the circle oab the carriers defined by $\{o, t\}$ are either a', a'' or b', b''. Interchanging roles of b and c we see that the carriers defined by $\{o, t\}$ are either a', a'' or c', c''. Hence the carriers defined by $\{o, t\}$ are a' and a''. If the flock defined by $\{o, t\}$ is linear, then clearly otb contains a' and a''. Since oab is the unique circle through o, b, a', a'', we have $t \in oab$, a contradiction. So the flock defined by $\{o, t\}$ is always a Thas flock. Let $(d'd''b'b'') = -1$ with d' and d'' conjugate on L and $\{d', d''\} \neq \{a', a''\}$, and let C be the circle through o, b, d', d''. The pair $\{d', d''\}$ can always be chosen in such a way that C intersects T in two distinct points o and u. Since $u \notin oab$, the pair $\{o, u\}$ defines the carriers a' and a''. Hence $(a'a''d'd'') = -1$, contradicting Lemma 3. Consequently, for each point c of $AG(2, q)$ not on oa and not on ob, we have either $\{c', c''\} = \{a', a''\}$ or $\{c', c''\} = \{b', b''\}$.

Let d', d'' be conjugate points of L with $(d'd''b'b'') = -1$ and $\{d', d''\} \neq \{a', a''\}$. Let again c be a point of $AG(2, q)$ which is neither on oa nor on ob. Let C be the unique circle through o, b, d', d''. The pair $\{d', d''\}$ can always be chosen in such a way that C intersects oc in two distinct points o and u. Since $(a'a''d'd'') \neq -1$, the carriers defined by $\{o, u\}$ are b' and b''. By Part 2 the pair $\{o, u\}$ defines a Thas flock. Interchanging the roles of u and b we see that for each affine point v of $ob - \{o\}$ the pair $\{o, v\}$ defines carriers a', a'' or b', b''. Consequently, for each point e of $AG(2, q)$ not on oa, we have either $\{e', e''\} = \{a', a''\}$ or $\{e', e''\} = \{b', b''\}$. Suppose that the pair $\{o, m\}$, with m an affine point of $ob - \{b\}$, defines a linear flock. Since the circle oam contains a' and a'', we have $\{m', m''\} = \{a', a''\}$. The carriers defined by $\{o, u\}$ are b' and b'', so there is a unique circle through o, u, a', a''. Hence the circles oua and oum coincide, so $u \in oam$. Now consider a line U through o, with

$U \notin \{oc, oa, ob\}$ and U not tangent to C at o. Let $U \cap C = \{o, u'\}$. Then again $u' \in oam$. Hence the distinct circles C and oam have at least three points of $AG(2, q)$ in common, a contradiction. It follows that all affine points of $ob - \{o\}$ define a Thas flock. Interchanging roles of b and u we immediately see that each pair $\{o, n\}$, with n an affine point not on oa and ob, defines a Thas flock. Hence o together with any point of $AG(2, q) - oa$, defines a Thas flock. The corresponding carriers are either a', a'' or b', b''.

Let again c be a point of $AG(2, q)$ which is neither on oa nor on ob, and let $oc = N$. The affine points of N are denoted by $o, c = c_1, c_2, \ldots, c_{q-1}$. By way of contradiction assume that at least $(q+1)/2$ pairs $\{o, c_i\}$ define the carriers a', a'', say the pairs $\{o, c_1\}, \{o, c_2\}, \ldots, \{o, c_{(q+1)/2}\}$. Let d', d'' be conjugate points on L with $(a'a''d'd'') = -1$ and $\{d', d''\} \neq \{b', b''\}$. The circle through o, c_i, d', d'' is denoted by C_i, $i = 1, 2, \ldots, (q+1)/2$. Such as in previous cases we see that $C_1, C_2, \ldots, C_{(q+1)/2}$ are mutually tangent at o, or that $C_1, C_2, \ldots, C_{(q+1)/2}$ all contain a common affine point d of $oa - \{o, a\}$.

First, let $C_1, C_2, \ldots, C_{(q+1)/2}$ all contain a common affine point d of $oa - \{o, a\}$. Then $\{o, d\}$ defines a linear flock F with carriers d' and d''. This contradicts Part 2.

Hence $C_1, C_2, \ldots, C_{(q+1)/2}$ are mutually tangent at o. Such as in Part 2, Case (2), we see that oa is the common tangent of $C_1, C_2, \ldots, C_{(q+1)/2}$ at o. Again such as in Part 2, Case (2), this leads to a contradiction.

If follows that at most $(q-1)/2$ pairs $\{o, c_i\}$ define the carriers a', a''. Analogously the carriers b', b'' are defined by at most $(q-1)/2$ of the points $\{o, c_j\}$. So we may assume that a', a'' correspond to $\{o, c_1\}, \ldots, \{o, c_{(q-1)/2}\}$, and that b', b'' correspond to $\{o, c_{(q+1)/2}\}, \ldots, \{o, c_{q-1}\}$.

Let e', e'' be conjugate points on L with $(b'b''e'e'') = -1$ and $\{e', e''\} \neq \{a', a''\}$. The circle through o, c_j, e', e'' is denoted by D_j, $j = (q + 1)/2, \ldots, q - 1$. Such as in Part 2, Case (2), we see that either the $(q-1)/2$ circles D_j all contain a common point r of $oa - \{o, a\}$, or that the $(q-1)/2$ circles D_j are mutually tangent at o with common tangent oa.

First, suppose that the circles D_j are mutually tangent at o with common tangent oa. Let $D'_1, D'_2, \ldots, D'_{(q-1)/2}$ be the remaining circles which are tangent to oa at o. Again L together with the conics $D'_1, \ldots, D'_{(q-1)/2}$, $D_{(q+1)/2}, \ldots, D_{q-1}$ define a linear flock F of the quadratic cone K of $PG(3, q)$. So also $D'_1, D'_2, \ldots, D'_{(q-1)/2}$ contain e' and e''. The point c_1 is on one of the circles $D'_1, D'_2, \ldots, D'_{(q-1)/2}$. Since $(a'a''e'e'') \neq -1$ the carriers defined by the pair $\{o, c_1\}$ are b', b'', a contradiction.

Hence the $(q-1)/2$ circles D_j all contain a common point r on $oa - \{o, a\}$. It follows that with the $(q-1)/2$ pairs $\{e', e''\}$ there correspond

$(q-1)/2$ distinct points of $oa - \{o, a\}$. Interchanging roles of a', a'' and b', b'' we again consider $(q-1)/2$ pairs $\{\bar{e}', \bar{e}''\}$ on L. By Lemma 3 any pair $\{e', e''\}$ is distinct from any pair $\{\bar{e}', \bar{e}''\}$. Hence we find $(q-1)/2$ new points of $AG(2, q)$ on $oa - \{o, a\}$. Hence $oa - \{o, a\}$ contains at least $q+1$ affine points, a final contradiction.

Conclusion of this Part 3. If o, a, b are non-collinear points of $AG(2, q)$, with $\{o, a\}$ defining a linear flock and $\{o, b\}$ defining a Thas flock, then $\{a', a''\} = \{b', b''\}$.

Part 4. *Assume that each point of $AG(2, q) - \{o\}$ defines a Thas flock.* Suppose that we can find non-collinear points o, a, b in $AG(2, q)$, with $\{a', a''\} \neq \{b', b''\}$ and $(a'a''b'b'') \neq -1$. If $oab \cap L = \{d', d''\}$, then $\{d', d''\}$ is the unique pair of L defined by $(a'a''d'd'') = (b'b''d'd'') = -1$. Let D be one of the circles through o, b, b', b''. Intersecting D with the $(q-1)/2$ circles through o, a, a', a'', we find at least $(q-3)/2$ affine points $d_1, d_2, \ldots, d_{(q-3)/2}$ on D. The carriers of each pair $\{o, d_i\}$ are clearly d' and d'', $i = 1, 2, \ldots, (q-3)/2$. Let c' and c'' be conjugate on L with $(b'b''c'c'') = -1$ and $\{c', c''\} \neq \{d', d''\}$. Further, let C be the circle through o, b, c', c''. Let p be a point of $C - \{o, b\}$. At least $(q-5)/2$ of the points $d_1, d_2, \ldots, d_{(q-3)/2}$ are not on the line op, say $d_1, d_2, \ldots, d_{(q-5)/2}$ are not on the line op. Let C_i be the circle opd_i, with $i = 1, 2, \ldots, (q-5)/2$. Since $q \geq 9$ there are at least two circles C_i. If d', $d'' \in C_i$ for at least one i, then, since $(b'b''d'd'') = (b'b''c'c'') = -1$, it follows that the carriers defined by $\{o, p\}$ are b' and b''. Now suppose that no C_i contains d', d''. If C_1 and C_2 both contain b' and b'', then again the carriers defined by $\{o, p\}$ are b' and b''. Next, suppose that C_1 and C_2 both contain r', r'' on L, with $\{r', r''\} \neq \{b', b''\}$ (since $(d'd''c'c'') \neq -1$ we have $\{r', r''\} \neq \{c', c''\}$). Then $(c'c''r'r'') = (d'd''r'r'') = -1$, implying $\{r', r''\} = \{b', b''\}$, a contradiction. Finally, suppose that $C_1 \cap L = \{s', s''\}$ and $C_2 \cap L = \{t', t''\}$, with $\{s', s''\} \neq \{t', t''\}$. Then $(d'd''s's'') = (d'd''t't'') = -1$. It follows that d', d'' are the carriers defined by $\{o, p\}$, hence $(c'c''d'd'') = -1$, a contradiction. We conclude that in each possible case the points b', b'' are the carriers defined by $\{o, p\}$. Now let u', u'' be conjugate on L with $(a'a''u'u'') = -1$ and $\{u', u''\} \neq \{d', d''\}$. Let E be the circle through o, a, u', u''. Interchanging roles of a and b, then by the previous argument the points a', a'' are the carriers defined by any point of $E - \{o\}$. Consequently C and E are tangent at o. Let $\bar{u}'$, $\bar{u}''$ be conjugate on L with $(a'a''\bar{u}'\bar{u}'') = -1$, $\{d', d''\} \neq \{\bar{u}', \bar{u}''\} \neq \{u', u''\}$, and let $\bar{E}$ be the circle through o, a, $\bar{u}'$, $\bar{u}''$. Then C and $\bar{E}$ are tangent at o. Since $a \in E \cap \bar{E}$, we have $E = \bar{E}$, a contradiction. We conclude that there do not exist non-collinear points o, a, b in $AG(2, q)$, with $\{a', a''\} \neq \{b', b''\}$ and $(a'a''b'b'') \neq -1$.

Next, suppose that we can find non-collinear points o, a, b in $AG(2, q)$, with $\{a', a''\} \neq \{b', b''\}$ and $(a'a''b'b'') = -1$. By Lemma 3 the circle

oab contains a', a'' or b', b'', say a', a''. Let c be an affine point which is neither on oa nor on ob. If $\{a', a''\} \neq \{c', c''\} \neq \{b', b''\}$, then we have one of $(a'a''c'c'') \neq -1$ or $(b'b''c'c'') \neq -1$, contradicting the previous section. Hence either $\{c', c''\} = \{a', a''\}$ or $\{c', c''\} = \{b', b''\}$. Let $o, c = c_1, c_2, \ldots, c_{q-1}$ be the affine points of the line oc. First, suppose that a', a'' are the carriers defined by $\{o, c_i\}$ and that b', b'' are the carriers defined by $\{o, c_j\}$, for at least one $i \neq j$. Interchanging roles of a and c_i (resp. b and c_j) we see that for each affine point of $oa - \{o\}$ (resp. $ob - \{o\}$) the corresponding carriers are a', a'' or b', b''. It follows that for any point of $AG(2, q) - \{o\}$ the corresponding carriers are a', a'' or b', b''. Assume that at least $(q-1)/2$ pairs $\{o, c_k\}$ have a', a'' as corresponding carriers, say $\{o, c_1\}, \{o, c_2\}, \ldots, \{o, c_{(q-1)/2}\}$. Let t', t'' be conjugate points on L with $(a'a''t't'') = -1$ and $\{t', t''\} \neq \{b', b''\}$. The circle through o, c_k, t', t'' will be denoted by C_k, $k = 1, 2, \ldots, (q-1)/2$. Since t', t'' can never act as carriers, the circles $C_1, C_2, \ldots, C_{(q-1)/2}$ are mutually tangent at o. The common tangent at o will be denoted by N. Let $C'_{(q+1)/2}, C'_{(q+3)/2}, \ldots, C'_{q-1}$ be the remaining circles which are tangent to N at o. Since $C_1, C_2, \ldots, C_{(q-1)/2}$ all contain t', t'', the line L together with the conics $C_1, C_2, \ldots, C_{(q-1)/2}, C'_{(q+1)/2}, \ldots, C'_{q-1}$ define a linear flock F of the quadratic cone K of $PG(3, q)$. By assumption there is a pair $\{o, c_j\}$ with b', b'' as corresponding carriers. One of the circles $C'_{(q+1)/2}, C'_{(q+3)/2}, \ldots, C'_{q-1}$ contains c_j. Hence $(b'b''t't'') = -1$, contradicting Lemma 3. Consequently at most $(q-3)/2$ pairs $\{o, c_k\}$ have a', a'' as corresponding carriers. Analogously, at most $(q-3)/2$ pairs $\{o, c_l\}$ have b', b'' as corresponding carriers. Since the number of affine points on $oc - \{o\}$ is equal to $q - 1$ we have a contradiction. Hence for each line $od \notin \{oa, ob\}$ all affine points of $od - \{o\}$ define the same carriers. Suppose e.g. that a', a'' are the carriers defined by the line $oc \notin \{oa, ob\}$. Interchanging roles of a and c, we see that all affine points of $oa - \{o\}$ define the same carriers. It follows that a', a'' are the carriers defined by any affine point of $oa - \{o\}$. Let t', t'' be conjugate on L with $(a'a''t't'') = -1$ and $\{t', t''\} \neq \{b', b''\}$. Let $o, c = c_1, c_2, \ldots, c_{q-1}$ be the affine points of oc and let C_i be the circle containing o, c_i, t', t'' with $i = 1, 2, \ldots, q - 1$. Since t', t'' can never act as carriers, the circles $C_1, C_2, \ldots, C_{q-1}$ are mutually tangent at o. If the common tangent at o is not ob then one of the circles C_i contains b, implying $(b'b''t't'') = -1$, a contradiction. Consequently the common tangent at o of the circles $C_1, C_2, \ldots, C_{q-1}$ is the line ob. Finally, consider conjugate points $\bar{t}', \bar{t}''$ on L, with $(a'a''\bar{t}'\bar{t}'') = -1$ and $\{b', b''\} \neq \{\bar{t}', \bar{t}''\} \neq \{t', t''\}$. Again there arise $q - 1$ circles $\bar{C}_1, \bar{C}_2, \ldots, \bar{C}_{q-1}$ which are mutually tangent at o with common tangent ob. Hence $\{C_1, C_2, \ldots, C_{q-1}\} = \{\bar{C}_1, \bar{C}_2, \ldots, \bar{C}_{q-1}\}$, and so $C_i \cap L = \{t', t''\} = \{\bar{t}', \bar{t}''\}$, a final contradiction.

Hence for non-collinear points o, a, b in $AG(2, q)$ we always have $\{a', a''\} = \{b', b''\}$. It immediately follows that all points of $AG(2, q) - \{o\}$ define the same carriers a', a''. Let r', r'' be conjugate points on L with $(a'a''r'r'') = -1$. For any point d of $AG(2, q) - \{o\}$ the points o, d, r', r'' are contained in a unique circle. It follows that the set $AG(2, q) - \{o\}$ of order $q^2 - 1$ can be partitioned into sets of order q, a contradiction.

Conclusion of this Part 4. At least one point of $AG(2, q) - \{o\}$ defines a linear flock.

Part 5. *Assume that a and b are distinct points of $AG(2, q) - \{o\}$, with $\{o, a\}$ defining a linear flock and with $\{o, b\}$ defining a Thas flock.* First, suppose that o, a, b are not collinear. By the conclusion of Part 3 we have $\{a', a''\} = \{b', b''\}$. Assume that for each affine point of $ob - \{o\}$ the corresponding flock is of Thas type. Let d', d'' be conjugate on L with $(a'a''d'd'') = -1$. For each $b_i \in ob - \{o\}$, with $i = 1, 2, \ldots, q-1$, there is a unique circle C_i through o, b_i, d', d''. By the conclusions of Parts 2 and 3, either $C_1, C_2, \ldots, C_{q-1}$ are mutually tangent at o or $C_1, C_2, \ldots, C_{q-1}$ all contain a common affine point on $oa - \{o, a\}$. If $C_1, C_2, \ldots, C_{q-1}$ all contain a common affine point r on $oa - \{o, a\}$, then $\{o, r\}$ defines a linear flock with carriers $\{d', d''\}$, contradicting the conclusion of Part 2. So $C_1, C_2, \ldots, C_{q-1}$ are mutually tangent at o. The common tangent line at o will be denoted by N. If $N \neq oa$, then one of the circles C_i, say C_1, contains a, so $C_1 \cap L = \{a', a''\}$, a contradiction. Hence the common tangent at o of $C_1, C_2, \ldots, C_{q-1}$ is the line oa. Now we consider conjugate points $\bar{d}', \bar{d}''$ on L with $(a'a''\bar{d}'\bar{d}'') = -1$ and $\{d', d''\} \neq \{\bar{d}', \bar{d}''\}$. Again there arise $q - 1$ circles $\bar{C}_1, \bar{C}_2, \ldots, \bar{C}_{q-1}$ which are mutually tangent at o with common tangent oa. Hence $\{C_1, C_2, \ldots, C_{q-1}\} = \{\bar{C}_1, \bar{C}_2, \ldots, \bar{C}_{q-1}\}$, so $C_1 \cap L = \{d', d''\} = \{\bar{d}', \bar{d}''\}$, a contradiction. It follows that on $ob - \{o\}$ there is at least one affine point c such that the flock defined by $\{o, c\}$ is linear. Now by the conclusions of Parts 2 and 3 it is clear that for each point $d \in AG(2, q) - \{o\}$, the flock defined by $\{o, d\}$ has a', a'' as carriers. Let d', d'' be conjugate on L with $(a'a''d'd'') = -1$ and let C be the circle through o, b, d', d''. Clearly all points of $C - \{o\}$ define a flock of Thas type. Let $C - \{o\} = \{d_1, d_2, \ldots, d_q\}$ and let $\bar{d}', \bar{d}''$ be conjugate on L with $(a'a''\bar{d}'\bar{d}'') = -1$ and $\{d', d''\} \neq \{\bar{d}', \bar{d}''\}$. Let D_i be the circle through $o, d_i, \bar{d}', \bar{d}''$, with $i = 1, 2, \ldots, q$. The circles $D_1, D_2, \ldots, D_q$ are mutually distinct and mutually tangent at o. So there arise $\left| D_1 \cup D_2 \cup \cdots \cup D_q \right| = q^2 + 1$ affine points, a contradiction.

Next, suppose that o, a, b are collinear, with $\{o, a\}$ defining a linear flock and $\{o, b\}$ defining a Thas flock. Let $c \in AG(2, q) - oa$. By the previous section $\{o, c\}$ defines a linear flock. Now again by the previous section, we see that $\{o, b\}$ defines a linear flock, a final contradiction.

Conclusion of this Part 5. Either all points of $AG(2, q) - \{o\}$ define a linear flock or all points of $AG(2, q) - \{o\}$ define a Thas flock.

Part 6. By the conclusions of Parts 4 and 5 all points of $AG(2, q) - \{o\}$ define a linear flock. By the conclusion of Part 2 the carriers defined by any point of $AG(2, q) - \{o\}$ are independent of the choice of that point. Let these common carriers be denoted by o', o''. Clearly each conical circle through o (i.e., a circle through o which is not a line of $AG(2, q)$) contains o' and o''.

Let $o_1 \in AG(2, q) - \{o\}$. Interchanging roles of o and o_1, the point o_1 defines carriers o_1' and o_1''. Considering a conical circle through o and o_1, it follows that $\{o', o''\} = \{o_1', o_1''\}$. Consequently each conical circle through o_1 also contains o' and o''.

It follows that any of the $q^3 - q^2$ conical circles contains o' and o''. Now by 4.1 I is the Miquelian inversive plane.

Theorem 12 can also be formulated as follows.

THEOREM $12'$. *For $q \equiv 1$ (mod 4) the Desarguesian affine plane of order q has a unique extension. This extension is the Miquelian inversive plane of order q.*

5. Recent results

L. Bader and N. L. Johnson constructed their flocks via the exceptional nearfields. For this reason L. Bader calls these flocks the *exceptional flocks*.

Relying on the crucial Theorem 2 and on some theorems on Bol planes by M. Kallaher, L. Bader and G. Lunardon [30] recently proved the following theorem.

THEOREM A. *Any flock of the hyperbolic quadric $Q^+(3, q)$, q odd, either is a linear flock, or a Thas flock, or an exceptional flock.*

A direct corollary of Theorem A and Corollary 5 is the following theorem.

THEOREM B. *The hyperbolic quadric $Q^+(2n - 1, q)$, with $n \geq 3$, q odd and $q \notin \{11, 23, 59\}$, has no MES.*

Relying on Theorem A we also proved the following generalization of Theorem 12 [31].

THEOREM C. *Let I be an inversive plane of odd order q, $q \notin \{11, 23, 59\}$. If for at least one point p of I the internal plane I_p is Desarguesian, then I is Miquelian.*

Theorem C also can be formulated as follows.

THEOREM C'. *For $q \notin \{11, 23, 59\}$ the Desarguesian affine plane of order q, q odd, has a unique extension. This extension is the Miquelian inversive plane of order q.*

In the proof of Theorem C we assume the uniqueness of the inversive plane of order q, $q \in \{3, 5\}$. By Theorem C and the uniqueness of the affine plane of order 7, we have the following corollary.

COROLLARY. *Up to isomorphism there is a unique inversive plane of order 7.*

This provides a computer-free proof of the uniqueness of the inversive plane of order 7.

REFERENCES

1. L. Bader, *Some new examples of flocks of* $Q^+(3, q)$, to appear.
2. R.D. Baker and G.L. Ebert, *A nonlinear flock in the Minkowski plane of order* 11, to appear.
3. Y. Chen, *The Steiner system* $S(3, 6, 26)$, J. Geometry 2(1972) 7-28.
4. Y. Chen and G. Kaerlein, *Eine Bemerkung über endliche Laguerre- und Minkowski-Ebenen*, Geometriae Ded. 2 (1973) 193-194.
5. F. De Clerck and J.A. Thas, *Exterior sets with respect to the hyperbolic quadric in* $PG(2n-1, q)$ in "Finite Geometries" (ed. by C. Baker and L. Batten), Lecture Notes in Pure and Applied Mathematics #103,Marcel Dekker 1985, 83-91.
6. F. De Clerck, H. Gevaert and J.A. Thas, *Flocks of a quadratic cone in* $PG(3, q)$, $q \leq 8$, Geometriae Ded., **26**(1988) 215-230.
7. P. Dembowski, *Finite Geometries*, Springer-Verlag 1968.
8. R.H.F. Denniston, *Uniqueness of the inversive plane of order* 5, Manuscr. Math. **8** (1973) 11-19.
9. _____, *Uniqueness of the inversive plane of order* 7, Manuscr. Math. **8**(1973) 21-26.
10. J.C. Fisher and J.A. Thas, *Flocks in* $PG(3, q)$, Math. Z. **169**(1979) 1-11.
11. H.-R. Halder and W. Heise, *Einführung in die Kombinatorik*, Carl Hanser Verlag 1976.
12. J.W.P. Hirschfeld, *Projective Geometries over Finite Fields*, Clarendon Press, Oxford 1979.
13. _____, *Finite Projective Spaces of Three Dimensions*, Clarendon Press, Oxford 1985.
14. N.L. Johnson, *Flocks of hyperbolic quadrics and translation planes admitting affine homologies*, J. Geometry, to appear.
15. _____, *The flocks of hyperbolic quadrics in* $PG(3, 5)$, to appear.
16. W.M. Kantor, *Ovoids and translation planes*, Can. J. Math. **34** (1982) 1195-1207.
17. A. Neumaier, *Some sporadic geometries related to* $PG(3, 2)$, Arch. Math. **42** (1984) 89-96.
18. D.J. Oakden, "Spreads in Three-Dimensional Projective Space," Ph.D. Thesis, Univ. Toronto 1973.
19. S.E. Payne and J.A. Thas, *Generalized quadrangles with symmetry, Part II*, Simon Stevin **49** (1976) 81-103.
20. S. Rees, C_2-*geometries arising from the Klein quadric*, Geometriae Ded. **18** (1985) 67-85.
21. B. Sègre and G. Korchmáros, *Una proprietà degli insieme di punti di un piano di Galois caratterizante quelli formati dai punti delle singole rette esterne ad una conica*, Rend. Accad. Naz. Lincei **62** (1977) 613-619.
22. J.A. Thas, *Flocks of finite egglike inversive planes*, in "Finite Geometric Structures and Their Applications," (ed. by A. Barlotti), Ed. Cremonese Roma (1973) 189-191.
23. _____, *Flocks of nonsingular ruled quadrics in* $PG(3, q)$, Rend. Accad. Naz. Lincei **59** (1975) 83-85.
24. _____, *Some results on quadrics and a new class of partial geometries*, Simon Stevin **55** (1981) 129-139.
25. _____, *Generalized quadrangles and flocks of cones*, European J. Comb., **18** (1987) 441-452.
26. M. Walker, *A class of translation planes*, Geometriae Ded. **5** (1976) 135-146.
27. E. Witt, *Uber Steinersche Systeme*, Abh. Math. Sem. Univ. Hamb. **12** (1938) 265-275.

28. P.-H. Zieschang, *Maximal exterior sets with respect to the hyperbolic quadric in PG(3,q)*, to appear.

29. _____, *On alternating bilinear forms on a 4-dimensional vector space over $GF(5)$*, to appear.

30. L. Bader and G. Lunardon, *On the flocks of $Q^+(3, q)$*, to appear.

31. J.A. Thas, *Solution of a classical problem on finite inversive planes*, Proc. Conf. "Finite Buildings, Related Geometries and Applications," Colorado State Univ., Pingree Park 1988, to appear.

32. G.M. Conwell, *The 3-space $PG(3, 2)$ and its group*, Ann. of Math. **11** (1910) 60–76.

33. J.A. Thas, *Recent results on flocks, maximal exterior sets and inversive planes*, Ann. Discrete Math., to appear.

Seminar of Geometry and Combinatorics, State University of Ghent, Krijgslaan 281, B-9000 Gent, Belgium

Contemporary Mathematics
Volume **111**, 1990

Self-Orthogonal Designs

VLADIMIR D. TONCHEV

ABSTRACT. The class of self-orthogonal designs provides a useful link between the theory of designs and self-orthogonal codes allowing the application of results and techniques from coding theory as a tool for the investigation of designs. On the other hand, self-orthogonal designs are used for
construction of extremal self-dual codes. This paper is a survey of recent
results and development in the theory of self-orthogonal designs and their
applications.

1. Self-orthogonal designs and codes. We assume familiarity with the basic
facts and notions from design and coding theory (cf., e.g. [**4**], [**19**], [**22**] for
designs, [**26**], [**28**], [**35**] for codes, and [**48**] both for designs and codes).

A $t - (v, k, \lambda)$ *design* (with $v > k > t > 0$ and $\lambda > 0$) is a collection D
of k-subsets (called *blocks*) of a set X of v *points*, such that any t-subset
of X is contained in exactly λ blocks of D. Two designs having the same
parameters t, v, k, λ are *isomorphic* if there is a bijection between their
point sets mapping the blocks of the first design to blocks of the second. A
$t - (v, k, \lambda)$ design is also an $s - (v, k, \lambda_s)$ design for any $s \leq t$, where
$\lambda_s = \lambda \binom{v-s}{t-s} / \binom{k-s}{t-s}$. Given a $t - (v, k, \lambda)$ design D, the *derived* design D_p
with respect to a point p is the $(t - 1) - (v - 1, k - 1, \lambda)$ design whose
points are the points of D different from p, and whose blocks are the sets
$B - \{p\}$ for each block B which contains p. The *residual* design D^p with
respect to p has the same point set as D_p, but its blocks are the blocks of
D not containing p; D^p is a $(t - 1) - (v - 1, k, \lambda_{t-1} - \lambda)$ design. We shall
say that D_p and D^p are obtained by derivation from D, and that D is an
extension of D_p. A $t - (v, k, 1)$ design, i.e. a t-design with $\lambda = 1$, is called
a Steiner system, and the notation S(t, k, v) is more customary in this case.
A projective plane of order n is simply an S(2, $n + 1$, $n^2 + n + 1$).

1980 *Mathematics Subject Classification* (1985 *Revision*). Primary 05B05.
Research partially supported by NCSCMB contract 37/1987 and a Research fellowship from
the Eindhoven University of Technology.
This paper is in final form and no version of it will be submitted for publication elsewhere.

A *self-orthogonal Steiner system* in the sense of Assmus, Mattson and Guza
[3] is a Steiner system S(t, k, v) with even k such that any two blocks intersect in an even number of points. As proved in [3] the only self-orthogonal
Steiner systems with $t > 2$ are the unique S(3,4,8), S(3,6,22), S(5,8,24), and
possibly an extension of a projective plane of order 10, S(3,12,112). The
nonexistence of a system with the last parameters follows by a result of Lam
et al. [25] obtained with the aid of computer running over 183 days.[1]

Generalizing the concept of a self-orthogonal Steiner system, we call a
$t - (v, k, \lambda)$ design *self-orthogonal* if the cardinality of the intersection of
any two blocks has the same parity (mod 2) as the block size k. More
generally, one may replace the condition (mod 2) by (mod p) where p is
a prime, although we restrict ourselves mainly to the case mod 2.

The term "self-orthogonal" is due to a natural connection between such
designs and self-orthogonal codes. By a (binary) code of length n and dimension k (or an (n, k) code) we mean a k-dimensional subspace of the
n-dimensional vector space V_n over $GF(2)$. Two (n, k) codes are equivalent if one can be obtained from the other by permuting the n coordinates.
If C is an (n, k) code, the *dual* code is defined to be the $(n, n - k)$ code
$C' = \{y \in V_n : xy = 0$ for each $x \in C\}$. C is *self-orthogonal*, if $C \subset C'$,
and *self-dual*, if $C = C'$. A matrix with the property that the linear span
of its rows generates the code C, is a *generator matrix* of C. The generator
matrices of the dual code C' are called parity check matrices of C. We
shall often refer to the elements of a code as *codewords*, or simply *words*.
The *weight* of a codeword is the number of its nonzero positions, and the
minimum weight of a code is the weight of a lightest nonzero codeword. An
(n, k, d) code is an (n, k) code with minimum weight d. A code has minimum weight at least d if and only if every $d - 1$ columns in a parity check
matrix are linearly independent. The *dual distance* of a code is the minimum weight of the dual code. The weights of all words in a self-orthogonal
code are even. If in addition all weights are divisible by 4, the code is called
doubly-even. The minimum weight in a doubly-even self-dual $(n, n/2)$ code
is less or equal to $4[n/24] + 4$; and the codes attaining this bound are called
extremal. By a famous theorem of Assmus and Mattson [2], the minimum weight codewords in an extremal self-dual doubly-even code form an
incidence matrix of an 1-, 3-, or 5-design according to $n \equiv 16, 8$, or 0
(mod 24).

For convenience, we shall further denote the total number of blocks λ_0 in
a $t-(v, k, \lambda)$ design by b, and the number λ_1 of blocks containing a given
point by r. The *incidence matrix* of a design is a b by v matrix $A = (a_{ij})$
with rows and columns indexed by the blocks and points respectively, where
$a_{ij} = 1$ if the $i-$th block contains the $j-$th point and $a_{ij} = 0$ otherwise.

---

[1]In late 1988 Lam, et al announced that their search for projective planes of order 10 was
completed.

The following theorems describe a connection between self-orthogonal 2– designs and self-orthogonal codes.

THEOREM 1.1. *Let A be a $b \times v$ incidence matrix of a self-orthogonal $2 - (v, k, \lambda)$ design with even k.*

(i) *If v is even, then the rows of A generate a self-orthogonal code of length v with dual distance $d \geq (v - 1)/(k - 1) + 1$.*

(ii) *If v is odd, then the rows of the matrix*

$$(1) \qquad A^* = \begin{bmatrix} 1 & \cdots & 1 & 1 \\ & & & 0 \\ & A & & \vdots \\ & & & 0 \end{bmatrix}$$

generate a self-orthogonal code of length $v + 1$ with dual distance $d \geq (v - 1)/(k - 1) + 1$.

PROOF. In both cases the weights of all rows of the generator matrix are even and the scalar product of any pair of rows is even, i.e. zero modulo 2. Hence the generated code is self-orthogonal. Suppose that S is a minimal set of linearly dependent (over $GF(2)$) columns of A. Then every row of A must intersect an even number of these columns in ones. Let n_i denote the number of rows intersecting exactly i columns from S in ones, and let $|S| = d$. Since every column of A contains $r = \lambda(v - 1)/(k - 1)$ ones and the scalar product (over the reals) of any two columns is λ, we have:

$$\sum_i 2i \cdot n_{2i} = rd,$$

$$\sum_i 2i(2i - 1)n_{2i} = d(d - 1)\lambda,$$

whence

$$\sum_i 2i(2i - 2)n_{2i} = d((d - 1)\lambda - r) \geq 0,$$

i.e. $d \geq (r + \lambda)/\lambda = r/\lambda + 1 = (v - 1)/(k - 1) + 1$.

To prove (ii) it is enough to mention that the dual of the code generated by A^* is obtained from the dual of the code generated by A by adding a new position equal to 1 for the words of odd weight and 0 for those of even weight.

The following theorem is proved in a similar way.

THEOREM 1.2. *Let A be a $b \times v$ incidence matrix of a self-orthogonal $2 - (v, k, \lambda)$ design with odd k. Then the rows of the matrix*

$$(2) \qquad \begin{bmatrix} 1 \\ \vdots & A \\ 1 \end{bmatrix}$$

generate a self-orthogonal code of length $v+1$ with dual distance $d \geq v/k+1$.

Let us note that the self-orthogonality of the design is essential only for the self-orthogonality of the related code, while the estimate for the dual distance depends only on the assumption that A is an incidence matrix of a 2-design (not necessarily self-orthogonal).

Every self-orthogonal code of even length is contained in a certain number of self-dual codes of the same length, and every doubly-even self-orthogonal code of length divisible by 8 is contained in doubly-even self-dual codes of the same length [28, Chapter 19, Section 6]. Combining the results of the previous theorems, one obtains the following

THEOREM 1.3 [44]. *If E is a self-dual code containing the code from Theorem* 1.1 *(resp.* 1.2*) then the minimum weight of E is at least $(v-1)/(k-1)+1$ (resp.* $v/k+1$*).*

PROOF. If C is a code defined as in Theorem 1.1 (or 1.2) and C is contained in E, then $E = E'$ is contained in C'. Hence the minimum weight of C' does not exceed that of E.

The relation between self-orthogonal designs and codes can be used to investigate designs on the basis of classification of self-dual codes. Using the knowledge of the total number of distinct self-orthogonal codes of a given length and dimension [32], techniques for construction of such codes and computing their automorphism groups, all binary self-dual codes of length up to 30, and all doubly-even self-dual codes of length 32 are classified up to equivalence [33], [36], [34], [13].

2. Self-orthogonal designs and a revision of Hamada's conjecture. A well known application of block designs to coding theory is the construction of majority-logic decodable codes having as parity check matrices incidence matrices of designs. The best codes constructed in this way are obtained from designs whose incidence matrices have minimum rank over the considered finite field. Most of the best majority-logic decodable codes are based on designs arising from finite geometries. The ranks of the incidence matrices of such designs were determined by MacWilliams and Mann [27], Goethals and Delsarte [16], Smith [37], and Hamada [20]. It was conjectured by Hamada [20] that a design arising from a finite affine or projective geometry has minimum rank among all designs with the same parameters. More precisely, if D is a design with the same parameters as those of a design G defined by the m-dimensional subspaces in $AG(t, q)$ or $PG(t, q)$, then $\operatorname{rank}_q D \geq \operatorname{rank}_q G$, with equality if and only if D is isomorphic to G. (Here by $\operatorname{rank}_q D$ (resp. $\operatorname{rank}_q G$) we mean the rank of any incidence matrix of the corresponding design over $GF(q)$). This conjecture has been proved by Hamada and Ohmori [21] in the case $q = 2$ and $m = t - 1$, i.e. for the designs defined by the hyperplanes in a binary affine or projective geometry. Doyen, Hubaut, and Vandensavel [15] proved that the conjecture is true also for designs formed by the lines in a binary projective or a ternary affine geometry, and Teirlink [40] (cf. Dehon [14]) proved the conjecture for the design of the planes in a binary affine geometry. In particular, Hamada's conjecture holds true for all designs derived from $PG(t, 2)$ or $AG(t + 1, 2)$ for $t \leq 3$. The first two open cases concern the designs of the planes in $PG(4, 2)$ with parameters $2 - (31, 7, 7)$, and the 3-dimensional subspaces in $AG(5, 2)$, with parameters $3 - (32, 8, 7)$, or $2 - (32, 8, 35)$ more generally.

Using Theorems 1.1–1.3 and the classification of the doubly-even (32,16) codes, one can show that in both cases there exist at least five nonisomorphic designs with the same parameters and rank over $GF(2)$. Therefore, the "only if" part of the Hamada's conjecture is not true in general.

The $3 - (32, 8, 7)$ design having as blocks the 3-dimensional subspaces in $AG(5, 2)$ is clearly self-orthogonal since the intersection of affine subspaces is again a subspace. Let A be an incidence matrix of a self-orthogonal $2 - (32, 8, 35)$ design D. By Theorems 1.1 and 1.3 any doubly-even self-dual (32,16) code E containing the code generated by A has minimum weight $d \geq 31/7 + 1 > 5$ and hence $d = 8$. Therefore E must be an extremal $(32, 16, 8)$ code and D must coincide with the $3-(32, 8, 7)$ design formed by the minimum weight words of E. Up to equivalence, there are exactly 5 doubly-even $(32, 16, 8)$ codes and each of them is generated by the vectors of minimum weight. Thus we have the following

THEOREM 2.1 [43]. *There are precisely* 5 *nonisomorphic self-orthogonal* $3 - (32, 8, 7)$ *designs, all having rank* 16 *over* $GF(2)$.

Thus, in addition to the design arising from $AG(5, 2)$ there are four other designs with the same parameters and the same rank over $GF(2)$, so the designs arising from a finite affine geometry are not characterized by their ranks in general.

REMARK. Since every extremal doubly-even $(32, 16)$ code is generated by the words of weight 8, the automorphism group of a self-orthogonal $3 - (32, 8, 7)$ design coincides with the group of the corresponding code. Thus by the results of [13] the five self-orthogonal $3 - (32, 8, 7)$ designs have transitive automorphism groups. The design obtained from the quadratic-residue code has a doubly-transitive group, and only the geometric design of the Reed-Muller code has a triply-transitive automorphism group.

The $2 - (31, 7, 7)$ design defined by the planes in $PG(4, 2)$ has the property that any two distinct blocks intersect in either 1 or 3 points, thus this design is self-orthogonal. Using Theorem 1.2 and arguments as in Theorem 2.1, the following can be proved.

THEOREM 2.2. *There are precisely* 5 *nonisomorphic self-orthogonal* $2 - (31, 7, 7)$ *designs, and all of them have rank* 16 *over* $GF(2)$.

Thus the designs arising from a finite projective geometry also are not characterized by their ranks in general.

3. Quasi-symmetric designs. A $2 - (v, k, \lambda)$ design is *quasi-symmetric* with intersection numbers x, y $(x < y)$ if the cardinality of the intersection of any two blocks is x or y. Quasi-symmetric designs are extensively studied because of their connections with strongly regular graphs [6]. Well known examples of quasi-symmetric designs are: the Steiner systems $S(2, k, v)$ which are not projective planes; the multiples of symmetric designs; the quasi-residual $2 - (v, k, 2)$ designs; and the strongly resolvable 2-designs. Following Neumaier [30], we call a quasi-symmetric design *exceptional*, if it does not belong to any of the above four classes. Neumaier [30] has investigated quasi-symmetric designs by use of various necessary conditions for strongly regular graphs. In particular, this condition left 23 sets of parameters for exceptional $2 - (v, k, \lambda)$ designs in the range $2k \leq v \leq 40$. For 8 of these 23 parameter sets quasi-symmetric designs are known to exist, the remaining cases are still open.

A quasi-symmetric design with $x \equiv y \equiv k \pmod 2$ is self-orthogonal. Thus the technique from Section 1 applies for such designs. Combined with the classification of certain self-dual codes, this technique allows to enumerate all non-isomorphic quasi-symmetric $2 - (31, 7, 7)$ designs, to establish the uniqueness of the classical quasi-symmetric designs arising from the Witt system $S(5, 8, 24)$, to prove the nonexistence of quasi-symmetric $2 - (28, 7, 16)$ or $2 - (29, 7, 12)$ designs; and to construct a new quasi-symmetric $2 - (45, 9, 8)$ design from an extremal doubly-even $(48, 24)$ code.

Let us begin with the self-orthogonal $2 - (31, 7, 7)$ design derived from $PG(4, 2)$ which is quasi-symmetric with $x = 1$, $y = 3$. In fact, every self-orthogonal $2 - (31, 7, 7)$ design is quasi-symmetric. For, let n_i denote the number of blocks having exactly i common points with a given block in a self-orthogonal $2 - (31, 7, 7)$ design. Then

$$n_1 + n_3 + n_5 + n_7 = 154,$$
$$n_1 + 3n_3 + 5n_5 + 7n_7 = 7.34,$$
$$3n_3 + 10n_5 + 21n_7 = 21.6,$$

whence $n_1 = 112$, $n_3 = 42$, $n_5 = n_7 = 0$.

On the other hand, suppose D is a quasi-symmetric $2 - (31, 7, 7)$ design with intersection numbers $x, y \, (x < y)$. Defining n_x, n_y as above, one has

$$n_x + n_y = 154,$$
(3)
$$xn_x + yn_y = 7.34,$$
$$x(x - 1)n_x + y(y - 1)n_y = 7.6.6,$$

whence $y(y - x)n_y = 14(18 - 17(x - 1)) > 0$, which is possible only if $x \leq 2$. It is easily checked that the system (3) has no solutions for $x = 0$ or 2, and if $x = 1$ then $y = 3$.

Hence, as a corollary of Theorem 2.2 we obtain the following

THEOREM 3.1 [**43**]. *There are exactly* 5 *nonisomorphic quasi-symmetric* $2 - (31, 7, 7)$ *designs.*

Let us mention that the automorphism group of a quasi-symmetric $2 - (31, 7, 7)$ design is isomorphic with the stabilizer of a coordinate in the automorphism group of the underlying doubly-even code. The design obtained from the Reed-Muller code is isomorphic with the design of the planes in $PG(4, 2)$ and has a doubly transitive group. The design derived from the extended quadratic-residue code has a transitive automorphism group, while the groups of the remaining three designs are not transitive.

If one considers the blocks of a quasi-symmetric $2 - (31, 7, 7)$ design as vertices of a graph, where two blocks are adjacent iff they have three common points, the resulting graph is strongly regular with parameters $v = 155$, $n_1 = 42$, $p_{11}^1 = 17$, $p_{11}^2 = 9$ (in the notation of Bose [**6**]). It was recently shown by Stoichev [**38**] that the five quasi-symmetric $2 - (31, 7, 7)$ designs produce nonisomorphic strongly regular graphs.

In the Neumaier's table [**30**] five of the eight parameter sets of known exceptional quasi-symmetric designs belong to designs related to the unique Steiner system $S(5, 8, 24)$. Two of them, $2-(22, 6, 5)$ and $2-(23, 7, 21)$ are those of the unique Steiner systems $S(3, 6, 22)$ and $S(4, 7, 23)$ constructed by Witt [**51**]. The remaining three are $2 - (21, 6, 4)(x = 0, y = 2)$, $2 - (21, 7, 12)(x = 1, y = 3)$, and $2 - (22, 7, 16)(x = 1, y = 3)$. All these quasi-symmetric designs were constructed by Goethals and Seidel [**17**] by derivation from the unique $S(5, 8, 24)$. It can be shown that the last three parameter sets also determine uniquely the corresponding quasi-symmetric designs.

THEOREM 3.2 [**44**]. *The quasi-symmetric* $2 - (21, 6, 4)$, $2 - (21, 7, 12)$ *and* $2 - (22, 7, 16)$ *designs are unique up to isomorphism.*

PROOF. We consider only the first parameter set. The remaining are treated similarly. Let A be a 56 by 21 incidence matrix of a quasi-symmetric $2-(21, 6, 4)$ design, and let C be the code generated by the matrix (1). By Theorems 1.1, 1.3 C is contained in a self-dual $(22, 11, 6)$ code. The only self-dual $(22, 11, 6)$ code (up to permutation of the positions) is the shortened Golay code G_{22} [**36**]. The code G_{22} contains exactly 77 words of weight 6 forming an incidence matrix of the unique Steiner system $S(3, 6, 22)$. Each point in $S(3, 6, 22)$ is contained in $21 = 77 - 56$ blocks, hence a quasi-symmetric $2 - (21, 6, 4)$ design must be isomorphic with a residual of the unique $3 - (22, 6, 1)$ design.

In a similar fashion, using the classification of self-dual $(30, 15)$ codes, one can prove the following

THEOREM 3.3 [**44**]. *Quasi-symmetric* $2 - (28, 7, 16)$ *or* $2 - (29, 7, 12)$ *designs do not exist.*

REMARK. The second parameter set, namely $2 - (29, 7, 12)$, has been previously eliminated by Haemers [18] using eigenvalue techniques, but not the first parameter set $2 - (28, 7, 16)$, as incorrectly mentioned in [10].

It is possible to construct strongly regular graphs from doubly-even codes in at least one other case, e.g. from extremal $(48, 24)$ codes [43]. Only one such code is known; this is the extended quadratic-residue code of length 48, and it has been proved by Huffman [23] that this is the only extremal doubly-even $(48, 24)$ code possessing automorphisms of odd order. The minimum weight words in this code form a $5 - (48, 12, 8)$ design D such that any two blocks have 0, 2, 4 or 6 common points. Let D^* be a derived $2 - (45, 9, 8)$ design obtained by the blocks of D containing three given points. Considered as blocks of D they must intersect each other in 4 or 6 points, hence the corresponding blocks of D^* must intersect in 1 or 3 points. Thus D^* is a quasi-symmetric design. If we call two blocks of D^* adjacent if they have exactly one common point, we get a strongly regular graph with parameters $v = 220$, $n_1 = 135$, $p_{11}^1 = 78$, $p_{11}^2 = 90$, i.e. a pseudo-geometric graph with $r = 15$, $k = 10$, $t = 6$. It is worth noting that the line graph of such a partial geometry would be a strongly regular graph with $v = 330$, $n_1 = 140$, $p_{11}^2 = p_{11}^1 + 2 = 60$, yielding a symmetric $2 - (330, 141, 60)$ design with a null polarity.[2]

REMARK. By construction, the design D^* is invariant under the stablizer of a set of three coordinates of the underlying code, which is a cyclic group of order 3. It has been shown recently by Stoichev [39] that this is in fact the full automorphism group of the related graph, and consequently, of D^*.

In the rest of this Section we shall sketch briefly the recent development in the theory of self-orthogonal designs due to Calderbank, Brouwer and Frankl [8], [9], [10], [12] giving new necessary conditions for the existence of quasi-symmetric designs.

Using the characterization of weight enumerators of maximal self-orthogonal doubly-even code, Calderbank [9] (cf. [12]) proved the following

THEOREM 3.3. *Let D be a $2-(v, k, \lambda)$ design where the block intersection sizes $s_1, s_2, \ldots, s_n$ satisfy $s_1 \equiv s_2 \equiv \ldots \equiv s_n \pmod 2$. If $r \not\equiv \lambda \pmod 4$, then after possibly taking complements either*

   (i) $k \equiv 0 \pmod 4$, $v \equiv 1 \pmod 8$, $r \equiv 0 \pmod 8$, *or*
   (ii) $k \equiv 0 \pmod 4$, $v \equiv -1 \pmod 8$, $2\lambda + r \equiv 0 \pmod 8$.

Another useful theorem of Calderbank and Frankl [12] is the following

THEOREM 3.4. *Let D be a $2 - (v, k, \lambda)$ design with block intersection numbers $s_1 \equiv s_2 \equiv \ldots \equiv s_n \pmod 2$ and an incidence matrix A. Let C be the code spanned by the rows of A.*

   (i) *If $r \equiv \lambda \pmod 4$ then either $r \equiv 0 \pmod 8$ or C contains the all-one vector.*

---

[2]W. Haemers (private communication) has shown that this graph on 220 vertices is not geometric.

(ii) *If $k \equiv 2$ (mod 4) then there exists a vector z such that $(z, a_i) \equiv 1$ (mod 2) for all rows a_i of A.*

For example, the non-existence of quasi-symmetric $2 - (21, 8, 14)$, $2 - (21, 9, 12)$ and $2 - (35, 7, 3)$ designs follows directly from the theorem.

An interesting parameter set from the Neumaier's table of exceptional quasi-symmetric designs is $2 - (24, 8, 7)$. It was shown in [44] that the block set of such a design must be a subset of the block set of the unique $S(5, 8, 24)$. Using the fact that the extended Golay $(24, 12, 8)$ code has covering radius 4 and a connection between quasi-symmetric $2 - (24, 8, 7)$ designs and the non-existent uniformly packed $(70, 58, 5)$ code [11], Brouwer and Calderbank [8] proved the non-existence of such a design.

In [10] Calderbank generalized the technique from [9] to designs with block intersection numbers $s_1, \ldots, s_n$ satisfying $s_1 \equiv s_2 \equiv \ldots \equiv s_n$ (mod p), p a prime, by use of self-orthogonal codes over $FG(p)$.

In Table 1 the updated Neumaier's list of exceptional quasi-symmetric designs with $2k \leq v \leq 70$ is given.

TABLE 1. Neumaier's list of exceptional quasi-symmetric 2-designs with $2k \leq v \leq 70$

No.	v	k	λ	r	b	x	y	EXISTENCE	REFERENCE
1	19	9	16	36	76	3	5	No	Calderbank[9]
2	20	10	18	38	76	4	6	No	Calderbank, Frankl[12]
3	20	8	14	38	95	2	4	No	Calderbank, Frankl[12]
4	21	9	12	30	70	3	5	No	Calderbank[10]
5	21	8	14	40	105	2	4	No	Calderbank[10]
6	21	6	4	16	56	0	2	Yes(1)	Goethals, Seidel[17] Tonchev[44]
7	21	7	12	40	120	1	3	Yes(1)	Goethals, Seidel[17] Tonchev[44]
8	22	8	12	36	99	2	4	No	Calderbank[9]
9	22	6	5	21	77	0	2	Yes(1)	Witt[51]
10	22	7	16	56	176	1	3	Yes(1)	Goethals, Seidel[17] Tonchev[44]
11	23	7	21	77	253	1	3	Yes(1)	Witt[51]
12	24	8	7	23	69	2	4	No	Brouwer, Calderbank[8]
13	28	7	16	72	288	1	3	No	Tonchev[44]
14	28	12	11	27	63	4	6	Yes	Cameron[30]
15	29	7	12	56	232	1	3	No	Haemers[18], Tonchev[44]
16	31	7	7	35	155	1	3	Yes(5)	Brouwer[30] Tonchev[43]
17	33	15	35	80	176	6	9		
18	33	9	6	24	88	1	3	No	Calderbank[9]
19	35	7	3	17	85	1	3	No	Calderbank[10]
20	35	14	13	34	85	5	8		

No.	v	k	λ	r	b	x	y	EXISTENCE	REFERENCE
21	36	16	12	28	63	6	8	Yes	Cameron[10]
22	37	9	8	36	148	1	3		
23	39	12	22	76	247	3	6		
24	41	9	9	45	205	1	3		
25	41	20	57	120	246	8	11		
26	41	17	34	85	205	5	8	No	Calderbank[10]
27	42	21	60	123	246	9	12		
28	42	18	51	123	287	6	9		
29	43	18	51	126	301	6	9	No	Calderbank[10]
30	43	16	40	112	301	4	7		
31	45	21	70	154	330	9	13		
32	45	9	8	44	220	1	3	Yes	Tonchev[43]
33	45	18	34	88	220	6	9		
34	45	15	42	132	396	3	6		
35	46	16	72	216	621	4	7		
36	46	16	8	24	69	4	6		
37	49	9	6	36	196	1	3		
38	49	16	45	144	441	4	7		
39	49	13	13	52	196	1	4		
40	51	21	14	35	85	6	9	No	Calderbank[10]
41	51	15	7	25	85	3	5	No	Calderbank[10]
42	52	16	20	68	221	4	7		
43	55	16	40	144	495	4	8		
44	55	15	63	243	891	3	6		
45	55	15	7	27	99	3	5		
46	56	16	18	66	231	4	8		
47	56	15	42	165	616	3	6		
48	56	12	9	45	210	0	3		
49	56	21	24	66	176	6	9		
50	56	20	19	55	154	5	8		
51	56	16	6	22	77	4	6	Yes	Tonchev[46]
52	57	27	117	252	532	12	17	No	Calderbank[10]
53	57	9	3	21	133	1	3		
54	57	15	30	120	456	3	6		
55	57	12	11	56	266	0	3		
56	57	24	23	56	133	9	12	No	Claderbank[10]
57	57	21	25	70	190	6	9		
58	57	21	10	28	76	7	9		
59	60	15	14	59	236	3	6		
60	60	30	58	118	236	14	18	No	Calderbank, Frankl[12]
61	61	25	160	400	976	9	13		
62	61	21	21	63	183	6	9		
63	63	15	35	155	651	3	7	Yes	Brouwer[10]
64	63	24	92	248	651	8	12		
65	63	18	17	62	217	3	6		
66	64	24	46	126	336	8	12		
67	65	20	19	64	208	5	8		
68	66	30	29	65	143	12	15		
69	69	33	176	374	782	15	21	No	Calderbank[10]
70	69	18	30	120	460	3	6		
71	70	10	6	46	322	0	2		
72	70	30	58	138	322	10	14		

REMARK. When a design exists the number in brackets in the column "Existence" indicates the number of nonisomorphic solutions if this is known.

For 19 of the total number of 72 parameter sets passing the Neumaier test non-existence is proved by applying techniques based on codes and self-orthogonality. For 11 parameter sets designs are known to exist: 5 of those (No. 6, 7, 9, 10, 11) arise from Witt designs; the designs No. 14, 21, 63, as well as the geometric design with parameters No. 16 belong to infinite series derived from finite geometries; the designs No. 32 and 51 seem to be "sporadic". The existence of designs for each of the remaining 42 cases is still undecided.

4. Characterization of designs related to the Witt system $S(5, 8, 24)$. One of the most important examples of Steiner systems is the celebrated $S(5, 8, 24)$ constructed by Witt in 1938 using the 5-transitive Mathieu group M_{24}. Applying consecutive derivation to an $S(5, 8, 24)$ one obtains a number of t-designs with $2 \leq t \leq 4$. The parameters of these designs are given in Table 2, where the last column contains the cardinalities of intersections of pairs of blocks. Since $S(5, 8, 24)$ is self-orthogonal, all its derived and residuals are also self-orthogonal. The uniqueness of the Steiner systems $S(2, 5, 21)$, $S(3, 6, 22)$, $S(4, 7, 23)$, $S(5, 8, 24)$ goes back to Witt **[51]**. In the previous Section we proved the uniqueness of the quasi-symmetric designs No. 5, 8, 9. In fact, our proof uses only the self-orthogonality rather than the quasi-symmetry of these designs. Using the same method, all remaining designs from Table 2 can be characterized as follows.

THEOREM 4.1 **[42]**. *All t-designs* $(t \geq 2)$ *derived from* $S(5, 8, 24)$ *are the unique self-orthogonal designs with the given parameters.*

A *dodecad* in an $S(5, 8, 24)$ is the symmetric difference of two blocks intersecting in two points. The set of all 2576 dodecads forms a $5-(24, 12, 48)$ design D. Both $S(5, 8, 24)$ and D arise as sets of supports of the words of weight 8 and 12 respectively, in the binary Golay code G_{24}. Removing one, two, or three points from D and considering the blocks containing (resp. not containing) them, one obtains a number of designs whose parameters are listed in Table 3.

The dodecad design and consequently the designs derived from it, are all self-orthogonal. Applying the method of the previous Section, one obtains the following characterization of these designs.

THEOREM 4.2. **[45]**. *The dodecad design* D, *as well as all t-designs* $(t \geq 2)$ *derived from* D *are the unique self-orthogonal designs with the given parameters.*

The proofs of Theorems 4.1, 4.2 use heavily the uniqueness of the Golay codes and the classification of self-dual codes of length 22.

5. Extremal self-dual codes constructed from self-orthogonal designs. Let

TABLE 2. Designs derived from $S(5, 8, 24)$

No.	Design	λ_5	λ_4	λ_3	λ_2	λ_1	λ_0	Block intersection numbers
1	$5 - (24, 8, 1)$	1	5	21	77	253	759	0, 2, 4
2	$4 - (23, 7, 1)$		1	5	21	77	253	1, 3
3	$4 - (23, 8, 4)$		4	16	56	176	506	0, 2, 4
4	$3 - (22, 6, 1)$			1	5	21	77	0, 2
5	$3 - (22, 7, 4)$			4	16	56	176	1, 3
6	$3 - (22, 8, 12)$			12	40	120	330	0, 2, 4
7	$2 - (21, 5, 1)$				1	5	21	1
8	$2 - (21, 6, 4)$				4	16	56	0, 2
9	$2 - (21, 7, 12)$				12	40	120	1, 3
10	$2 - (21, 8, 28)$				28	80	210	0, 2, 4

TABLE 3. Designs derived from dodecads in $S(5, 8, 24)$

No.	Design	λ_5	λ_4	λ_3	λ_2	λ_1	λ_0	Block intersection numbers
1	$5 - (24, 12, 48)$	48	120	280	616	1288	2576	0, 2, 4, 6, 8
2	$4 - (23, 11.48)$		48	120	280	616	1288	1, 3, 5, 7
3	$4 - (23, 12, 72)$		72	160	336	672	1288	0, 2, 4, 6, 8
4	$3 - (22, 10, 48)$			48	120	280	616	0, 2, 4, 6
5	$3 - (22, 11, 72)$			72	160	336	672	1, 3, 5, 7
6	$3 - (22, 12, 88)$			88	176	336	616	0, 2, 4, 6, 8
7	$2 - (21, 9, 48)$				48	120	280	1, 3, 5
8	$2 - (21, 10, 72)$				72	160	336	0, 2, 4, 6
9	$2 - (21, 11, 88)$				88	176	336	1, 3, 5, 7
10	$2 - (21, 12, 88)$				88	160	280	0, 2, 4, 6, 8

$X = \{x_1, \ldots, x_v\}$ be a set of "points". We generalize our initial definition of a self-orthogonal design as follows. We call a family $B = \{B_j\}_{j=1}^{b}$ of subsets of X a *weakly self-orthogonal* design if the following conditions are satisfied:

(a) $|B_i| \equiv |B_j|$ (mod 2) for $i, j \in \{1, \ldots, b\}$.

(b) $\left|B_i \cap B_j\right| \equiv \left|B_k \cap B_m\right|$ (mod 2) for $i, j, k, m \in \{1, \ldots, b\}$, $i \neq j$, $k \neq m$.

There are four possible types of weakly self-orthogonal designs according to the parity of the block size and the cardinality of the intersection of pairs of blocks:

(i) $\left|B_i \cap B_j\right| \equiv |B_k| \equiv 0$ (mod 2).

(ii) $\left|B_i \cap B_j\right| \equiv |B_k| \equiv 1$ (mod 2)).

(iii) $\left|B_i \cap B_j\right| \equiv 1$ (mod 2), $|B_k| \equiv 0$ (mod 2).

(iv) $\left|B_i \cap B_j\right| \equiv 0$ (mod 2), $|B_k| \equiv 1$ (mod 2).

A design of type (i) is called *properly self-orthogonal*, or simply *self-orthogonal*.

Evidently one has the following

THEOREM 5.1. *If A is a block-point $b \times v$ incidence matrix of a self-orthogonal design, then A generates a self-orthogonal binary code of length v.*

Designs of type (ii), (iii), or (iv) are easily extendable to self-orthogonal designs of type (i) by adding one or two new points to each block in an appropriate way. Suppose that $D = (X, B)$ is a weakly self-orthogonal design. Define the sets X', X'', X''' and the families of blocks B', B'', B''' as follows: $X' = X \cup \{x_{v+1}\}$, $X'' = X \cup \{x_{v+1}, \ldots, x_{v+b}, x_{v+b+1}\}$, $X''' = X'' - \{x_{v+b+1}\}$; $B' = \{B_j \cup \{x_{v+1}\} : j = 1, \ldots, b\}$, $B'' = \{B_j \cup \{x_{v+j}, x_{v+b+1}\} : j = 1, \ldots, b\}$, $B''' = \{B_j \cup \{x_{v+j}\} : j = 1, \ldots, b\}$. Then the following assertions are easily verified.

5.2. If D is of type (ii) then $D' = (X', B')$ is self-orthogonal, i.e. of type (i).

5.3. If D is of type (iii) then $D'' = (X'', B'')$ is of type (i).

5.4. If D is of type (iv) then $D''' = (X''', B''')$ is of type (i).

In other words, if A is a $b \times v$ incidence matrix of a weakly self-orthogonal design D, then the following matrix

(4)
$$\left[\begin{array}{cc} A & \begin{array}{c} 1 \\ \vdots \\ 1 \end{array} \end{array} \right]$$

generates a binary self-orthogonal code of length $v + 1$ provided that D is of type (ii). The following matrix

(5)
$$\left[\begin{array}{ccc} I_b & A & \begin{array}{c} 1 \\ \vdots \\ 1 \end{array} \end{array} \right]$$

generates a self-orthogonal code of length $b + v + 1$ provided that D is of type (iii). Let us mention that if v is odd, then one can add one more block consisting of all old points plus one new point x_{v+b+2}, so that a self-orthogonal design with $v + b + 2$ points and $b + 1$ blocks is obtained. In coding terms, the following matrix

(6)
$$\left[\begin{array}{ccc} I_{b+1} & A & \begin{array}{c} 1 \ldots 10 \\ 1 \\ \vdots \\ 1 \end{array} \end{array} \right]$$

generates a self-orthogonal code of length $v + b + 2$. Finally, the matrix

(7)
$$(I_b, A)$$

generates a self-orthogonal code of length $b + v$ provided that D is of type (iv).

Familiar examples of weakly self-orthogonal designs are provided by symmetric 2-designs. In such a case the corresponding codes generated by matrices of the form (6) or (7) are in fact self-dual, and the construction is well known [1],[5], [24], [41], [47], [48, pp. 108, 109], [49], [50]. As pointed out in [49], the absence of ovals, i.e maximal sets of points meeting each block in at most 2 points [1], is usually a necessary, and sometimes even a sufficient condition for extremality of the related code. For instance, four new extremal doubly-even codes of length 64 have been constructed (cf. [24, 47]) from the known symmetric $2 - (31, 10, 3)$ designs without ovals [29].

If A is an incidence matrix of a Hadamard symmetric $2 - (4t - 1, 2t, t)$ design with odd t then the matrix (6) generates a doubly-even self-dual code of length $8t$. As proved in [41], any such code has minimum weight at least 8 provided that $t > 0$; in particular, the codes derived from Hadamard designs with $t \leq 5$ are extremal. The last result has been recently "rediscovered" by Ozeki [31] (without the bound $d \geq 8$; the extremality has been checked by computer in [31]). Hadamard 2-designs arising from equivalent Hadamard matrices yield equivalent codes [31].

Exploring the concept of a self-orthogonal design, we generalize the construction of self-dual codes based on Hadamard 2-designs to a construction using a wider class of (0,1)-Hadamard matrices [50]. This general construction can produce inequivalent codes from equivalent Hadamard matrices. Given a Hadamard matrix H of order $n = 4t$, let us define incidence between rows and columns of H, a row and a column being incident if they intersect in $+1$. We call the incidence structure thus defined *the design of H*. An incidence matrix of the design of H is $(H + J)/2$, where J is the all-one matrix.

An essential property of a Hadamard matrix of order n is that the Hamming distance between each pair of rows is $n/2$. Consequently, the parity of the number of $+1$'s in a row is the same for all rows if $n > 2$. Using this, it is easy to prove the following theorem.

THEOREM 5.5 [50]. *The design of a Hadamard matrix H of order $n > 2$ such that the numbers of $+1$'s in all rows of H are congruent modulo 4, is weakly self-orthogonal.*

Suppose now that H is a Hadamard matrix of order $n = 8t + 4$ with some row (and hence all rows) containing an odd number of $+1$'s. Note that if the number of $+1$'s in a row is $\equiv 1 \pmod 4$, then multiplying that row by -1 transforms it in a row containing a number of $+1$'s $\equiv 3 \pmod 4$.

THEOREM 5.6 [50]. *Let H be a Hadamard matrix of order $n = 8t + 4$ such that the number of $+1$'s in each row is $\equiv 3 \pmod 4$. Then the following matrix*

$$(8) \qquad\qquad\qquad (I, A),$$

where $A = (H + J)/2$ is the incidence matrix of the design of H, generates a *self-dual doubly-even code C of length $2n$. The minimum distance of C is at least 8 if and only if each row and column of H contains at least 7 $+1$'s.*

PROOF. The self-orthogonality of the code follows from the fact that the design of H is self-orthogonal of type (iv). The code is doubly-even since the weights of all rows of the generator matrix (8) are divisible by 4.

Suppose that the minimum weight d is less than 8, i.e. $d = 4$.

Since the matrix A is non-singular over $GF(2)$, a codeword of weight 4 must be a sum of at most 3 rows of (8). A row of (8) can be of weight 4 only if some row of H contains exactly 3 $+1$'s. Since H is a Hadamard matrix, the weight of the sum of any two rows of (8) is $2 + n/2 > 4$ for $n > 4$. If there is a codeword of weight 4 being a sum of 3 rows of (8), then this word must be a row of the following matrix

$$(9) \qquad\qquad (A^T, I),$$

which is, due to the self-duality of C, both parity check and generator matrix of C. However, (9) can have a row of weight 4 only if H contains a column with exactly 3 $+1$'s.

Let us mention that if H is of the form

$$(10) \qquad\qquad H = \begin{bmatrix} -1 & 1 & \cdots & 1 \\ 1 & & & \\ \vdots & & H' & \\ 1 & & & \end{bmatrix},$$

then H' is an $(+1,-1)$-incidence matrix of a symmetric Hadamard $2 - (n - 1, n/2, n/4)$ design, thus the condition of Theorem 5.6 is fulfilled if $n > 4$. Hence Theorem 5.6 generalizes the result from [41].

Theorem 5.6 gives a simple criterion for extremality of codes arising from Hadamard matrices of order 8, 12, or 20. Starting from a particular Hadamard matrix, one can transform it into many different (but equivalent) matrices by multiplying rows and columns with -1 so that all rows and columns contain a number of $+1$'s $\equiv 3 \pmod 4$. It has been checked by an incomplete computer search [50] that at least 79 inequivalent extremal $(40, 20, 8)$ codes arise from the three Hadamard matrices of order 20.

REMARK. An automorphism of a code is any equivalence of the code to itself. A powerful method for the construction of self-dual codes that has recently been considerably developed, is based on consideration of automorphisms (cf., e.g. [7], [23], [52]). An essential feature of this method is that it is applicable only for automorphism groups of order not divisible by the characteristic of the underlying field. In particular, for binary codes only automorphisms of an odd order are handled. In this respect, it is worth noting that many of the extremal codes derived from Hadamard matrices of order 20 do not possess any nontrivial automorphisms of odd order, and there is at least one code with trivial full automorphism group [50].

Another extremal code with a very small automorphism group, namely a doubly-even $(64, 32, 12)$ code with full automorphism group of order 2, has been encountered in [24] by use of a symmetric $2 - (31, 10, 3)$ design.

REFERENCES

1. E. F. Assmus, Jr., and J. H. van Lint, *Ovals in projective designs,* J. Combin. Theory, A **27** (1979), 307–324.
2. E. F. Assmus, Jr., and H. F. Mattson, Jr., *New 5-designs,* J. Combin. Theory, **6** (1969), 122–151.
3. E. F. Assmus, Jr., H. F. Mattson, Jr., and M. Guza, *Self-orthogonal Steiner systems and projective planes,* Math. Z. **138** (1974), 89–96.
4. Th. Beth, D. Jungnickel, H. Lenz, "Design Theory," B. I. Wissenschaftsverlag, Zurich 1985.
5. V. K. Bhargava, J. M. Stein, (v, k, λ) *configurations and self-dual codes,* Information and Control, **28** (1975), 352–355.
6. R. C. Bose, *Strongly regular graphs, partial geometries, and partially balanced designs,* Pacific J. Math. **13** (1963), 389–419.
7. W. G. Bridges, M. Hall, Jr., and J. L. Hayden, *Codes and designs,* J. Combin. Theory, A **31** (1981), 155–174.
8. A. E. Brouwer, A. R. Calderbank, *An Erdös-Ko-Rado theorem for regular intersecting families of octads,* Graphs and Combinatorics, **2** (1986), 309–316.
9. A. R. Calderbank, *The application of invariant theory to the existence of quasi-symmetric designs,* J. Combin. Theory, Ser. A, **44**(1987), 94–109.
10. ____, *Geometric invariants for quasi-symmetric designs,* J. Combin. Theory, Ser. A **47** (1988), 101–110.
11. ____, *Nonexistence of uniformly packed* $(70, 58, 5)$ *code,* IEEE Trans. Info. Theory **32** (1986), 828–833.
12. A. R. Calderbank and P. Frankl, *Binary codes and quasi-symmetric designs,* Discr. Math. (to appear).
13. J. H. Conway and V. Pless, *On the enumeration of self-dual codes,* J. Combin. Theory, Ser. A, **28** (1980), 26–53.
14. M. Dehon, *Ranks of incidence matrices of t-design* $S_\lambda(t, t+1, v)$, Europ. J. Combin., **1** (1980), 97–100.
15. J. Doyen, X. Hubaut, and M. Vandensavel, *Ranks of incidence matrices of Steiner triple systems,* Math. Z., **163**(1978), 251–259.
16. J. -M. Goethals and P. Delsarte, *On a class of majority logic cyclic codes,* IEEE Trans. Inform. Theory, **14** (1968), 182–188.
17. J.-M. Goethals, J. J. Seidel, *Strongly regular graphs derived from combinatorial designs,* Canad. J. Math., **22** (1970), 597–614.
18. W. Haemers, "Sterke grafen en block designs", Thesis, Technische Hogeschool Eindhoven, October 1975 (in Dutch).
19. M. Hall, Jr., "Combinatorial Theory," Sec. Ed., Wiley–Interscience Publ., New York, 1986.
20. N. Hamada, *On the p-rank of the incidence matrix of a balanced or partially balanced incomplete block design and its applications to error correcting codes,* Hiroshima Math. J., **3** (1973), 153–226.
21. N. Hamada and H. Ohmori, *On the BIB design having minimum p-rank,* J. Combin. Theory, Ser. A, **18** (1975), 131–140.
22. D. R. Hughes and F. C. Piper, "Design Theory," Cambridge University Press, Cambridge 1985.
23. W. C. Huffman, *Automorphisms of codes with applications to extre nal doubly even codes of length* 48, IEEE Trans. Info. Theory, **28** (1982), 511–521.
24. S. N. Kapralov and V. D. Tonchev, *Extremal doubly even code of length* 64 *derived from symmetric designs,* Discrete Math. (submitted).
25. C. W. H. Lam, L. Thiel, S. Swiercz and J. McKay, *The nonexistence of ovals in a projective*

plane of order 10, Discrete Math., **45** (1983), 319–321.

26. J. H. van Lint, "Introduction to Coding Theory," Springer-Verlag, Berlin/Heidelberg/New York 1982.

27. F. J. MacWilliams and H. B. Mann, *On the p-rank of the design matrix of a difference set,* Inform. and Control, **12** (1968), 474–488.

28. F. J. MacWilliams and N. J. A. Sloane, "The Theory of Error-Correcting Codes," North-Holland, Amsterdam, 1977.

29. R. Mathon, *Symmetric* (31, 10, 3) *design with non-trivial automorphism group,* Ars Combinatoria **25** (1988), 171–183.

30. A. Neumaier, *Regular sets and quasi-symmetric 2-designs,* in: Combinatorial Theory (D. Jungnickel and K. Vedder, eds.) Lecture Notes in Math. **969** (1982), 258–275.

31. M. Ozeki, *Hadamard matrices and doubly-even self-dual error-correcting codes,* J. Combin. Theory, A **44** (1987), 274–281.

32. V. Pless, *The number of isotropic subspaces in a finite geometry,* Accad. Naz. Lincei., Rend. Cl. Sci. Fiz., Mat. e Nat. (8) **39** (1965), 418–421.

33. ____, *A classification of self-orthogonal codes over* $GF(2)$, Discrete Math., **3** (1972), 209–246.

34. ____, *The children of the* (32, 16) *doubly even codes,* IEEE Trans. Info. Theory, **24** (1978), 738–746.

35. ____, "Introduction to the Theory of Error-Correcting Codes," Wiley-Interscience, New York, 1982.

36. V. Pless and N. J. A. Sloane, *On the classification and enumeration of self-dual codes,* J. Combin. Theory, Ser. A, **18** (1975), 313–335.

37. K. J. C. Smith, *On the p-rank of the incidence matrix of points and hyperplanes in a finite projective geometry,* J. Combin. Theory **7** (1969), 122–129.

38. S. D. Stoichev, *The nonisomorphism of the strongly regular graphs derived from the quasi-symmetric* $2-(31, 7, 7)$ *designs,* Compt. rend. Acad. bulg. Sci. **40** (1987), No. 5, 33–35.

39. ____, *The automorphism group of a strongly regular graph on* 220 *vertices,* Compt. rend. Acad. bulg. Sci. (to appear).

40. L. Teirlink, *On projective and affine hyperplanes,* J. Combin. Theory, Ser. A, **28** (1980), 290–306.

41. V. D. Tonchev, *Block designs of Hadamard type and self-dual codes,* Problems of Information Transmission, April 1984, 270–275.

42. ____, *A characterization of designs related to the Witt system* $S(5, 8, 24)$, Math. Z. **191** (1986), 225–230.

43. ____, *Quasi-symmetric* $2-(31, 7, 7)$ *designs and a revision of Hamada's conjecture,* J. Combin. Theory, Ser. A, **42** (1986), 104–110.

44. ____, *Quasi-symmetric designs and self-dual codes,* Europ. J. Combinatorics, **7** (1986), 67–73.

45. ____, *A characterization of designs related to dodecads in the Witt system* $S(5, 8, 24)$, J. Combin. Theory, Ser. A, **43** (1986), 219–227.

46. ____, *Embedding of the Witt-Mathieu system* $S(3, 6, 22)$ *in a symmetric* $2-(78, 22, 6)$ *design,* Geometriae Dedicata, **22**(1987), 49–75.

47. ____, *Symmetric* $2-(31, 10, 3)$ *designs with automorphisms of order* 7, Ann. Discrete Math. **34**(1987), 461–464.

48. ____, "Combinatorial Configurations," Longman Scientific & Technical, J. Wiley & Sons, New York 1988.

49. ____, *Symmetric designs without ovals and extremal self-dual codes,* Ann. Discr. Math. **37** (1988), 451–458.

50. ____, *Self-orthogonal designs and extremal doubly-even codes,* J. Combin. Theory, Ser. A **52** (1989), 197–205.

51. E. Witt, *Über Steinersche Systeme,* Abh. Math. Sem. Hamburg Univ. **12** (1938), 265–275.

52. V. Y. Yorgov, *Binary self-dual codes with an automorphism of an odd order,* Problems of Information Transmission, April 1984, 260–270.

DEPARTMENT OF MATHEMATICS AND COMPUTING SCIENCE, EINDHOVEN UNIVERSITY OF TECHNOLOGY, 5600 MB EINDHOVEN, THE NETHERLANDS

Contemporary Mathematics
Volume **111**, 1990

Nonembeddable Quasi-Residual Designs

TRAN van TRUNG

1. Introduction

We assume that the reader is familiar with the basic notions about theory of block designs (cf. [**1, 5, 8, 10**]). However, we shall recall some definitions concerning the subject studied in this paper. We shall specify a 2-design by naming its parameter set $2 - (v, b, r, k, \lambda)$; such a 2-design consists of b blocks of size k from a set of v points, such that each point occurs in r blocks and each pair of points occurs in λ blocks. The number $n = r - \lambda$ is called the order of the design. It is well-known that

$$bk = vr,$$

$$\lambda(v - 1) = r(k - 1).$$

We also have the famous Fisher's Inequality

$$b \geq v.$$

If in a 2-design a given block is repeated m times (i.e. there is a set of $m + 1$ identical blocks), then we have the inequality

$$b \geq (m + 1)v - (m - 1),$$

(cf.[**19**]). In particular, it implies that if $b < 2v$ then there are no repeated blocks. We call a design simple if there are no repeated blocks. Throughout the paper we are only interested in simple 2-designs. If $b = v$ (and hence $r = k$) we speak of a symmetric 2-design or symmetric design and in this case we write $2 - (v, k, \lambda)$ for the parameter set. For such a 2-design any two distinct blocks intersect in λ points. Let $\mathscr{D}^*$ be a $2 - (v^*, k^*, \lambda^*)$ symmetric design. A residual design $\mathscr{D}$ of $\mathscr{D}^*$ is obtained by deleting a single block of $\mathscr{D}^*$ and all points in it. This residual design has parameter set

$$2 - (v = v^* - k^*, b = v^* - 1, r = k^*, k = k^* - \lambda^*, \lambda = \lambda^*).$$

1980 *Mathematics Subject Classification* (1985 *Revision*). Primary 05B05.
This paper is in final form and no version of it will be submitted for publication elsewhere.

A $2 - (v, b, r, k, \lambda)$ design $\mathcal{D}$ is called a quasi-residual design if $r = k + \lambda$. Obviously, any residual design is a quasi-residual design. Now given a $2 - (v, b, k + \lambda, k, \lambda)$ quasi-residual design $\mathcal{D}$, if there exists a $2 - (v^* = v + k + \lambda, k^* = k + \lambda, \lambda^* = \lambda)$ symmetric design $\mathcal{D}^*$ such that $\mathcal{D}$ is a residual design of $\mathcal{D}^*$, then $\mathcal{D}$ is called embeddable or residual; otherwise $\mathcal{D}$ is called nonembeddable. In fact if $\lambda = 1$ or 2, then a quasi-residual design is always embeddable, and even uniquely. A quasi-residual design with $\lambda = 1$ is an affine plane which is known to be embeddable in a projective plane as we can easily see it. The case for $\lambda = 2$ is more difficult and the proof is due to M. Hall and W. S. Connor (cf.[7]). When $\lambda > 2$, R. C. Bose, S. S. Shrikhande and N. M. Singhi (cf.[3]) show that there is a function $f(\lambda)$ such that, if $k > f(\lambda)$, then a quasi-residual design can always be embedded in a unique symmetric design. It was observed for the first time by K. N. Bhattacharya [3] that there exists a 2-(16,24,9,6,3) quasi-residual design which is not residual. The fact that his example is not embeddable in a 2-(25,9,3) symmetric design is obvious since it has a pair of blocks meeting in 4 points.

The purpose of this paper is to construct nonembeddable quasi-residual designs; we are mainly interested in case $k < \frac{1}{2}v$. However we shall discuss nonembeddable quasi-residual designs with $k \geq \frac{1}{2}v$ in Chapter 3.

Some investigations for the case $k \geq \frac{1}{2}v$ were made by E. T. Parker [18], J. H. van Lint and V. D. Tonchev [15]; but very little is known about the case $k < \frac{1}{2}v$. J. F. Lawless [13], R. B. Brown [4] and J. H. van Lint [14] constructed further nonembeddable quasi-residual designs for 2-(16,24,9,6,3). This parameter set had been, for a long time, the unique one for which nonembeddable quasi-residual designs were known to exist. Recently, Tran van Trung [21] found two nonembeddable quasi-residual designs for the new parameter sets 2-(25,40,16,10,6) and 2-(36,63,28,16,12).

We say a $2 - (v, b, k + \lambda, k, \lambda)$ quasi-residual design $\mathcal{D}$ is of Bhattacharya type if there are two blocks B_1 and B_2, $B_1 \neq B_2$, in $\mathcal{D}$ with $|B_1 \cap B_2| = \Lambda > \lambda$. Evidently, a quasi-residual design of Bhattacharya type is nonembeddable.

The methods of constructing nonembeddable quasi-residual designs in this paper are either direct or recursive. One of the direct methods used is constructing a design with an automorphism group, in which we first study the orbit structures of the group, and then try to extend them to a design. The other direct method is to construct incidence matrices of designs from Hadamard designs or incidence structures which are derived from the Galois fields.

2. A recursive construction

Let $\mathcal{C}$ be a $2 - (v, b, r, \kappa, \lambda)$ design admitting a parallelism, i.e. there is a partition of the blocks of $\mathcal{C}$ into classes such that each point is on exactly one block of each class (and hence two blocks of each class have no points in common). There are then $\mu = \frac{v}{k}$ blocks in each class and $r = \frac{\lambda(v-1)}{(k-1)}$ parallel classes. Suppose there exist $\mathcal{D}_1, \mathcal{D}_2, \ldots, \mathcal{D}_r$, not necessarily isomorphic,

$2 - (\mu, \beta, \rho, \kappa, \alpha)$ designs. Then we can construct a new 2-design $\mathscr{B}$ in the following way: Let $\mathscr{P}_1, \mathscr{P}_2, \ldots, \mathscr{P}_r$ be r parallel classes of $\mathscr{C}$. Let π_i be any bijection from the points of the design $\mathscr{D}_i$ onto the parallel class $\mathscr{P}_i$. We define a structure $\mathscr{B}$ as follows. The points of $\mathscr{B}$ are the points of $\mathscr{C}$. For any block $B = x_1, x_2, \ldots, x_\kappa$ in $\mathscr{D}_i$ we define a block $B^{\pi i}$ of $\mathscr{B}$ by $B^{\pi i} = x_1^{\pi i}, x_2^{\pi i}, \ldots, x_\kappa^{\pi i}$.

From the definition each block $B^{\pi i}$ of $\mathscr{B}$ is the union of κ parallel block of $\mathscr{C}$; since these blocks are parallel they have no common points and since the points of $\mathscr{C}$ are the points of $\mathscr{B}$, so each block of $\mathscr{B}$ contains $k\kappa$ points.

Let P, Q be any two points of $\mathscr{B}$ (and hence also of $\mathscr{C}$). They occur together in λ blocks of $\mathscr{C}$. But, since there are ρ blocks incident with each point of $\mathscr{D}_i$, thus these λ blocks of $\mathscr{C}$ give rise to $\rho\lambda$ blocks of $\mathscr{B}$ containing P and Q. On the other hand, there are also $r - \lambda$ blocks of $\mathscr{C}$ which are incident with P but not Q. For each of these blocks of $\mathscr{C}$ there is one parallel block incident with Q, thus this pair of blocks occurs in exactly α blocks of $\mathscr{B}$. Hence any two points P and Q are on exactly $\rho\lambda + (r - \lambda)\alpha$ blocks. We have proved that the structure $\mathscr{B}$ is a 2-design. Computing the other parameters of $\mathscr{B}$ we see that each point of $\mathscr{B}$ is on $r\rho$ blocks and there are $r\beta$ blocks in $\mathscr{B}$. Thus we have the following

THEOREM 2.1. *If there is a $2 - (v, b, r, k, \lambda)$ design admitting a parallelism and if there are, not necessarily isomorphic, $2-(\mu, \beta, \rho, \kappa, \alpha)$ designs $\mathscr{D}_1, \mathscr{D}_2, \ldots, \mathscr{D}_r$, where $\mu = \frac{v}{k}$, then there exists a 2-design with parameters $2 - (v, r\beta, r\rho, k\kappa, \rho\lambda + (r - \lambda)\alpha)$.*

We now observe that the obtained design $\mathscr{B}$ in Theorem 2.1 can be a quasi-residual design provided that $r\rho = k\kappa + \rho\lambda + (r - \lambda)\alpha$ holds, but this condition is equivalent to $\frac{r-\lambda}{k} = \frac{\kappa}{\rho-\alpha}$, hence we have

COROLLARY 2.1. *If the parameters of $\mathscr{C}$ and $\mathscr{D}_i$ in Theorem 2.1 satisfy*

$$(*) \qquad \frac{r - \lambda}{k} = \frac{\kappa}{\rho - \alpha},$$

then $\mathscr{B}$ is a quasi-residual design.

COROLLARY 2.2. *If there is a $2 - (v, b, k + \lambda, k, \lambda)$ quasi-residual design $\mathscr{C}$ having a parallelism and if there are, not necessarily isomorphic, $2 - (\mu, \beta, \kappa + \alpha, \kappa, \alpha)$ quasi-residual designs $\mathscr{D}_1, \mathscr{D}_2, \ldots, \mathscr{D}_{k+\lambda}$, where $\mu = v/k$, then there exists a*

$$2 - (v, (k + \lambda)\beta, (k + \lambda)(\kappa + \alpha), k\kappa, (\kappa + \alpha)\lambda + k\alpha)$$

quasi-residual design $\mathscr{B}$. Moreover if there are two blocks in some $\mathscr{D}_i$ meeting in γ points then there are two blocks in $\mathscr{B}$ having $k\gamma$ common points. In particular, if $\gamma > \alpha$ then $k\gamma > (\kappa + \alpha)\lambda + k\alpha$; hence $\mathscr{B}$ is a quasi-residual design of Bhattacharya type.

PROOF. It is clear by computing the parameter set for $\mathscr{B}$ that if $\mathscr{C}$ and $\mathscr{D}_i$ are both quasi-residual, then so is $\mathscr{B}$. If there are two blocks B_1 and B_2 in some $\mathscr{D}_i$ having γ common points then from the construction of the blocks $B_1^{\pi i}$ and $B_2^{\pi i}$ of $\mathscr{B}$ corresponding to the parallel class $\mathscr{P}_i$ we see that $B_1^{\pi i}$ and $B_2^{\pi i}$ have $k\gamma$ common points. We have to show that if $\gamma > \alpha$ then $(\kappa + \alpha) + k\alpha < k\gamma$ or equivalently

$$(1) \qquad\qquad (\kappa + \alpha) - k(\alpha - \gamma) < 0.$$

Further $(\kappa + \alpha)\lambda = \rho\lambda = \frac{\alpha(\mu-1)}{\kappa-1}\lambda = \frac{\alpha(\nu-k)}{k(\kappa-1)}\lambda$. Thus (1) is equivalent to

$$(2) \qquad\qquad \alpha\lambda\nu - k(\alpha\lambda + k(\gamma - \alpha)(\kappa - 1)) < 0,$$

but, since $\lambda\nu = k(k + \lambda - 1)$, hence (2) is equivalent to

$$(3) \qquad\qquad \alpha(k - 1) - k(\gamma - \alpha)(\kappa - 1) < 0.$$

But (3) is always satisfied, since we have $\alpha < \kappa - 1$ and $k - 1 < k < k(\gamma - \alpha)$. Thus Corollary 2.2 is proved. $\diamond$

Let $\mathscr{C}$ be the affine geometry $\mathscr{A}\mathscr{G}$ (m,q) over GF(q), i. e. $\mathscr{C}$ is a

$$2 - (q^m, q\frac{q^m - 1}{q - 1}, \frac{q^m - 1}{q - 1}, q^{m-1}, \frac{q^{m-1} - 1}{q - 1})$$

residual design having a parallelism with $\mu = q$ blocks for each class. Then we have the following result as a special case of Corollary 2.2.

COROLLARY 2.3. *If there is a* $2 - (q, \beta, \kappa + \alpha, \kappa, \alpha)$ *quasi-residual design* $\mathscr{D}$, *where q is a prime power, then there exists a*

$$2 - (q^m, \frac{q^m - 1}{q - 1}\beta, \frac{q^m - 1}{q - 1}(\kappa + a), q^{m-1}\kappa, (\kappa + \alpha)\frac{q^{m-1} - 1}{q - 1} + q^{m-1}\alpha)$$

quasi-residual design $\mathscr{B}$.

The next Lemma gives a condition related to the number of blocks of certain type of a residual design.

LEMMA 2.1. *Let $\mathscr{B}$ be a* $2 - (v, b, k + \lambda, k, \lambda)$ *quasi-residual design. Assume that there are x blocks in $\mathscr{B}$ having pairwise $(\lambda - 1)$ common points. If $x(2\lambda + 1 - x) > 2(k + \lambda)$, then $\mathscr{B}$ is nonembeddable.*

PROOF. In order to extend $\mathscr{B}$ to a symmetric design we need a set of $k + \lambda$ new points. Let $B_1, B_2, \ldots, B_x$ be x blocks of $\mathscr{B}$ meeting pairwise in $(\lambda - 1)$ points. For extended blocks B_1^* and B_2^* we need $\lambda + (\lambda - 1) = 2\lambda - 1$ new points; for an extended block B_3^*, by comparing it with B_1^* and B_2^*, we must have at least $(\lambda - 2)$ more new points. If we continue to count the least number of new points needed for constructing an extended block B_i^*, comparing it with $B_1^*, \ldots, B_{i-1}^*$, we see that the least number of new points used for $B_1^*, \ldots, B_i^*$ is $\lambda + (\lambda - 1) + (\lambda - 2) + \cdots + (\lambda - i + 1)$. Thus for x blocks $B_1^*, \ldots, B_x^*$ we need at least $\lambda + (\lambda - 1) + (\lambda - 2) + \cdots + (\lambda - x + 1)$

new points; of course, this number is at most $(k + \lambda)$, the total number of new points. Hence we have $\lambda + (\lambda - 1) + (\lambda - 2) + \cdots + (\lambda - x + 1) \leq (k + \lambda)$ or

$$x\lambda - \frac{x(x-1)}{2} \leq (k + \lambda),$$

and hence

$$x(2\lambda + 1 - x) \leq 2(k + \lambda).$$

Thus Lemma 2.1 is proved. $\diamond$

3. Nonembeddable quasi-residual designs with $k \geq \frac{1}{2}v$

Before we deal with the constructions of quasi-residual designs with $k \leq \frac{1}{2}v$ in Chapter 4 we first discuss the existence problem for quasi-residual designs with large k, namely with $k \geq \frac{1}{2}v$. In general, quasi-residual designs with $k \geq \frac{1}{2}v$ are rather simple to construct as we shall see in the following observation. There are examples of nonembeddable quasi-residual designs for $k \geq \frac{1}{2}v$ (cf. [15],[18]).

Let $\mathscr{B}$ be a $2-(v, k, \lambda)$ symmetric design. A derived design is obtained by removing a single block of $\mathscr{B}$ and all the points not incident with this block. In this way we get a $2 - (k, v - 1, k - 1, \lambda, \lambda - 1)$ design. A $2 - (v, b, r, k, (\lambda - 1))$ design $\mathscr{D}$ is called a *quasi-derived design* if $k = \lambda$ and therefore the parameter set of $\mathscr{D}$ will be $2 - (v, v(v - 1)/\lambda, v - 1, \lambda, v - 1, \lambda, \lambda - 1)$. Evidently, any derived design is a quasi-derived design. We say that a quasi-derived design $\mathscr{D}$ is embeddable if there exists a symmetric design S such that $\mathscr{D}$ is a derived design of S; otherwise $\mathscr{D}$ is called nonembeddable. The complement of a $2 - (v, v(v - 1)/\lambda, v - 1, \lambda, \lambda - 1)$ quasi-derived design $\mathscr{D}$ is a design $\mathscr{R}$ with parameters

$$2 - (v, \frac{v(v - 1)}{\lambda}, \frac{(v - 1)(v - \lambda)}{\lambda}, v - \lambda, \frac{(v - 1)(v - 2\lambda)}{\lambda} + (\lambda - 1)).$$

It turns out that $\mathscr{R}$ is a quasi-residual design. Hence we have

LEMMA 3.1. *The complement of a* $2 - (v, v(v - 1)/\lambda, v - 1, \lambda, \lambda - 1)$ *quasi-derived design* $\mathscr{D}$ *is a*

$$2 - (v, \frac{v(v - 1)}{\lambda}, \frac{(v - 1)(v - \lambda)}{\lambda}, v - \lambda, \frac{(v - 1)(v - 2\lambda)}{\lambda} + (\lambda - 1))$$

quasi-residual design. Conversely, the complement of a quasi-residual design is a quasi-derived design.

Lemma 3.1 shows us the reason for the existence of nonembeddable quasi-residual designs with very large block size. For instance, any trivial $2 - (k, k(k-1), k-1, 1, 0)$ design is a quasi-derived design, in fact its complement is a $2-(k, k(k-1), (k-1)^2, k-1, (k-1)(k-2))$ quasi-residual design and the corresponding symmetric design has parameters $2-(k(k-1)+1, (k-1)^2, (k-1)(k-2))$, which is in turn the complement of a $2-(k(k-1), k, 1)$ symmetric design, namely a projective plane of order $(k - 1)$; in particular, if the later does not satisfy the condition of Bruck-Ryser-Chowla Theorem, then there does not exist a projective plane of that order, though a corresponding quasi-residual design exists, hence it is nonembeddable. Similarly, any $2 - (k, k(k - 1)/2, k - 1, 2, 1)$ design is a quasi-derived design and its complement is a $2 - (k, k(k - 1)/2, (k - 1)(k - 2)/2, (k - 1), (k - 1)(k - 4)/2 + 1)$ nonembeddable quasi-residual design, if a corresponding $2 - (k(k - 1)/2 + 1, (k - 1)(k - 2)/2, (k - 1)(k - 4)/2 + 1)$ symmetric design does not satisfy the condition of Bruck-Ryser-Chowla Theorem; moreover the complement of this symmetric design is a biplane of order $(k - 2)$, i.e. a $2 - (k(k - 1)/2 + 1, k, 2)$ design.

From a result of Hanani [9] we have that any $2 - (v, b, r, k, k - 1)$ quasi-derived design exists for $k \leq 6$, therefore its complement, a quasi-residual design, does as well. Moreover if there is no symmetric design corresponding to this quasi-derived design (or quasi-residual design), then this quasi-residual design is nonembeddable. Therefore infinitely many nonembeddable quasi-residual designs with large k are known.

Our aim in this discussion is to show the effect of Corollary 2.1 in constructing quasi-residual designs. We shall give the existence of new nonembeddable quasi-residual designs with $k \geq \frac{1}{2} v$ by using this Corollary.

First of all we remember that the condition

$$(*) \qquad\qquad \frac{r - \lambda}{k} = \frac{\kappa}{\rho - \alpha}$$

is crucial for constructing quasi-residual designs. The problem is to find a pair of designs $\mathscr{C}$ and $\mathscr{D}$ in Corollary 2.1, where the former is a resolvable design (a design with parallelism) satisfying $(*)$. In fact there is a large number for such a pair of designs, though resolvable designs are not so much known when k is big.

Now let $\mathscr{C}$ be the classical unital of order q, i.e. a resolvable $2 - (q^3 + 1, q^2(q^2 - q + 1), q^2, q + 1, 1)$ design, where q is a prime power. Each class of $\mathscr{C}$ has $\mu = q^2 - q + 1$ blocks. We take $\mathscr{D}$ to be the complement of a projective plane of order $(q - 1)$, i.e. a $2 - (q^2 - q + 1, (q - 1)^2, (q - 1)(q - 2))$ symmetric design; then the equation $(*)$ is satisfied for this pair $\mathscr{C}$ and $\mathscr{D}$. Hence if there is a projective plane of order $(q - 1)$, then there exists a

$$2 - (q^3 + 1, q^2(q^2 - q + 1), q^2(q - 1)^2, (q + 1)(q - 1)^2, (q - 1)^2(q^2 - q - 1))$$

quasi-residual design. In particular, if $(q - 1)$ is a prime power (i.e. $(q - 1)$ is q Mersenne prime or q is a Fermat prime or $q = 9$) then a quasi-residual design with this parameter set exists. Moreover a corresponding symmetric design has parameters

$$2 - (q^2(q^2 - q + 1) + 1, q^2(q - 1)^2, (q - 1)^2(q^2 - q - 1)),$$

which does not exist if $q \neq 3, 8$, for the number of points of this symmetric design is even, while the order is $(q - 1)^2(q + 1)$ which is not a square. This means in this case that the obtained quasi-residual designs are nonembeddable. Thus we have

THEOREM 3.1. *Let* $q = 9$, *or* q *be a Fermat prime, or* $(q - 1)$ *be a Mersenne prime distinct from* 2 *and* 7. *Then there exists a*

$$2 - (q^3 + 1, q^2(q^2 - q + 1), q^2(q - 1)^2, (q + 1)(q - 1)^2, (q - 1)^2(q^2 - q - 1))$$

nonembeddable quasi-residual design.

For instance, if $q = 5, 7, 9, 17$ the corresponding quasi-residual designs in Theorem 3.1 have parameters $2 - (126, 525, 400, 96, 304)$, $2 - (344, 2107, 1764, 288, 1476)$, $2 - (730, 5913, 5184, 640, 4544)$, $2 - (4914, 78897, 73984, 4608, 69376)$, respectively.

We consider the resolvable Steiner system (a Steiner system is a $2 - (v, b, r, k, 1)$ design) $\mathscr{C}$ with parameters

$$2 - (v, b, r, k, \lambda) =$$
$$2 - (2^{n+l} - 2^n + 2^l, (2^n + 1)(2^n - 2^{n-l} + 1), 2^n + 1, 2^l, 1),$$

where $n > l$, constructed from maximal n-arcs in the Desarguesian plane of order 2^n by Denniston [6]. Each parallel class of $\mathscr{C}$ has $(2^n - 2^{n-l} + 1)$ blocks.

Suppose there exists a

$$2 - (2^n - 2^{n-l} + 1, (2^n - 2^{n-l} + 1)(2^l - 1)/(2^{n-l} + 1), 2^l - 1, 2^{n-l} + 1, 1)$$

Steiner system then its complement is a

$$2 - (\mu, \beta, \rho, \kappa, \alpha) =$$
$$2 - (2^n - 2^{n-l} + 1, (2^n - 2^{n-l} + 1)(2^l - 1)/(2^{n-l} + 1), 2^l - 1)$$
$$(2^n - 2^{n-l+1}/(2^{n-l} + 1), (2^n - 2^{n-l+1}), (2^l - 1)(2^n - 3.2^{n-l} - 1)/(2^{n-l} + 1) + 1)$$

Moreover, we have $(r - \lambda)/k + \kappa/(\rho - \alpha) + 2^{n-l}$. Thus by Corollary 2.1 we have a

$$2 - (2^{n+l} - 2^n + 2^l, (2^n - 2^{n-l} + 1)(2^l - 1)(2^n + 1)/(2^{n-l} + 1),$$
$$(2^n - 2^{n-l+1})(2^l - 1)(2^n + 1)/(2^{n-l} + 1), 2^{n+1}(2^{l-1} - 1),$$
$$(2^l - 1)(2^{2n} - 3.2^{2n-l} - 2^{n-l+1})/(2^{n-l} + 1) + 2^n)$$

quasi-residual design $\mathscr{B}$. A symmetric design $\mathscr{S}$ corresponding to $\mathscr{B}$ has parameters

$$2 - ((2^n - 2^{n-l} + 1)(2^l - 1)(2^n + 1)/(2^{n-l} + 1) + 1,$$
$$(2^n - 2^{n-l+1})(2^l - 1)(2^n + 1)/(2^{n-l} + 1),$$
$$(2^l - 1)(2^{2n} - 3.2^{2n-l} - 2^{n-l+1})/(2^{n-l} + 1) + 2^n).$$

Note that $\mathscr{S}$ has an even number of points and its order is $2^{n+1}(2^{l-1} - 1)$. Thus if n is even or if n is odd and $(2^{l-1} - 1)$ is not a square, then $\mathscr{S}$ does not exist. This implies that $\mathscr{B}$ is nonembeddable and we have the following result.

THEOREM 3.2. *Suppose there exists a*

$$2 - (2^n - 2^{n-l} + 1, (2^n - 2^{n-l} + 1)(2^l - 1)/(2^{n-l} + 1), 2^l - 1, 2^{n-l} + 1, 1)$$

Steiner system, where n, l *are integers and* $n > l$. *If* n *is even or if* n *is odd and* $(2^{l-1} - 1)$ *is not a square, then there exists a*

$$2 - (2^{n+l} - 2^n + 2^l, (2^n - 2^{n-l} + 1)(2^l - 1)(2^n + 1)/(2^{n-l} + 1),$$
$$(2^n - 2^{n-l+1})(2^l - 1)(2^n + 1)/(2^{n-l} + 1), 2^{n+1}(2^{l-1} - 1),$$
$$(2^l - 1)(2^{2n} - 3.2^{2n-1} - 2^{n-l+1})/(2^{n-l} + 1) + 2^n)$$

nonembeddable quasi-residual design.

In particular, if $l = n - 1$, then each parallel class of $\mathscr{C}$ has $\mu = 2^n - 1$ blocks. As $2^n - 1 \equiv 1$ *or* 3 (mod 6) there always exists a $2 - (2^n - 1, (2^n - 1)(2^{n-1} - 1)/3, (2^{n-1}), 3, 1)$ Steiner system.

If $l = n - 2$ for the Denniston's series, then each parallel class of $\mathscr{C}$ has $\mu = 2^n - 3$ blocks. Further, if $n \equiv 2$ *or* 3 (mod 4), then $2^n - 3 \equiv 1$ *or* 5 (mod 20) and there exists a $2 - (2^n - 3, (2^n - 3)(2^{n-2} - 1)/5, (2^{n-2} - 1), 5, 1)$ Steiner system [9]. Thus as a special case of Theorem 3.3 we have

COROLLARY 3.1. (i) *Let* $n \geq 3$ *be an integer. If* n *is even or if* n *is odd and* $(2^{n-2} - 1)$ *is not a square, then there exists a*

$$2 - (2^{2n-1} - 2^{n-1}, (2^{2n} - 1)(2^{n-1} - 1)/3, (2^n + 1)(2^{n-1} - 1)(2^n - 4)/3,$$
$$2^{n+1}(2^{n-2} - 1), 4(2^{3n-3} - 2^{2n} + 7.2^{n-2} + 1)/3))$$

nonembeddable quasi-residual design.

(ii) *Let* $n \equiv 2$ *or* 3 (mod 4) ≥ 3 *be an integer. If* n *is even or if* n *is odd and* $(2^{n-3} - 1)$ *is not a square, then there exists a*

$$2 - (2^{2n-2} - 2^n + 2^{n-2}, (2^n + 1)(2^n - 3)(2^{n-2} - 1)/5, (2^n + 1)(2^{n-2} - 1)(2^n - 8)/5,$$
$$2^{n+1}(2^{n-3} - 1), (2^{n-2} - 1)(2^{2n} - 12.2^n - 8)/5 + 2^n)$$

nonembeddable quasi-residual design.

For instance, if $n = 4, 5$ then Corollary 3.1.(i) gives $2 - (120, 595, 476, 96, 380)$ and $2 - (496, 5115, 5620, 448, 4172)$ nonembeddable quasi-residual designs; or if $n = 6$, (ii) yields a $2 - (976, 11895, 10920, 896, 10024)$ nonembeddable design.

Before we continue our topic we make a digression about the existence of disjoint designs. Let $\mathscr{A}$ be a $2 - (v, b, r, k, \lambda)$ design and S be the point set of $\mathscr{A}$. Let $\mathscr{B}$ be another $2 - (v, b, r, k, \lambda)$ design having the same point set S. We say that $\mathscr{A}$ and $\mathscr{B}$ are disjoint if any block of $\mathscr{A}$ is distinct from any block of $\mathscr{B}$. In general, if $\mathscr{A}_1, \ldots, \mathscr{A}_l$ are $2 - (v, b, r, k, \lambda)$ designs having a same point set, then $\mathscr{A}_i$, $i = 1, 2, \ldots, l$, are called mutually disjoint if they are pairwise disjoint. The next Theorem gives us an upper bound for the number of mutually disjoint designs.

THEOREM 3.3. *If there is a* $2 - (v, b, r, k, \lambda)$ *design, then there are at least* $\lceil \frac{1}{b^2} \binom{v}{k} \rceil$ *mutually disjoint designs, where* $\lceil x \rceil$ *denotes the least integer* $\geq x$.

PROOF. Let S be the point set of a $2 - (v, b, r, k, \lambda)$ design. Suppose there are l mutually disjoint $\mathscr{A}_1, \ldots, \mathscr{A}_l$ $2 - (v, b, r, k, \lambda)$ designs. Let $\mathscr{A}$ be the union of $\mathscr{A}_i$, $i = 1, \ldots, l$, then $\mathscr{A}$ is a $2 - (v, l.b, l.r, k, l\lambda)$ design. Let $\mathscr{D}$ be a $2 - (v, b, r, k, \lambda)$ design. We now find an upper bound of the number of permutations h on S for which there is a block A in $\mathscr{A}$ and a block D in $\mathscr{D}$ such that $A^h = D$. We see that for each block D in $\mathscr{D}$ and each block A in $\mathscr{A}$ there are $k!(v-k)!$ permutations h on S such that $D^h = A$. Thus for each block D in $\mathscr{D}$ there are $b.l.k!(v-k)!$ such permutations sending D onto a block of $\mathscr{A}$. Since there are b blocks in $\mathscr{D}$, the total number of permutations h sending a block of $\mathscr{D}$ onto a block of $\mathscr{A}$ is $b.b.l.k!(v-k)!$. But the symmetric group on S has $v!$ permutations, hence if $b.b.l - k!(v-k)! < v!$ then there exists a permutation h on S such that the image of $\mathscr{D}$ under h is disjoint from $\mathscr{A}$. But this inequality is equivalent to $b^2.l < \binom{v}{k}$ or $l < \frac{1}{b^2} \binom{v}{k}$. ◇

Applying Theorem 3.3 shows that some simple resolvable 2-designs exist. For instance, if we take two copies of the affine geometry $\mathscr{A} \mathscr{G} (m, q)$ over $GF(q)$ we have a resolvable

$$2 - (q^m, 2q(q^m - 1)/(q - 1), 2(q^m - 1)/(q - 1), q^{m-1}, 2(q^{m-1} - 1)/(q - 1))$$

design, however this design has repeated blocks. But now if we use Theorem 3.3 we see that there always exists a simple resolvable design with this parameter set. In general, applying Theorem 3.3 to the affine geometry $\mathscr{A} \mathscr{G}$ (m, q) over $GF(q)$ we have

COROLLARY 3.2. *If q is a prime power then there exists a simple resolvable*

$$2 - (q^m, lq(q^m - 1)/(q - 1), l(q^m - 1)/(q - 1), q^{m-1}, l(q^{m-1} - 1)/(q - 1))$$

design $\mathscr{C}$, *where* $l < \frac{(q-1)^2}{q^2(q^m-1)^2} \binom{q^m}{q^{m-1}}$.

Next we use the resolvable designs in Corollary 3.2 for constructing new nonembeddable quasi-residual designs. At first we observe that $(r - \lambda)/k = l$ for $\mathscr{C}$. Suppose there is a $2 - (q, \beta, \rho, \kappa, \alpha)$ design with $\kappa/(\rho - \alpha) = l$ then we have a

$$2 - (q^m, \beta l(q^m - 1)/(q - 1), \rho l(q^m - 1)/(q - 1), \kappa q^{m-1},$$
$$\rho l(q^{m-1} - 1)/(q - 1) + \alpha l q^{m-1})$$

quasi residual design $\mathscr{B}$

A symmetric design corresponding to $\mathscr{B}$ has parameters

$$2 - (\beta l(q^m - 1)/(q - 1) + 1, \rho l(q^m - 1)/(q - 1), \rho l(q^{m-1} - 1)/(q - 1) + \alpha l q^{m-1}).$$

The order of $\mathscr{S}$ is $q^{m-1}l(\rho - \alpha)$. In particular, if $\beta l(q^m - 1)/(q - 1) + 1$ is even and $q^{m-1}l(\rho - \alpha)$ is not a square then $\mathscr{B}$ is nonembeddable. Hence we have

THEOREM 3.4. *Suppose there exists a* $2 - (q, \beta, \rho, \kappa, \alpha)$ *design* $\mathscr{D}$ *with* $\kappa/(\rho - \alpha) = l < \frac{(q-1)^2}{q^2(q^m-1)^2}\binom{q^m}{q^{m-1}}$, *where* q *is a prime power, then there exists a*

$$2 - (q^m, \beta l(q^m - 1)/(q - 1), \rho l(q^m - 1)/(q - 1), \kappa q^{m-1},$$
$$\rho l(q^{m-1} - 1)/(q - 1) + \alpha l q^{m-1})$$

quasi-residual design $\mathscr{B}$. *Furthermore, if* $\beta l(q^m - 1)/(q - 1) + 1$ *is even and* $q^{m-1}l(\rho - \alpha)$ *is not a square, then* $\mathscr{B}$ *is nonembeddable.*

The following infinite series of nonembeddable quasi-residual designs are obtained by using Theorem 3.4.

(1) $q = 31, l = 3, m \equiv 1 \pmod 2, \mathscr{D} = 2 - (31, 31, 21, 21, 14)$ and
$\mathscr{B} = 2 - (31^m, 93.(31^m - 1)/32, 63.(31^m - 1)/32, 21.31^{m-1},$
$\qquad\qquad\qquad 63.(31^{m-1} - 1)/32 + 42.31^{m-1}).$

(2) $q = 49, l = 3, m \equiv 1 \pmod 2, \mathscr{D} = 2 - (49, 49, 33, 33, 22)$ and
$\mathscr{B} = 2 - (49^m, 147.(49^m - 1)/48, 99.(49^m - 1)/48,$
$\qquad\qquad\qquad 33.49^{m-1}, 99.(49^{m-1} - 1)/48 + 66.49^{m-1}).$

(3) $q = 37, l = 3, m \equiv 1 \pmod 2, \mathscr{D} = 2 - (37, 111, 99, 33, 88)$ and
$\mathscr{B} = 2 - (37^m, 333.(37^m - 1)/36, 297.(37^m - 1)/36,$
$\qquad\qquad\qquad 33.37^{m-1}, 297.(37^{m-1} - 1)/36 + 264.37^{m-1}).$

(4) $q = 61, l = 3, m \equiv 1 \pmod 2, \mathscr{D} = 2 - (61, 305, 285, 57, 266)$ and
$\mathscr{B} = 2 - (61^m, 915.(61^m - 1)/60, 855.(61^m - 1)/60,$
$\qquad\qquad\qquad 57.61^{m-1}, 855.(61^{m-1} - 1)/60 + 798.61^{m-1}).$

(5)
$q = 151, l = 5, m \equiv 1 \pmod 2, \mathscr{D} = 2 - (151, 755, 725, 145, 696)$ and
$\mathscr{B} = 2 - (151^m, 3775.(151^m - 1)/150, 3625.(151^m - 1)/150,$
$\qquad\qquad\qquad 145.151^{m-1}, 3625.(151^{m-1} - 1)/150 + 3480.151^{m-1}).$

(6)
$q = 109, l = 3, m \equiv 1 \pmod 2, \mathscr{D} = 2 - (109, 981, 945, 105, 910)$ and
$\mathscr{B} = 2 - (109^m, 2943.(109^m - 1)/108, 2835.(109^m - 1)/108,$
$\qquad\qquad\qquad 105.109^{m-1}, 2835.(109^{m-1} - 1)/108 + 2730.109^{m-1}).$

(7)
$$q = 739, l = 9, m \equiv 1 \pmod 2, \mathscr{D} = 2 - (739, 739, 657, 657, 584) \text{ and}$$
$$\mathscr{B} = 2 - (739^m, 6651.(739^m - 1)/738, 5913.(739^m - 1)/738,$$
$$657.739^{m-1}, 5913.(739^{m-1} - 1)/738 + 5256.739^{m-1}).$$

For the existence of $\mathscr{D}$ in (1) to (7) see [1, 16].

4. Nonembeddable quasi-residual designs with $k < \frac{1}{2}v$

As mentioned in the introduction our main interest is to investigate non-embeddable quasi-residual designs with $k < \frac{1}{2}v$. The reason is that very few examples are known and that it is more difficult to construct such a design than one with $k \geq \frac{1}{2}v$. In fact, in some sense a quasi-residual design with $k < \frac{1}{2}v$ is closer to a corresponding symmetric design, this means that if the former is embeddable then we already have most "part" of the latter. Especially, the result of Bose, Singhi and Shrikhande [3] shows that when this "part" is large enough then a symmetric design can be constructed; this also implies that for a fixed λ the number of nonembeddable quasi-residual designs is *finite*. Most of nonembeddable quasi-residual designs in this chapter are constructed by using an automorphism group and the remainder by incidence matrices of smaller structures. Throughout the paper we do not go into details to study how to use an automorphism group to construct a design, we only give the group and its action on designs. (For a more detailed construction, and related matters, see for instance [5], [11]. However we show the nonembedding of designs, which is in general a very difficult problem.

4.1. A nonembeddable quasi-residual Design for 2-(27, 39, 13, 9, 4). In this section we are dealing with the existence of a nonembeddable quasi-residual design which has the same parameters as the affine geometry $\mathscr{A}\mathscr{G}(3,3)$. It is worth noting here that quasi-residual designs which have the parameters of $\mathscr{A}\mathscr{G}(3,2)$ are all embeddable, see for instance [14].

The quasi-residual design $\mathscr{D}$ for $2 - (27, 39, 13, 9, 4)$ presented below is constructed with a cyclic automorphism group $C = \langle \tau\rho \mid \tau^2 = 1, \rho^3 = 1, \tau\rho = \rho\tau \rangle$ of order 6. If we denote the points of $\mathscr{D}$ by the numbers $1, 2, 3, \ldots,$ 27, then the action of τ and ρ on the points of $\mathscr{D}$ are as follows:

$$\tau = (1, 4)(2, 5)(3, 6)(7, 10)(8, 11)(9, 12)(13, 16)(14, 17)(15, 18)$$
$$(19)(29)(21)(22)(23)(24)(25)(26)(27).$$

$$\rho = (1, 2, 3)(4, 5, 6)(7, 8, 9)(10, 11, 12)(13, 14, 15)(16, 17, 18)$$
$$(19, 20, 21)(22, 23, 24)(25, 26, 27).$$

Next we show that $\mathscr{D}$ is not embedded in a $2 - (40, 13, 4)$ symmetric design. Let $28, 29, 30, 31, 32, 33, 34, 35, 36, 37, 38, 39, 40$ be thirteen new points. Let us consider the blocks $(1), (4), (7), (10), (16), (34), (37),$

The blocks of $\mathscr{D}$ are the following 39 sets:

(1) = 2	3	8	9	14	16	19	22	26
(2) = 3	1	9	7	15	17	20	23	27
(3) = 1	2	7	8	13	18	21	24	25
(4) = 5	6	11	12	17	13	19	22	26
(5) = 6	4	12	10	18	14	20	23	27
(6) = 4	5	10	11	16	15	21	24	25
(7) = 8	11	14	15	17	18	19	24	27
(8) = 9	12	15	13	18	16	20	22	25
(9) = 7	10	13	14	16	17	21	23	26
(10) = 1	2	4	5	9	12	19	23	25
(11) = 2	3	5	6	7	10	20	24	26
(12) = 3	1	6	4	8	11	21	22	27
(13) = 1	2	9	10	11	13	18	26	27
(14) = 2	3	7	11	12	14	16	27	25
(15) = 3	1	8	12	10	15	17	25	26
(16) = 4	5	12	7	8	16	15	26	27
(17) = 5	6	10	8	9	17	13	27	25
(18) = 6	4	11	9	7	18	14	25	26
(19) = 3	4	5	7	8	13	18	19	23
(20) = 1	5	6	8	9	14	16	20	24
(21) = 2	6	4	9	7	15	17	21	22
(22) = 6	1	2	10	11	16	15	19	23
(23) = 4	2	3	11	12	17	13	20	24
(24) = 5	3	1	12	10	18	14	21	22
(25) = 1	5	7	11	13	14	15	20	22
(26) = 2	6	8	12	14	15	13	21	23
(27) = 3	4	9	10	15	13	14	19	24
(28) = 4	2	8	10	16	17	19	20	22
(29) = 5	3	9	11	17	18	16	21	23
(30) = 6	1	7	12	18	16	17	19	24
(31) = 8	9	11	12	20	21	23	24	26
(32) = 9	7	12	10	21	19	24	22	27
(33) = 7	8	10	11	19	20	22	23	25
(34) = 1	4	13	16	22	23	24	26	27
(35) = 2	5	14	17	23	24	22	27	25
(36) = 3	6	15	18	24	22	23	25	26
(37) = 2	5	15	18	19	20	21	26	27
(38) = 3	6	13	16	20	21	19	27	25
(39) = 1	4	14	17	21	19	20	25	26

(38) and (39) of $\mathscr{D}$ and let $[(a), (b)]$ denote the number of points common

to the blocks (a) and (b). Since $[(37), (38)] = [(37), (39)] = [(38), (39)] = 4$ we may choose $(37)^* = (37) \cup \{28\,29\,30\,31\}$, $(38)^* = (38) \cup \{32, 33\,34\,35\}$, $(39)* = (39) \cup \{36\,37\,38\,39\}$. Since $[(10, 37)] = 3$, $[(10), (38)] = 2$, $[(10), (39)] = 4$, we can choose $(10)* = (10) \cup \{28\,32\,33\,40\}$.

Further since $[(7), (37)] = 4$, $[(7), (38)] = 2$, $[(7), (39)] = 3$, $[(7), (10)] = 1$ we see that $28 \notin (7)^*$ so we may choose $36 \in (7)^*$ and then $(7)^* = (7) \cup \{36\,32\,33\,40\}$.

Now since $[(34), (10)] = 3$ and $[(34), (7)] = 2$ we have two possibilities: either $40 \in (34)^*$ or $40 \notin (34)^*$. If $40 \in (34)^*$ then $28, 32, 33 \notin (34)^*$ and $36 \in (34)^*$ but since $[(34), (37)] = 2$ we can write $(34)^* = (34) \cup \{40\,36\,29\,30\}$, which is a contradiction with $[(34), (38)] = 3$ (i.e., $(34)^*$ contains exactly one of the new point set $\{32,33,34,35\}$). Therefore we have $40 \notin (34)^*$, this means we may choose $32, 36 \in (34)^*$; further since $[(34), (37)] = 2$ we may put $(34)^* = (34) \cup \{28\,29\,32\,36\}$.

Now since $[(16), (34)] = 4$, so $28, 29, 30, 31, 32, 36 \notin (16)^*$; on the other hand $[(16), (7)] = [(16), (10)] = 3$ we have either $40 \in (16)^*$ or $40 \notin (16)^*$. If $40 \in (16)^*$ then $33 \notin (16)^*$ and then $34, 35 \in (16)^*$ since $[(16), (38)] = 2$, further as $[(16), (39)] = 2$ two of $\{37, 38, 39\}$ occur in $(16)^*$ which is a contradiction with the fact the $(16)^*$ contains five instead of four new points. Hence $33 \in (16)^*$ and we may choose $(16)^* = (16) \cup \{33\,34\,37\,38\}$.

We sum up here the extended blocks $(37)^*$, $(38)^*$, $(39)^*$, $(10)^*$, $(7)^*$, $(34)^*$, $(16)^*$:

$$(37)^* = (37) \cup \{28\,29\,30\,31\},$$
$$(38)^* = (38) \cup \{32\,33\,34\,35\},$$
$$(39)^* = (39) \cup \{36\,37\,38\,39\},$$
$$(10)^* = (10) \cup \{28\,32\,33\,40\},$$
$$(7)^* = (7) \cup \{36\,32\,33\,40\},$$
$$(34)^* = (34) \cup \{28\,29\,32\,36\},$$
$$(16)^* = (16) \cup \{33\,34\,37\,38\}.$$

We now look at an extended block$(1)*$. First of all we remark that $[(1), (37)] = [(1), (38)] = [(1), (39)] = [(1), (10)] = [(1), (7)] = [(1), (34)] = [(1), (16)] = 3$. We have the following possibilities: (i) $28, 36 \in (1)^*$, (ii) $32 \in (1)^*$, (iii) $33 \in (1)^*$ or (iv) $40 \in (1)^*$. The case (i) cannot occur since $[(1), (34)] = 3$, i.e., $(1)^*$ contains exactly one of $\{28, 29, 32, 36\}$. If we are in case (ii): $(32) \in (1)^*$ then $28, 33, 34, 35, 36, 40, 29 \notin (1)^*$ and we can choose $30 \in (1)^*$ and then $31 \in (1)^*$, but since $[(1), (16)] = 3$ we may choose $37 \in (1)^*$ i.e., $38, 39 \notin (1)^*$ so $(1)^* \ni 32, 30, 37$, but then we have no new points to fill up $(1)^*$, which is a contradiction. If we are in case (iii): $33 \in (1)^*$, then $28, 32, 36, 40, 34, 35, 37, 38 \notin (1)^*$, therefore $39 \in (1)^*$ and we can choose $29 \in (1)^*$, and then $30, 31 \notin (1)^*$, so $(1)^* \in 33, 39, 29$, we have again a contradiction because there are no more new points to fill up $(1)^*$. Hence we are in case (iv): $40 \in (1)^*$. From the fact that $(10)^* = (10) \cup \{28\ 36\ 33\ 40\}$, $(7)^* = (7) \cup \{36\ 32\ 33\ 40\}$, and $(34)^* = (34) \cup \{28\ 29\ 32\ 36\}$, we see that $28, 32, 36, 33 \notin (1)^*$ and $29 \in (1)^*$, so we have $(1)^* \in 40, 29$.

The fact that $[(4), (37)] = [(4), (38)] = [(4), (39)] = [(4), (10)] = [(4), (7)] = [(4), (34)] = [(4), (16)] = 3$ leads to the result that $(4)^* \in 40, 29$, by the same way as for $(1)^*$. On the other hand $[(4), (1)] = 3$ i.e. $(1)^*$ and $(4)^*$ have exactly one new point in common, which is a contradiction with the fact that both $(1)^*$ and $(4)^*$ contain the new points 40 and 29. We have proved that $\mathscr{D}$ is not embeddable.

THEOREM 4.1. *There is a nonembeddable quasi-residual design for* $2 - (27, 39, 13, 9, 4)$.

From Corollary 2.3 and Theorem 4.1 we get a family of quasi-residual designs having parameters:

$$2 - (27^m, 39.(27^m - 1)/26, 13.(27^m - 1)/26, 9.27^{m-1},$$

$$13.(27^{m-1} - 1)/26 + 4.27^{m-1}),$$

where $m \geq 1$ is an integer.

We conjecture that the designs of this family are nonembeddable.

4.2 A nonembeddable quasi-residual design for 2-(21, 35, 15, 9, 6). Here we present a nonembeddable quasi-residual design $\mathscr{D}$ for 2-(21,35,15,9,6) which admits an automorphism group $H = C \times F$, where $C = \langle \tau \mid \tau^2 = 1 \rangle$ is a cyclic group of order 2 and $F = \langle \rho, \sigma \mid \rho^5 = 1, \sigma^2 = 1, \sigma^{-1}\rho\sigma = \rho^{-1} \rangle$ is a Frobenius group of order 10. Denoting by $\infty, 1, 2, 3, \ldots, 20$ the points of $\mathscr{D}$, the action of τ, ρ, σ on the points of $\mathscr{D}$ should be as follows:

$\tau = (\infty)(1)(2)(3)(4)(5)(6)(7)(8)(9)(10)(11, 16)(12, 17)(13, 18)(14, 19)(15, 20).$

$\rho = (\infty)(1, 2, 3, 4, 5)(6, 7, 8, 9, 10)(11, 12, 13, 14, 15)(16, 17, 18, 19, 20).$

$\sigma = (\infty)(1)(2, 5)(3, 4)(6)(7, 10)(8, 9)(11)(12, 15)(13, 14)(16)(17, 20)(18, 19).$

Next we show that $\mathscr{D}$ is not embeddable. Let 22, 23, 24, 25, 26, 27, 28, 29, 30, 31, 32, 33, 34, 35, 36 be a set of fifteen new points; and let $(a)^*$ be an extended block of a block (a) in a corresponding $2 - (36, 25, 6)$ symmetric design $\mathscr{D}^*$ of $\mathscr{D}$. Consider the blocks (1),(6),(11),(14) and (27) of $\mathscr{D}$. Since $[(1), (6)] = [(1), (11)] = [(6), (11)] = 5$, we may choose

$$(1)^* = (1) \cup \{22\ 23\ 24\ 25\ 26\ 27\},$$

$$(6)^* = (6) \cup \{22\ 28\ 29\ 30\ 31\ 32\},$$

$$(11)^* = (11) \cup \{23\ 28\ 33\ 34\ 35\ 36\}.$$

Further, since $[(14), (1)] = [(14), (6)] = [(14), (11)] = 3$ we have the following possibilities: either none, one ,two or three of the point set {22,23, 28} are contained in (14)*. If none of {22,23,28} is in (14)* then (14)* contains three of {24,25,26,27}, three of {29,30,31,32} and three of {33,34,

The 35 blocks of $\mathscr{D}$ are:

$(1) =$	∞	2	5	8	9	12	15	18	19
$(2) =$	∞	3	1	9	10	13	11	19	20
$(3) =$	∞	4	2	10	6	14	12	20	16
$(4) =$	∞	5	3	6	7	15	13	16	17
$(5) =$	∞	1	4	7	8	11	14	17	18
$(6) =$	∞	2	5	8	9	13	14	17	20
$(7) =$	∞	3	1	9	10	14	15	18	16
$(8) =$	∞	4	2	10	6	15	11	19	17
$(9) =$	∞	5	3	6	7	11	12	20	18
$(10) =$	∞	1	4	7	8	12	13	16	19
$(11) =$	∞	3	4	8	9	12	15	17	20
$(12) =$	∞	4	5	9	10	13	11	18	16
$(13) =$	∞	5	1	10	6	14	12	19	17
$(14) =$	∞	1	2	6	7	15	13	20	18
$(15) =$	∞	2	3	7	8	11	14	16	19
$(16) =$	1	6	8	9	11	17	18	19	20
$(17) =$	2	7	9	10	12	18	19	20	16
$(18) =$	3	8	10	6	13	19	20	16	17
$(19) =$	4	9	6	7	14	20	16	17	18
$(20) =$	5	10	7	8	15	16	17	18	19
$(21) =$	1	6	8	9	12	13	14	15	16
$(22) =$	2	7	9	10	13	14	15	11	17
$(23) =$	3	8	10	6	14	15	11	12	18
$(24) =$	4	9	6	7	15	11	12	13	19
$(25) =$	5	10	7	8	11	12	13	14	20
$(26) =$	1	2	5	11	12	15	16	17	20
$(27) =$	2	3	1	12	13	11	17	18	16
$(28) =$	3	4	2	13	14	12	18	19	17
$(29) =$	4	5	3	14	15	13	19	20	18
$(30) =$	5	1	4	15	11	14	20	16	19
$(31) =$	2	3	4	5	6	8	9	11	16
$(32) =$	3	4	5	1	7	8	10	12	17
$(33) =$	4	5	1	2	8	10	6	13	18
$(34) =$	5	1	2	3	9	6	7	14	19
$(35) =$	1	2	3	4	10	7	8	15	20.

35,36}, which is impossible since $(14)^*$ would have nine instead of six new points. If one of $\{22,23,28\}$ is in $(14)^*$, so by permuting the blocks $(1)^*,(6)^*$, and $(11)^*$ we may suppose that $22 \in (14)_*$ and then $(14)^*$ contains two of $\{24,25,26,27\}$, two of $\{29,30,31,32\}$ and three of $\{33,34,35,36\}$, which is impossible since $(14)^*$ would contain more then six new points. If two of $\{22,23,28\}$ are contained in $(14)^*$ then we may choose that $22,23 \in (14)^*$;

this means that $(14)^*$ contains one of $\{24,25,26,27\}$, two of $\{29,30,31,32\}$ and two of $\{33,34,35,36\}$, which is again contrary to the fact that $(14)^*$ has exactly six new points. Thus we have shown that $(14)^*$ must contain $22,23,28$.

Now since $[(27),(1)] = [(27),(6)] = [(27),(11)] = 3$ we see with the same argument that $(27)^*$ contains $22,23,28$, which contradicts the fact that $[(27),(14)] = 4$. Hence we have proved that $\mathscr{D}$ is not embeddable.

THEOREM 4.2. *There is a nonembeddable quasi-residual design for* $2-(21,35,15,9,6)$.

4.3 Three infinite families of quasi-residual designs of Bhattacharya type with parameters.

$$2-\left(49^m, \frac{7}{4}(49^m-1), \frac{3}{4}(49^m-1), 21.49^{m-1}, \frac{3}{4}(21.49^{m-1}-1)\right).$$

In this section we prove the existence of three non-isomorphic infinite families of quasi-residual designs of Bhattacharya type with given parameters. A large number of non-isomorphic families is constructed, but we only present here three of them with some characteristic properties.

In fact, we construct the first member of each family, namely a quasi-residual $2-(49,84,36,21,15)$ design. Then Corollary 2.2 and 2.3 provide new families of nonembeddable quasi-residual designs.

Since the upper bound of the block intersection numbers of a $2-(v,b,r,k,\lambda)$ design is $\frac{2k\lambda}{r}-(k-r+\lambda)$, (cf. [17]), the possible block intersection numbers greater than λ for an $2-(49,84,36,21,15)$ are 16 and 17. We give the existence of three non-isomorphic quasi-residual designs $\mathscr{D}_1$, $\mathscr{D}_2$ and $\mathscr{D}_3$ of Bhattacharya type for $2-(49,84,36,21,15)$ with the following properties: $\mathscr{D}_1$ contains pairs of block having 16 common points but not 17; $\mathscr{D}_2$ contains pairs of block having 17 common points but not 16; and $\mathscr{D}_3$ contains both pairs of block having 16 or 17 common points. It is worth mentioning that the designs in the family starting with $\mathscr{D}_3$ are the first ones known having more than one block intersection number greater than λ.

Let $\mathscr{D}$ be a $2-(49,84,36,21,15)$ design admitting a group $G = F \times H$, where $F = \langle \rho, \zeta \mid \rho^7 = 1, \zeta^3 = 1, \rho^\zeta = \rho^2 \rangle$ is a Frobenius group of order 21 with ρ acting fixed-point-free on the points of $\mathscr{D}$ and $H = \langle \tau \mid \tau^2 = 1 \rangle$ is a cyclic group of order 2. Since ρ has no fixed points we may put

$$\rho = (1_0, 1_1, \ldots, 1_6)(2_0, 2_1, \ldots, 2_6)\ldots(7_0, 7_1, \ldots, 7_6),$$

where $1_0, 1_1, \ldots, 7_6$ are all 49 points of $\mathscr{D}$. We prescribe the action of ζ and τ as follows:

$$\zeta = (I_0)(I_1, I_2, I_4)(I_3, I_6, I_5),$$

where $I = 1,2,3,4,5,6,7$ and

$$\tau = (1_i, 2_i)(3_i, 4_i)(5_i)(6_i)(7_i),$$

where $i = 0,1,2,3,4,5,6$.

Below are three quasi-residual designs of Bhattacharya type for $2-(49, 84, 36, 21, 15)$ $\mathscr{D}_1$, $\mathscr{D}_2$ and $\mathscr{D}_3$ constructed with the group G. We only present 12 base blocks for $\langle \rho \rangle$. The other blocks of $\mathscr{D}_i$ are obtained by applying $\langle \rho \rangle$ to these 12 base blocks. For short we shall write $I_{ab...x}$ instead of $I_a I_b ... I_x$ for the points $I_a, I_b, ..., I_x$ in the Ith ρ-point orbit, $I = 1, 2, 3, 4, 5, 6, 7$.

$$\mathscr{D}_1$$

$$B_1 = 2_{0365} \quad 3_{0365} \quad 4_{0124} \quad 5_{365} \quad 6_{365} \quad 7_{365}$$
$$B_2 = 1_{0365} \quad 3_{0124} \quad 4_{0365} \quad 5_{365} \quad 6_{365} \quad 7_{365}$$
$$B_3 = 1_{0124} \quad 2_{365} \quad 3_0 \quad 4_{124} \quad 5_{365} \quad 6_{0124} \quad 7_{365}$$
$$B_4 = 1_{365} \quad 2_{0124} \quad 3_{124} \quad 4_0 \quad 5_{365} \quad 6_{0124} \quad 7_{365}$$
$$B_5 = 1_{0365} \quad 2_{0365} \quad 3_{124} \quad 4_{124} \quad 5_{124} \quad 6_0 \quad 7_{124}$$
$$B_6 = 1_0 \quad 2_{365} \quad 3_{124} \quad 4_{0365} \quad 5_{0365} \quad 6_{365} \quad 7_{124}$$
$$B_7 = 1_{365} \quad 2_0 \quad 3_{0365} \quad 4_{124} \quad 5_{0365} \quad 6_{365} \quad 7_{124}$$
$$B_8 = 1_{124} \quad 2_{365} \quad 4_{0124} \quad 5_{0365} \quad 6_{0124} \quad 7_{124}$$
$$B_9 = 1_{365} \quad 2_{124} \quad 3_{0124} \quad 5_{0365} \quad 6_{0124} \quad 7_{124}$$
$$B_{10} = 1_{124} \quad 2_{124} \quad 3_{124} \quad 4_{124} \quad 6_{365} \quad 7_{124365}$$
$$B_{11} = 1_{0124} \quad 2_{0124} \quad 3_{365} \quad 4_{365} \quad 5_{0365} \quad 7_{124}$$
$$B_{12} = 1_{0365} \quad 2_{0365} \quad 3_{0365} \quad 4_{0365} \quad 5_0 \quad 6_{0124}$$

$$\mathscr{D}_2$$

$$B_1 = 2_{0124} \quad 3_{0124} \quad 4_{0365} \quad 5_{365} \quad 5_{365} \quad 7_{124}$$
$$B_2 = 1_{0124} \quad 3_{0365} \quad 4_{0124} \quad 5_{365} \quad 6_{365} \quad 7_{124}$$
$$B_3 = 1_{0124} \quad 2_{365} \quad 3_0 \quad 4_{365} \quad 5_{365} \quad 6_{0124} \quad 7_{124}$$
$$B_4 = 1_{365} \quad 2_{0124} \quad 3_{365} \quad 4_0 \quad 5_{365} \quad 6_{0124} \quad 7_{124}$$
$$B_5 = 1_{0124} \quad 2_{0124} \quad 3_{365} \quad 4_{365} \quad 5_{365} \quad 6_0 \quad 7_{365}$$
$$B_6 = 1_0 \quad 2_{124} \quad 3_{124} \quad 4_{0365} \quad 5_{0124} \quad 6_{124} \quad 7_{124}$$
$$B_7 = 1_{124} \quad 2_0 \quad 3_{0365} \quad 4_{124} \quad 5_{0124} \quad 6_{124} \quad 7_{124}$$
$$B_8 = 1_{365} \quad 2_{124} \quad 4_{0365} \quad 5_{0124} \quad 6_{0365} \quad 7_{124}$$
$$B_9 = 1_{124} \quad 2_{365} \quad 3_{0365} \quad 5_{0124} \quad 6_{0365} \quad 7_{124}$$
$$B_{10} = 1_{365} \quad 2_{365} \quad 3_{365} \quad 4_{365} \quad 6_{124} \quad 7_{124365}$$
$$B_{11} = 1_{0124} \quad 2_{0124} \quad 3_{365} \quad 4_{365} \quad 5_{0124} \quad 7_{365}$$
$$B_{12} = 1_{0365} \quad 2_{0365} \quad 3_{0365} \quad 4_{0365} \quad 5_0 \quad 6_{0365}$$

$$\mathscr{D}_3$$

$$B_1 = 2_{0365} \quad 3_{0365} \quad 4_{0124} \quad 5_{124} \quad 6_{124} \quad 7_{124}$$

$$B_2 = 1_{0365} \quad 3_{0124} \quad 4_{0365} \quad 5_{124} \quad 6_{124} \quad 7_{124}$$

$$B_3 = 1_{0124} \quad 2_{365} \quad 3_0 \quad 4_{124} \quad 5_{124} \quad 6_{0365} \quad 7_{124}$$

$$B_4 = 1_{365} \quad 2_{0124} \quad 3_{124} \quad 4_0 \quad 5_{124} \quad 6_{0365} \quad 7_{124}$$

$$B_5 = 1_{0124} \quad 2_{0124} \quad 3_{365} \quad 4_{365} \quad 5_{124} \quad 6_0 \quad 7_{124}$$

$$B_6 = 1_0 \quad 2_{365} \quad 3_{124} \quad 4_{0365} \quad 5_{0365} \quad 6_{124} \quad 7_{124}$$

$$B_7 = 1_{365} \quad 2_0 \quad 3_{0365} \quad 4_{124} \quad 5_{0365} \quad 6_{124} \quad 7_{124}$$

$$B_8 = 1_{124} \quad 2_{365} \quad 4_{0124} \quad 5_{0365} \quad 6_{0365} \quad 7_{124}$$

$$B_9 = 1_{365} \quad 2_{124} \quad 3_{0124} \quad 5_{0365} \quad 6_{0365} \quad 7_{124}$$

$$B_{10} = 1_{365} \quad 2_{365} \quad 3_{365} \quad 4_{365} \quad 6_{365} \quad 7_{124365}$$

$$B_{11} = 1_{0124} \quad 2_{0124} \quad 3_{365} \quad 4_{365} \quad 5_{0365} \quad 7_{124}$$

$$B_{12} = 1_{0124} \quad 2_{0124} \quad 3_{0124} \quad 4_{0124} \quad 5_0 \quad 6_{0124}$$

The table below gives the block intersection numbers and the corresponding numbers of pairs for $\mathscr{D}_1$, $\mathscr{D}_2$ and $\mathscr{D}_3$, and shows at the same time that they are pairwise not isomorphic.

Intersection numbers	$\mathscr{D}_1$ block pair numbers	$\mathscr{D}_2$ block pair numbers	$\mathscr{D}_3$ block pair numbers
0	7	0	7
1	0	0	0
2	0	0	0
3	0	0	0
4	14	0	14
5	7	14	0
6	0	49	0
7	196	182	210
8	777	945	721
9	1988	1701	2093
10	350	357	294
11	84	154	98
12	0	35	0
13	35	28	14
14	14	14	14
15	0	0	0
16	14	0	14
17	0	7	7

THEOREM 4.3. *There exist at least three infinite families of quasi-residual*

designs of Bhattacharya type with parameters:

$$2 - \left(49^m, \frac{7}{4}(49^m - 1), \frac{3}{4}(49^m - 1), 21.49^{m-1}, \frac{3}{4}(21.49^{m-1} - 1)\right),$$

for $m \geq 1$.

4.4 A nonembeddable quasi-residual design for 2-(33, 48, 16, 11, 5). In this section we shall present a $2 - (33, 48, 16, 11, 5)$ nonembeddable quasi-residual design $\mathcal{D}$ with a cyclic automorphism group $G = \langle \rho, \zeta \mid \rho^5 = 1, \zeta^3 = 1, \rho\zeta = \zeta\rho \rangle$ of order 15. The automorphism ζ acts fixed-point-free on $\mathcal{D}$ and ρ has three points $\infty_1, \infty_2, \infty_3$ and 6 non trivial point orbits denoted by 1,2,3,4,5,6. If we denote by I_0, I_1, I_2, I_3, I_4 the 5 points of $\mathcal{D}$ in the I-point orbit of ρ for $I = 1, 2, 3, 4, 5, 6$, we may put the action of ρ on the points of $\mathcal{D}$ as follows:

$$\rho = (\infty_1)(\infty_2)(\infty_3)(1_0, 1_1, 1_2, 1_3, 1_4)(2_0, 2_1, 2_2, 2_3, 2_4)$$
$$(3_0, 3_1, 3_2, 3_3, 3_4, 3_4)(4_0, 4_1, 4_2, 4_3, 4_4)(5_0, 5_1, 5_2, 5_3, 5_4)$$
$$(6_0, 6_1, 6_2, 6_3, 6_4).$$

And ζ should act on the points of $\mathcal{D}$ as follows:

$$\zeta = (\infty_1, \infty_2, \infty_3)(1_i, 2_i, 3_i)(4_i, 5_i, 6_i),$$

where $i = 0, 1, 2, 3, 4$.

The blocks of $\mathcal{D}$ are the following 48 sets:

$$
\begin{array}{llllll}
X = & \infty_1 & 1_{01234} & 3_{01234} & & \\
Y = & \infty_2 & 1_{01234} & 2_{01234} & & \\
Z = & \infty_3 & 2_{01234} & 3_{01234} & & \\
A_0 = & 1_{02} & 2_{01} & 3_0 & 4_{0123} & 5_4 & 6_0 \\
A_1 = & 1_{13} & 2_{12} & 3_1 & 4_{1234} & 5_0 & 6_1 \\
A_2 = & 1_{24} & 2_{23} & 3_2 & 4_{2340} & 5_1 & 6_2 \\
A_3 = & 1_{30} & 2_{34} & 3_3 & 4_{3401} & 5_2 & 6_3 \\
A_4 = & 1_{41} & 2_{40} & 3_4 & 4_{4012} & 5_3 & 6_4 \\
B_0 = & 2_{02} & 3_{01} & 1_0 & 5_{0123} & 6_4 & 4_0 \\
B_1 = & 2_{13} & 3_{12} & 1_1 & 5_{1234} & 6_0 & 4_1 \\
B_2 = & 2_{24} & 3_{23} & 1_2 & 5_{2340} & 6_1 & 4_2 \\
B_3 = & 2_{30} & 3_{34} & 1_3 & 5_{3401} & 6_2 & 4_3 \\
B_4 = & 2_{41} & 3_{40} & 1_4 & 5_{4012} & 6_3 & 4_4 \\
C_0 = & 3_{02} & 1_{01} & 2_0 & 6_{0123} & 4_4 & 5_0 \\
C_1 = & 3_{13} & 1_{12} & 2_1 & 6_{1234} & 4_0 & 5_1 \\
C_2 = & 3_{24} & 1_{23} & 2_2 & 6_{2340} & 4_1 & 5_2 \\
\end{array}
$$

$$C_3 = \quad 3_{30} \quad 1_{34} \quad 2_3 \quad 6_{3401} \quad 4_2 \quad 5_3$$
$$C_4 = \quad 3_{41} \quad 1_{40} \quad 2_4 \quad 6_{4012} \quad 4_3 \quad 5_4$$
$$D_0 = \quad \infty_1 \quad \infty_3 \quad 1_{13} \quad 2_{34} \quad 4_0 \quad 5_{24} \quad 6_{01}$$
$$D_1 = \quad \infty_1 \quad \infty_3 \quad 1_{24} \quad 2_{40} \quad 4_1 \quad 5_{30} \quad 6_{12}$$
$$D_2 = \quad \infty_1 \quad \infty_3 \quad 1_{30} \quad 2_{01} \quad 4_2 \quad 5_{41} \quad 6_{23}$$
$$D_3 = \quad \infty_1 \quad \infty_3 \quad 1_{41} \quad 2_{12} \quad 4_3 \quad 5_{02} \quad 6_{34}$$
$$D_4 = \quad \infty_1 \quad \infty_3 \quad 1_{02} \quad 2_{23} \quad 4_4 \quad 5_{13} \quad 6_{40}$$
$$E_0 = \quad \infty_2 \quad \infty_1 \quad 2_{13} \quad 3_{34} \quad 5_0 \quad 6_{24} \quad 4_{01}$$
$$E_1 = \quad \infty_2 \quad \infty_1 \quad 2_{24} \quad 3_{40} \quad 5_1 \quad 6_{30} \quad 4_{12}$$
$$E_2 = \quad \infty_2 \quad \infty_1 \quad 2_{30} \quad 3_{01} \quad 5_2 \quad 6_{41} \quad 4_{23}$$
$$E_3 = \quad \infty_2 \quad \infty_1 \quad 2_{41} \quad 3_{12} \quad 5_3 \quad 6_{02} \quad 4_{34}$$
$$E_4 = \quad \infty_2 \quad \infty_1 \quad 2_{02} \quad 3_{23} \quad 5_4 \quad 6_{13} \quad 4_{40}$$
$$F_0 = \quad \infty_3 \quad \infty_2 \quad 3_{13} \quad 1_{34} \quad 6_0 \quad 4_{24} \quad 5_{01}$$
$$F_1 = \quad \infty_3 \quad \infty_2 \quad 3_{24} \quad 1_{40} \quad 6_1 \quad 4_{30} \quad 5_{12}$$
$$F_2 = \quad \infty_3 \quad \infty_2 \quad 3_{30} \quad 1_{01} \quad 6_2 \quad 4_{41} \quad 5_{23}$$
$$F_3 = \quad \infty_3 \quad \infty_2 \quad 3_{41} \quad 1_{12} \quad 6_3 \quad 4_{02} \quad 5_{34}$$
$$F_4 = \quad \infty_3 \quad \infty_2 \quad 3_{02} \quad 1_{23} \quad 6_4 \quad 4_{13} \quad 5_{40}$$
$$G_0 = \quad \infty_2 \quad 1_{02} \quad 2_{34} \quad 3_1 \quad 4_1 \quad 5_{01} \quad 6_{13}$$
$$G_1 = \quad \infty_2 \quad 1_{13} \quad 2_{40} \quad 3_2 \quad 4_2 \quad 5_{12} \quad 6_{24}$$
$$G_2 = \quad \infty_2 \quad 1_{24} \quad 2_{01} \quad 3_3 \quad 4_3 \quad 5_{23} \quad 6_{30}$$
$$G_3 = \quad \infty_2 \quad 1_{30} \quad 2_{12} \quad 3_4 \quad 4_4 \quad 5_{34} \quad 6_{41}$$
$$G_4 = \quad \infty_2 \quad 1_{41} \quad 2_{23} \quad 3_0 \quad 4_0 \quad 5_{40} \quad 6_{02}$$
$$H_0 = \quad \infty_3 \quad 2_{02} \quad 3_{34} \quad 1_1 \quad 5_1 \quad 6_{01} \quad 4_{13}$$
$$H_1 = \quad \infty_3 \quad 2_{13} \quad 3_{40} \quad 1_2 \quad 5_2 \quad 6_{12} \quad 4_{24}$$
$$H_2 = \quad \infty_3 \quad 2_{24} \quad 3_{01} \quad 1_3 \quad 5_3 \quad 6_{23} \quad 4_{30}$$
$$H_3 = \quad \infty_3 \quad 2_{30} \quad 3_{12} \quad 1_4 \quad 5_4 \quad 6_{34} \quad 4_{41}$$
$$H_4 = \quad \infty_3 \quad 2_{41} \quad 3_{23} \quad 1_0 \quad 5_0 \quad 6_{40} \quad 4_{02}$$
$$I_0 = \quad \infty_1 \quad 3_{02} \quad 1_{34} \quad 2_1 \quad 6_1 \quad 4_{01} \quad 5_{13}$$
$$I_1 = \quad \infty_1 \quad 3_{13} \quad 1_{40} \quad 2_2 \quad 6_2 \quad 4_{12} \quad 5_{24}$$
$$I_2 = \quad \infty_1 \quad 3_{24} \quad 1_{01} \quad 2_3 \quad 6_3 \quad 4_{23} \quad 5_{30}$$
$$I_3 = \quad \infty_1 \quad 3_{30} \quad 1_{12} \quad 2_4 \quad 6_4 \quad 4_{34} \quad 5_{41}$$
$$I_4 = \quad \infty_1 \quad 3_{41} \quad 1_{23} \quad 2_0 \quad 6_0 \quad 4_{40} \quad 5_{02}$$

Next we show that $\mathscr{D}$ is not embeddable in a 2-(49,16,5) symmetric design.

Let 34, 35, 36, 37, 38, 39, 40, 41, 42, 43, 44, 45, 46, 47, 48, 49 be a set of sixteen new points. First of all we observe that $[X, Y] = [X, Z] =$

$[Y, Z] = 5$ implies that the new points in extended blocks X^*, Y^* and Z^* are pairwise disjoint, i.e., fifteen new points are used. Further $[G_i, X] = [G_i, Z] = 3$ and $[G_i, Y] = 5$ means that the still free 16th new point must appear in extended blocks G_i^*, $i = 0, 1, 2, 3, 4$. By a similar observation we see that the extended blocks H_j^*, I_k^* contain the 16th new point. Thus we have the following assertion:

(Δ) *The extended blocks* G_i^*, H_j^*, I_k^*, $i, j, k = 0, 1, 2, 3, 4$, *have a common new point.*

We now consider the blocks B_0, B_1, D_0, D_4. We have the following intersection numbers:

$$[B_0, B_1] = 4, \quad [D_0, B_0] = 2, \quad [D_0, B_1] = 5,$$
$$[D_4, B_0] = 5, \quad [D_4, B_1] = 4, \quad [D_4, D_0] = 4.$$

Thus for B_0^* and B_1^* and D_0^* we may choose

$$B_0^* = B_0 \cup \{34\,35\,36\,37\,38\},$$
$$B_1^* = B_1 \cup \{34\,39\,40\,41\,42\},$$
$$D_0^* = D_0 \cup \{35\,36\,37\,43\,44\},$$

and for D_4^* we may put

$$D_4^* = D_4 \cup \{39\,43\,45\,46\,47\}.$$

Now let us look at the block G_3, we have

$$[G_4, B_0] = 4, \quad [G_3, B_1] = 3, \quad [G_3, D_0] = 3, \quad [G_3, D_4] = 5.$$

Since $[G_3, D_4] = 5$, so 39, 43, 45, 46, 47 are not in G_3^*. Suppose 34 or 38 $\in G_3^*$, then 35, 36, 37, 38 $\notin G_3^*$ as $[G_3, B_0] = 4$; and then 43, 44 $\in G_3^*$, since $[G_3, D_0] = 2$, which is a contradiction. Thus 34, 38 $\notin G_3^*$. Hence one of $\{35, 36, 37\}$ is in G_3^*, and we may choose 35 $\in G_3^*$, and then we may put

$$G_3^* = G_3 \cup \{35\,44\,40\,41\,48\}.$$

Let us consider the block G_4. We have

$$[G_4, B_0] = 4, \ [G_4, B_1] = 4, \ [G_4, D_0] = 5, \ [G_4, D_4] = 3, \ [G_4, G_3] = 3.$$

Since $[G_4, D_0] = 5$, so 35, 36, 37, 43, 44 $\notin G_4^*$. Suppose 34 $\in G_4^*$, then 38, 39, 40, 41, 42 $\notin G_4^*$, thus $G_4^* = G_4 \cup \{34\,45\,46\,48\,49\}$, contradicting $[G_4, G_3] = 3$. Hence 34 $\notin G_4^*$. This forces 38 $\in G_4^*$. Suppose 39 $\in G_4^*$, then we may set $G_4^* = G_4 \cup \{38\,39\,45\,48\,49\}$, contradicting $[G_4, G_3] = 3$. Thus 39 $\notin G_4^*$. Suppose 42 $\in G_4^*$, then we may set $G_4^* = G_4 \cup \{38\,42\,45\,46^*\}$, where $^* = 48$ or 49, which contradicts again $[G_4, G_3] = 3$. Thus 42 $\notin G_4^*$. Hence we may choose 40 $\in G_4^*$ and then

$$G_4^* = G_4 \cup \{38\,40\,45\,46\,48\}.$$

Next we consider the block I_4. We have

$$[I_4, B_0] = 5, [I_4, B_1] = 3, [I_4, D_0] = 5, [I_4, D_4] = 4, [I_4, G_3] = 3,$$
$$[I_4, G_4] = 3.$$

By (Δ) we have either $40 \in I_4^*$ or $48 \in I_4^*$ or both 40 and $48 \in I_4^*$. Suppose $48 \notin I_4^*$ then $40 \in I_4^*$. From $[I_4, B_0] = 5$ and $[I_4, D_0] = 5$ we have 34, 35, 36, 37, 38, 43, 44 $\notin I_4^*$; this fact together with $[I_4, G_3] = 3$ yield $41 \in I_4^*$. Again $[I_4, B_1] = 3$ gives 39, 42 $\notin I_4^*$; on the other hand, since $[I_4, G_4] = 3$ we may choose $45 \in I_4^*$; it follows from $[I_4, D_4] = 4$ that 46, 47 $\notin I_4^*$. Thus we have shown that 40, 41, 45 $\in I_4^*$; in order to fill up I_4^* we must have two more new points, but we only have one free new point, namely 49; this fact leads to a contradiction. So we have

48 *is the common new point for* G_i^*, H_j^*, I_k^*, i, j, $k = 0, 1, 2, 3, 4.$

The fact that $[I_4, B_0] = [I_4, D_0] = 5$ gives 34, 35, 36, 37, 38, 43, 44 $\notin I_4^*$. Further, $[I_4, G_3] = 3$ implies that either $\alpha) : 40 \in I_4^*$ or $\beta) : 41 \in I_4^*$.

Case α. $40 \in I_4^*$. From $[I_4, G_3] = 3$ and $[I_4, G_4] = 3$ we have 41, 45, 46 $\notin I_4^*$. Moreover $[I_4, D_4] = 4$ yields that either $39 \in I_4^*$ or $47 \in I_4^*$. Suppose $39 \in I_4^*$, then 42, 47 $\notin I_4^*$, which leads to a contradiction, because we only have 49 as a free new point, so we cannot get an extended block I_4^*. Thereby $47 \in I_4^*$. Again $[I_4, B_1] = 3$ forces $42 \in I_4^*$. Thus we have

$$I_4^* = I_4 \cup \{48\ 40\ 47\ 42\ 49\}.$$

Case β. $41 \in I_4^*$. From $[I_4, G_4] = 3$ we see that we may choose $45 \in I_4^*$ and then $[I_4, D_4] = 4$ gives 39, 46, 47 $\notin I_4^*$. Further $[I_4, B_1] = 3$ forces $42 \in I_4^*$. Hence we have

$$I_4^* = I_4 \cup \{48\ 41\ 45\ 42\ 49\}.$$

Next we consider the block G_2. We have the following intersection numbers:

$$[G_2, B_0] = 3, [G_2, B_1] = 4, [G_2, D_0] = 2, [G_2, D_4] = 3,$$
$$[G_2, G_3] = 3, [G_2, G_4] = 3, [G_2, I_4] = 4.$$

We first deal with
Case β :
Here we summarize the extended blocks constructed so far:

$$B_0^* = B_0 \cup \{34\ 35\ 36\ 37\ 38\},$$
$$B_1^* = B_1 \cup \{34\ 39\ 40\ 41\ 42\},$$
$$D_0^* = D_0 \cup \{35\ 36\ 37\ 43\ 44\},$$
$$D_4^* = D_4 \cup \{39\ 43\ 45\ 46\ 47\},$$
$$G_3^* = G_3 \cup \{35\ 44\ 40\ 41\ 48\},$$
$$G_4^* = G_4 \cup \{38\ 40\ 45\ 46\ 48\},$$
$$I_4^* = I_4 \cup \{48\ 41\ 45\ 42\ 49\}.$$

First of all, remember that $48 \in G_2^*$, as already shown. Since $[G_2, I_4] = 4$ we have $41, 42, 45, 49 \notin G_2^*$. Moreover 35, 36 and 37 cannot occur at the same time in G_2^* since $[G_2, B_0] = 3$; this implies, by $[G_2, D_0] = 2$, that either 43 or 44 or both must occur in G_2^*.

Suppose 43 and 44 both appear in G_2. From $[G_2, G_3] = 3$ we see $35, 40, 41 \notin G_2^*$ and from $[G_2, D_0] = 2$ we can choose $36 \in G_2^*$. So we have $G_2^* = G_2 \cup \{48\,43\,44\,36^*\}$, where $^* = 34$ or 38, since $[G_2, B_0] = 3$, which contradicts $[G_2, D_4] = 2$. Thus both of 43 and 44 cannot appear in G_2^*.

Suppose $44 \in G_2^*$. Then $43 \notin G_2^*$. As $[G_2, G_3] = 3$ we have $35, 40 \notin G_2^*$. Further, $[G_2, D_0] = 2$ forces that $36, 37 \in G_2^*$ and then $[G_2, B_0] = 3$ gives $34, 38 \notin G_2^*$; comparing G_2 with B_1 yields $39 \in G_2^*$. Thus we have $G_2^* = G_2 \cup \{48\,44\,36\,37\,39\}$, contradicting $[G_2, D_4] = 3$.

Hence $43 \in G_2^*$ and $44 \notin G_2^*$. As $[G_2, G_3] = 3$, either $35 \in G_2^*$ or $40 \in G_2^*$.

(i) If $35 \in G_2^*$ then $40 \notin G_2^*$. And $[G_2, D_0] = 2$ implies that we may choose $36 \in G_2^*$. Moreover, $[G_2, B_0] = 2$ forces 34, 37, 38 $\notin G_2^*$; and $[G_2, G_4] = 3$ gives $46 \in G_2^*$. Thus we have $G_2^* = G_2 \cup \{48\,43\,35\,36\,46\}$, contradicting $[G_2, B_1] = 4$.

Hence we have

(ii) $40 \in G_2^*$ and $35 \notin G_2^*$. Now $[G_2, D_0] = 2$ gives $36, 37 \in G_2^*$. Thus $G_2^* = G_2 \cup \{48\,43\,40\,36\,37\}$, which is impossible since $[G_2, D_4] = 3$.

Hence *Case* β is ruled out.

Therefore we have

Case α:

In this case we have:

$$B_0^* = B_0 \cup \{34\,35\,36\,37\,38\},$$
$$B_1^* = B_1 \cup \{34\,39\,40\,41\,42\},$$
$$D_0^* = D_0 \cup \{35\,36\,37\,43\,44\},$$
$$D_4^* = D_4 \cup \{39\,43\,45\,46\,47\},$$
$$G_3^* = G_3 \cup \{35\,44\,40\,41\,48\},$$
$$G_4^* = G_4 \cup \{38\,40\,45\,46\,48\},$$
$$I_4^* = I_4 \cup \{48\,40\,47\,42\,49\}.$$

Of course, we have $48 \in G_2^*$. Since $[G_2, I_4] = 4$ we get $40, 42, 47, 49 \in G_2^*$. Further, $[G_2, B_0] = 3$ and $[G_2, D_0] = 2$ imply that at most two of $\{35,36,37\}$ can occur in G_2^* and at least one of $\{43,44\}$ must be in G_2^*.

Suppose both $43, 44 \in G_2^*$. Then $[G_2, G_3] = 3$ gives $35, 41 \notin G_2^*$. Hence we may choose $36 \in G_2^*$ and $37 \notin G_2^*$ by looking at $[G_2, D_0] = 2$. From $[G_2, B_0] = 2$ we have either $34 \in G_2^*$ or $38 \in G_2^*$. If $34 \in G_2^*$ then

$G_2^* = G_2 \cup \{48\,43\,44\,36\,34\}$, contradicting $[G_2, D_4] = 3$. Thus $38 \in G_2^*$ and then $G_2^* = G_2 \cup \{48\,43\,44\,36\,38\}$, contradicting again $[G_2, D_4] = 3$. This contradiction shows that both of 43 and 44 cannot appear in G_2^*.

Suppose $44 \in G_2^*$ and $43 \notin G_2^*$. Then $[G_2, G_3] = 3$ gives $35, 41 \notin G_2^*$ and then $[G_2, D_0] = 2$ yields $36, 37 \in G_2^*$. Further $[G_2, B_0] = 3$ forces $34, 38 \notin G_2^*$; and $[G_2, G_4] = 3$ allows us to choose $45 \in G_2^*$. So we have $G_2^* = G_2 \cup \{48\,44\,36\,37\,45\}$, contradicting $[G_2, B_1] = 4$.

Hence we finally have $43 \in G_2^*$ and $44 \notin G_2^*$. Further, $[G_2, D_4] = 3$ implies that either $39 \in G_2^*$ or $45 \in G_2^*$ (since 45 and 46 both appear only in D_4^* and G_4^*, i.e., if 45 or 46 appears in G_2^*, then we may choose $45 \in G_2^*$). Suppose $39 \in G_2^*$, then $34, 41, 45, 46 \notin G_2^*$ by reason of $[G_2, B_1] = 4$ and $[G_2, D_4] = 3$. Further, $[G_2, G_4] = 3$ yields $38 \in G_2^*$; and $[G_2, G_3] = 3$ shows that $35 \in G_2^*$. Thus we have $G_2^* = G_2 \cup \{48\,43\,39\,38\,45\}$, contradicting $[G_2, D_0] = 2$.

Therefore we must have $45 \in G_2^*$. As $[G_2, D_4] = 3$ and $[G_2, G_4] = 3$, we have that $38, 46, 39 \notin G_2^*$. Further, $[G_2, G_3] = 3$ means that either $35 \in G_2^*$ or $41 \in G_2^*$.

If $35 \in G_2^*$ then we may choose $36 \in G_2^*$ by considering $[G_2, D_0] = 2$. Thus in this case we get $G_2^* = G_2 \cup \{48\,43\,45\,35\,36\}$, contradicting $[G_2, B_1] = 4$.

Hence we have $41 \in G_2^*$ and $35 \notin G_2^*$. And then 36 and 37 must occur in G_2^*, as $[G_2, D_0] = 2$. This fact yields a contradiction, because G_2^* would contain more than five new points, namely 48, 43, 45, 41, 36, 37.

Hence we have proved that $\mathscr{D}$ is not embeddable.

THEOREM 4.4. *There exists a* $2 - (33, 48, 16, 11, 5)$ *nonembeddable quasi-residual design.*

4.5. Nonembeddable quasi-residual designs for 2-(40, 65, 26, 16, 10) and a new 2-(66, 26, 10) symmetric design. In this section we are dealing with the existence of nonembeddable quasi-residual designs for 2-(40,65,26,16,10) and their embedding in 2-(66,26,10) symmetric designs. In fact the quasi-residual designs are constructed via a Frobenius group of order 39. Let $\mathscr{D}$ be a quasi-residual design with parameters 2-(40,65,26,16,10). Also let $G = \langle \rho, \zeta \mid \rho^{13} = 1, \zeta^3 = 1, \zeta^{-1}\rho\zeta = \rho^3 \rangle$ be a Frobenius group of order 39 acting on $\mathscr{D}$. Obviously, ρ has precisely one fixed point on $\mathscr{D}$. Let $\infty, 1, 2, 3$ be four point orbits of ρ, where ∞ is its fixed point and 1,2,3 are non-trivial point orbits of length 13. So we may put the action of ρ on the points of $\mathscr{D}$ as follows:

$$\rho = (\infty)(I_0, I_1, I_2, I_3, I_4, I_5, I_6, I_7, I_8, I_9, I_{10}, I_{11}, I_{12}),$$

where $I = 1, 2, 3$; and ζ should act on the points of $\mathscr{D}$ as follows:

$$\zeta = (\infty)(I_0)(I_1, I_3, I_9)(I_2, I_6, I_5)(I_4, I_{12}, I_{10})(I_8, I_{11}, I_7),$$

where $I = 1, 2, 3$.

Below we present two nonembeddable quasi-residual designs for $2-(40, 65, 26, 16, 10)$ which are not isomorphic. We only write 5 base blocks for G.

Solution $\mathscr{D}_1$

$$A = \infty \; 1_{1\,3\,9} \; 2_{1\,3\,9\,2\,6\,5} \; 3_{4\,12\,10\,8\,11\,7},$$
$$B = \infty \; 1_{0\,1\,3\,9\,4\,12\,10} \; 2_{0\,1\,3\,9} \; 3_{0\,4\,12\,10},$$
$$C = 1_{2\,6\,5\,8\,11\,7} \; 2_{1\,3\,9} \; 3_{0\,4\,12\,10\,8\,11\,7},$$
$$D = 1_{2\,6\,5\,4\,12\,10} \; 2_{0\,4\,12\,10\,8\,11\,7} \; 3_{4\,12\,10},$$
$$E = 1_{0\,2\,6\,5} \; 2_{1\,3\,9\,4\,12\,10} \; 3_{1\,3\,9\,4\,12\,10}.$$

We next show that $\mathscr{D}_1$ is not embeddable in a $2-(66, 26, 10)$ symmetric design. We have the following intersection numbers between the blocks A, B, C, D and E : $[A, B] = 10$, $[A, C] = 9$, $[A, D] = 3$, $[A, E] = 6$, $[B, C] = 7$, $[B, D] = 7$, $[B, E] = 7$, $[C, D] = 6$, $[C, E] = 9$, $[D, E] = 9$. Let $1, 2, \ldots, 26$ be a set of 26 new points. For extended blocks A^* and B^* and D^* we may set

$$A^* = A \cup \{1\,2\,3\,4\,5\,6\,7\,8\,9\,10\},$$
$$B^* = B \cup \{11\,12\,13\,14\,15\,16\,17\,18\,19\,20\},$$
$$C^* = C \cup \{1\,11\,12\,13\,21\,22\,23\,24\,25\,26\}.$$

We see that $E^* = E \cup \{X \cup Y \cup Z\}$, where $X \subset \{1\,2\,3\,4\,5\,6\,7\,8\,9\,10\}$, $Y \subset \{11\,12\,13\,14\,15\,16\,17\,18\,19\,20\}$, $Z \subset \{1\,11\,12\,13\,21\,22\,23\,24\,25\,26\}$, $|X| = 4$, $|Y| = 3$, and $|Z| = 1$, but then $|X| + |Y| + |Z| \leq 8$ which is impossible because we must have ten new points in an extended block E^*. Hence $\mathscr{D}_1$ is not embeddable.

Solution $\mathscr{D}_2$

$$A = \infty \; 1_{1\,3\,9} \; 2_{1\,3\,9\,2\,6\,5} \; 3_{4\,12\,10\,8\,11\,7},$$
$$B = \infty \; 1_{0\,1\,3\,9\,2\,6\,5} \; 2_{0\,1\,3\,9} \; 3_{0\,4\,12\,10},$$
$$C = 1_{1\,3\,9\,8\,11\,7} \; 2_{4\,12\,10} \; 3_{0\,1\,3\,9\,4\,12\,10},$$
$$D = 1_{4\,12\,10\,8\,11\,7} \; 2_{0\,1\,3\,9\,2\,6\,5} \; 3_{4\,12\,10},$$
$$E = 1_{0\,4\,12\,10} \; 2_{1\,3\,9\,4\,12\,10} \; 3_{1\,3\,9\,4\,12\,10}.$$

The intersection numbers between the blocks A, B, C, D and E in $\mathscr{D}_2$ are: $[A, B] = 10$, $[A, C] = 6$, $[A, D] = 9$, $[A, E] = 6$, $[B, C] = 7$, $[B, D] = 7$, $[B, E] = 7$, $[C, D] = 6$, $[C, E] = 9$, $[D, E] = 9$. With the same argument as for $\mathscr{D}_1$, applying to the blocks A, B, D, E, we see that $\mathscr{D}_2$ is not embeddable in a 2-(66,26,10) symmetric design.

If we consider all $\binom{65}{2}$ intersection numbers between any two blocks in $\mathscr{D}_1$ and in $\mathscr{D}_2$ we see that they are of different patterns, hence $\mathscr{D}_1$ is not isomorphic to $\mathscr{D}_2$.

THEOREM 4.5. *There exist at least two nonembeddable quasi-residual designs for* $2 - (40, 65, 26, 16, 10)$.

By constructing quasi-residual designs for these parameters we have found that some of them have an extension to a $2 - (66, 26, 10)$ symmetric design $\mathscr{B}$ which admits G as an automorphism group. The six point orbits of ρ are denoted by $\infty, 1, 2, 3, 4, 5$, where ∞ is the fixed point of ρ. The action of ρ and ζ are given by:

$$\rho = (\infty)(I_0, I_1, I_2, I_3, I_4, I_5, I_6, I_7, I_8, I_9, I_{10}, I_{11}, I_{12}),$$
$$\zeta = (\infty)(I_0)(I_1, I_3, I_9)(I_2, I_6, I_5)(I_4, I_{12}, I_{10})(I_8, I_{11}, I_7),$$

where $I = 1, 2, 3, 4, 5$.
Below are six base blocks of $\mathscr{B}$:

$$A = 1_{0\ 1\ 2\ 3\ 4\ 5\ 6\ 7\ 8\ 9\ 10\ 11\ 12}\quad 2_{0\ 1\ 2\ 3\ 4\ 5\ 6\ 7\ 8\ 9\ 10\ 11\ 12},$$
$$B = \infty\quad 1_{2\ 6\ 5\ 8\ 11\ 7}\quad 2_{0\ 2\ 6\ 5}\quad 3_{4\ 12\ 10\ 8\ 11\ 7}\quad 4_{1\ 3\ 9\ 4\ 12\ 10}\quad 5_{1\ 3\ 9},$$
$$C = \infty\quad 1_{0\ 2\ 6\ 5}\quad 2_{1\ 3\ 9\ 8\ 11\ 7}\quad 3_{0\ 1\ 3\ 9}\quad 4_{0\ 1\ 3\ 9}\quad 5_{0\ 1\ 3\ 9\ 4\ 12\ 10},$$
$$D = 1_{1\ 3\ 9}\quad 2_{0\ 1\ 3\ 9\ 2\ 6\ 5}\quad 3_{1\ 3\ 9\ 4\ 12\ 10}\quad 4_{2\ 6\ 5\ 8\ 11\ 7}\quad 5_{0\ 1\ 3\ 9},$$
$$E = 1_{4\ 12\ 10\ 8\ 11\ 7}\quad 2_{0\ 4\ 12\ 10}\quad 3_{0\ 4\ 12\ 10\ 8\ 11\ 7}\quad 4_{1\ 3\ 9}\quad 5_{1\ 3\ 9\ 4\ 12\ 10},$$
$$F = 1_{2\ 6\ 5\ 4\ 12\ 10}\quad 2_{0\ 1\ 3\ 9}\quad 3_{0\ 1\ 3\ 9\ 4\ 12\ 10}\quad 4_{4\ 12\ 10}\quad 5_{2\ 6\ 5\ 8\ 11\ 7}.$$

We note that $\mathscr{B}$ is not isomorphic to the known design given in [20] and seems to be new as far as we know.

4.6. Nonembeddable quasi-residual designs for 2-(52, 78, 27, 18, 9). In this section we shall construct nonembeddable quasi-residual designs $\mathscr{D}$ with parameters $2 - (52, 78, 27, 18, 9)$ admitting a Frobenius automorphism group $G = \langle \rho, \zeta \mid \rho^{13} = 1, \zeta^3 = 1, \zeta^{-1}\rho\zeta = \rho^3 \rangle$ of order 39, on which ρ acts fixed point free and ζ has four fixed points. We denote by 0, 1, 2, 3 four point orbits of ρ, then we may put the action of ρ and ζ on the points of $\mathscr{D}$ as follows:

$$\rho = (I_0, I_1, I_2, I_3, I_4, I_5, I_6, I_7, I_8, I_9, I_{10}, I_{11}, I_{12}),$$
$$\zeta = (I_0)(I_1, I_3, I_9)(I_2, I_6, I_5)(I_4, I_{12}, I_{10})(I_8, I_{11}, I_7),$$

where $I = 0, 1, 2, 3$.

With this group G we have found a large number of solutions of nonembeddable quasi-residual designs and some of them are of Bhattacharya type and reach the Majumdar upper bound $\frac{2k\lambda}{r} - (k - r + \lambda)$ for block intersection numbers, which is 12 in our case. In the following we present five non-isomorphic solutions for $\mathscr{D}$. Of course, only 6 base blocks of $\mathscr{D}$ are displayed.

Solution $\mathcal{D}_1$

$A = 0_{4\ 12\ 10}\quad 1_{0\ 1\ 3\ 9}\quad 2_{0\ 4\ 12\ 10\ 8\ 11\ 7}\quad 3_{0\ 1\ 3\ 9},$

$B = 0_{4\ 12\ 10}\quad 1_{0\ 1\ 3\ 9}\quad 3_{0\ 4\ 12\ 10\ 8\ 11\ 7}\quad 2_{0\ 1\ 3\ 9},$

$C = 0_{1\ 3\ 9}\quad 1_{0\ 1\ 3\ 9\ 4\ 12\ 10}\quad 2_{0\ 2\ 6\ 5}\quad 3_{0\ 2\ 6\ 5},$

$D = 0_{1\ 3\ 9\ 2\ 6\ 5}\quad 1_{1\ 3\ 9}\quad 2_{1\ 3\ 9\ 4\ 12\ 10}\quad 3_{8\ 11\ 7},$

$E = 0_{1\ 3\ 9\ 2\ 6\ 5}\quad 1_{1\ 3\ 9}\quad 3_{1\ 3\ 9\ 4\ 12\ 10}\quad 2_{8\ 11\ 7},$

$F = 0_{1\ 3\ 9\ 2\ 6\ 5}\quad 1_{4\ 12\ 10\ 8\ 11\ 7}\quad 2_{1\ 3\ 9}\quad 3_{1\ 3\ 9}.$

It is easy to see that $\mathcal{D}_1$ is not embeddable in a $2-(79, 27, 9)$ symmetric design. Since $[A, B] = 9$, the new point sets occurring in extended blocks A^* and B^* are disjoint; further $[F, A] = [F, B] = 3$ implies that an extended block F^* contains six new points from A^* and six new points from B^*, which is impossible because F^* has precisely 9 new points. Hence $\mathcal{D}_1$ is not embeddable. Note that $\mathcal{D}_1$ is not of Bhattacharya type.

Solution $\mathcal{D}_2$

$A = 0_{8\ 11\ 7}\quad 1_{0\ 1\ 3\ 9}\quad 2_{0\ 2\ 6\ 5\ 8\ 11\ 7}\quad 3_{0\ 1\ 3\ 9},$

$B = 0_{8\ 11\ 7}\quad 1_{0\ 1\ 3\ 9}\quad 3_{0\ 2\ 6\ 5\ 8\ 11\ 7}\quad 2_{0\ 1\ 3\ 9},$

$C = 0_{2\ 6\ 5}\quad 1_{0\ 2\ 6\ 5\ 4\ 12\ 10}\quad 2_{0\ 8\ 11\ 7}\quad 3_{0\ 8\ 11\ 7},$

$D = 0_{1\ 3\ 9\ 8\ 11\ 7}\quad 1_{8\ 11\ 7}\quad 2_{2\ 6\ 5\ 8\ 11\ 7}\quad 3_{4\ 12\ 10},$

$E = 0_{1\ 3\ 9\ 8\ 11\ 7}\quad 1_{8\ 11\ 7}\quad 3_{2\ 6\ 5\ 8\ 11\ 7}\quad 2_{4\ 12\ 10},$

$F = 0_{1\ 3\ 9\ 8\ 11\ 7}\quad 1_{2\ 6\ 5\ 4\ 12\ 10}\quad 2_{8\ 11\ 7}\quad 3_{8\ 11\ 7}.$

Solution $\mathcal{D}_3$

$A = 0_{8\ 11\ 7}\quad 1_{0\ 1\ 3\ 9}\quad 2_{0\ 2\ 6\ 5\ 8\ 11\ 7}\quad 3_{0\ 1\ 3\ 9},$

$B = 0_{8\ 11\ 7}\quad 1_{0\ 1\ 3\ 9}\quad 3_{0\ 2\ 6\ 5\ 8\ 11\ 7}\quad 2_{0\ 1\ 3\ 9},$

$C = 0_{2\ 6\ 5}\quad 1_{0\ 2\ 6\ 5\ 4\ 12\ 10}\quad 2_{0\ 8\ 11\ 7}\quad 3_{0\ 8\ 11\ 7},$

$D = 0_{1\ 3\ 9\ 8\ 11\ 7}\quad 1_{8\ 11\ 7}\quad 2_{2\ 6\ 5\ 4\ 12\ 10}\quad 3_{8\ 11\ 7},$

$E = 0_{1\ 3\ 9\ 8\ 11\ 7}\quad 1_{8\ 11\ 7}\quad 3_{2\ 6\ 5\ 4\ 12\ 10}\quad 2_{8\ 11\ 7},$

$F = 0_{1\ 3\ 9\ 8\ 11\ 7}\quad 1_{2\ 6\ 5\ 4\ 12\ 10}\quad 2_{8\ 11\ 7}\quad 3_{8\ 11\ 7}.$

Solution $\mathscr{D}_4$

$$A = 0_{8\ 11\ 7} \quad 1_{0\ 2\ 6\ 5} \quad 2_{0\ 1\ 3\ 9\ 8\ 11\ 7} \quad 3_{0\ 2\ 6\ 5},$$
$$B = 0_{8\ 11\ 7} \quad 1_{0\ 2\ 6\ 5} \quad 3_{0\ 1\ 3\ 9\ 8\ 11\ 7} \quad 2_{0\ 2\ 6\ 5},$$
$$C = 0_{2\ 6\ 5} \quad 1_{0\ 2\ 6\ 5\ 8\ 11\ 7} \quad 2_{0\ 4\ 12\ 10} \quad 3_{0\ 4\ 12\ 10},$$
$$D = 0_{2\ 6\ 5\ 4\ 12\ 10} \quad 1_{2\ 6\ 5} \quad 2_{1\ 3\ 9\ 8\ 11\ 7} \quad 3_{2\ 6\ 5},$$
$$E = 0_{2\ 6\ 5\ 4\ 12\ 10} \quad 1_{2\ 6\ 5} \quad 3_{1\ 3\ 9\ 8\ 11\ 7} \quad 2_{2\ 6\ 5},$$
$$F = 0_{2\ 6\ 5\ 4\ 12\ 10} \quad 1_{1\ 3\ 9\ 11\ 7} \quad 2_{2\ 6\ 5} \quad 3_{2\ 6\ 5}.$$

Solution $\mathscr{D}_5$

$$A = 0_{4\ 12\ 10} \quad 1_{0\ 4\ 12\ 10} \quad 2_{0\ 1\ 3\ 9\ 4\ 12\ 10} \quad 3_{0\ 2\ 6\ 5},$$
$$B = 0_{4\ 12\ 10} \quad 1_{0\ 4\ 12\ 10} \quad 3_{0\ 1\ 3\ 9\ 4\ 12\ 10} \quad 2_{0\ 2\ 6\ 5},$$
$$C = 0_{4\ 12\ 10} \quad 1_{0\ 1\ 3\ 9\ 2\ 6\ 5} \quad 2_{0\ 4\ 12\ 10} \quad 3_{0\ 4\ 12\ 10},$$
$$D = 0_{2\ 6\ 5\ 8\ 11\ 7} \quad 1_{4\ 12\ 10} \quad 2_{1\ 3\ 9\ 4\ 12\ 10} \quad 3_{2\ 6\ 5},$$
$$E = 0_{2\ 6\ 5\ 8\ 11\ 7} \quad 1_{4\ 12\ 10} \quad 3_{1\ 3\ 9\ 4\ 12\ 10} \quad 2_{2\ 6\ 5},$$
$$F = 0_{2\ 6\ 5\ 8\ 11\ 7} \quad 1_{1\ 3\ 9\ 2\ 6\ 5} \quad 2_{4\ 12\ 10} \quad 3_{4\ 12\ 10}.$$

We see from the base blocks of $\mathscr{D}_2$, $\mathscr{D}_3$, $\mathscr{D}_4$, and $\mathscr{D}_5$ that they are all of Bhattacharya type and therefore not isomorphic to $\mathscr{D}_1$; moreover they are pairwise non-isomorphic as we see from the table below which gives the distribution of the block intersection numbers for $\mathscr{D}_i$, $i = 1, 2, 3, 4, 5$.

Intersection numbers	$\mathscr{D}_1$ block pair numbers	$\mathscr{D}_2$ block pair numbers	$\mathscr{D}_3$ block pair numbers	$\mathscr{D}_4$ block pair numbers	$\mathscr{D}_5$ block pair numbers
0	0	0	0	0	0
1	0	0	0	0	0
2	0	0	0	0	0
3	26	52	26	52	78
4	117	78	78	39	0
5	390	312	390	234	0
6	1846	1924	1976	2288	0
7	390	546	312	234	2808
8	156	0	156	78	0
9	78	78	52	52	78
10	0	0	0	0	0
11	0	0	0	0	0
12	0	13	13	26	39

Hence we have:

THEOREM 4.6. *There exist at least five non-isomorphic nonembeddable quasi-residual designs for* $2 - (52, 78, 27, 18, 9)$.

4.7 A nonembeddable quasi-residual design for 2-(78, 143, 66, 36, 30). In this section we shall construct a nonembeddable quasi-residual design $\mathscr{D}$ with parameters 2-(78,143,66,36,30) admitting an automorphism group $H = F \times G$, where $F = \langle \rho, \zeta \mid \rho^{11} = 1, \zeta^5 = 1, \zeta^{-1}\rho\zeta = \rho^3 \rangle$ is a Frobenius group of order 55 and $G = \langle \theta \mid \theta^6 = 1 \rangle$ is a cyclic group of order 6. Moreover ρ should act on $\mathscr{D}$ with one fixed point, ζ has 8 fixed points and θ has 12 fixed points. We denote by $\infty, 0, 1, 2, 3, 4, 5, 6$ eight point orbits of ρ, where ∞ is its fixed point; we may put the action of ρ, ζ and θ on the points of $\mathscr{D}$ as follows:

$$\rho = (\infty)(I_0, I_1, I_2, I_3, I_4, I_5, I_6, I_7, I_8, I_9, I_{10}),$$
$$\zeta = (\infty)(I_0)(I_1, I_3, I_9, I_5, I_4)(I_2, I_6, I_7, I_{10}, I_8),$$

where $I = 0, 1, 2, 3, 4, 5, 6$, and

$$\theta = (\infty)(0_i)(1_i, 2_i, 3_i, 4_i, 5_i, 6_i),$$

where $i = 0, 1, 2, 3, 4, 5, 6, 7, 8, 9, 10$.

Below are 13 base blocks of $\mathscr{D}$ with respect to $\langle \rho \rangle$:

$$(l_1) = \infty \quad 0_{1\,3\,9\,5\,4} \quad 1_{1\,3\,9\,5\,4} \quad 2_{2\,6\,7\,10\,8} \quad 3_{2\,6\,7\,10\,8}$$
$$4_{1\,3\,9\,5\,4} \quad 5_{1\,3\,9\,5\,4} \quad 6_{2\,6\,7\,10\,8},$$
$$(l_2) = \infty \quad 0_{1\,3\,9\,5\,4} \quad 2_{1\,3\,9\,5\,4} \quad 3_{2\,6\,7\,10\,8} \quad 4_{2\,6\,7\,10\,8}$$
$$5_{1\,3\,9\,5\,4} \quad 6_{1\,3\,9\,5\,4} \quad 1_{2\,6\,7\,10\,8}$$
$$(l_3) = \infty \quad 0_{1\,3\,9\,5\,4} \quad 3_{1\,3\,9\,5\,4} \quad 4_{2\,6\,7\,10\,8} \quad 5_{2\,6\,7\,10\,8}$$
$$6_{1\,3\,9\,5\,4} \quad 1_{1\,3\,9\,5\,4} \quad 2_{2\,6\,7\,10\,8}$$
$$(l_4) = \infty \quad 0_{1\,3\,9\,5\,4} \quad 4_{1\,3\,9\,5\,4} \quad 5_{2\,6\,7\,10\,8} \quad 6_{2\,6\,7\,10\,8}$$
$$1_{1\,3\,9\,5\,4} \quad 2_{1\,3\,9\,5\,4} \quad 3_{2\,6\,7\,10\,8},$$
$$(l_5) = \infty \quad 0_{1\,3\,9\,5\,4} \quad 5_{1\,3\,9\,5\,4} \quad 6_{2\,6\,7\,10\,8} \quad 1_{2\,6\,7\,10\,8}$$
$$2_{1\,3\,9\,5\,4} \quad 3_{1\,3\,9\,5\,4} \quad 4_{2\,6\,7\,10\,8},$$
$$(l_6) = \infty \quad 0_{1\,3\,9\,5\,4} \quad 6_{1\,3\,9\,5\,4} \quad 1_{2\,6\,7\,10\,8} \quad 2_{2\,6\,7\,10\,8}$$
$$3_{1\,3\,9\,5\,4} \quad 4_{1\,3\,9\,5\,4} \quad 5_{2\,6\,7\,10\,8}$$
$$(l_7) = 0_{0\,2\,6\,7\,10\,8} \quad 1_{0\,2\,6\,7\,10\,8} \quad 3_{0\,2\,6\,7\,10\,8} \quad 4_{0\,1\,3\,9\,5\,4}$$
$$5_{0\,1\,3\,9\,5\,4} \quad 6_{0\,1\,3\,9\,5\,4},$$
$$(l_8) = 0_{0\,2\,6\,7\,10\,8} \quad 2_{0\,2\,6\,7\,10\,8} \quad 4_{0\,2\,6\,7\,10\,8} \quad 5_{0\,1\,3\,9\,5\,4}$$
$$6_{0\,1\,3\,9\,5\,4} \quad 1_{0\,1\,3\,9\,5\,4},$$
$$(l_9) = 0_{0\,2\,6\,7\,10\,8} \quad 3_{0\,2\,6\,7\,10\,8} \quad 5_{0\,2\,6\,7\,10\,8} \quad 6_{0\,1\,3\,9\,5\,4}$$
$$1_{0\,1\,3\,9\,5\,4} \quad 2_{0\,1\,3\,9\,5\,4},$$
$$(l_{10}) = 0_{0\,2\,6\,7\,10\,8} \quad 4_{0\,2\,6\,7\,10\,8} \quad 6_{0\,2\,6\,7\,10\,8} \quad 1_{0\,1\,3\,9\,5\,4}$$
$$2_{0\,1\,3\,9\,5\,4} \quad 3_{0\,1\,3\,9\,5\,4},$$

$$(l_{11}) = 0_{0\ 2\ 6\ 7\ 10\ 8} \quad 5_{0\ 2\ 6\ 7\ 10\ 8} \quad 1_{0\ 2\ 6\ 7\ 10\ 8} \quad 2_{0\ 1\ 3\ 9\ 5\ 4}$$
$$3_{0\ 1\ 3\ 9\ 5\ 4} \quad 4_{0\ 1\ 3\ 9\ 5\ 4},$$
$$(l_{12}) = 0_{0\ 2\ 6\ 7\ 10\ 8} \quad 6_{0\ 2\ 6\ 7\ 10\ 8} \quad 2_{0\ 2\ 6\ 7\ 10\ 8} \quad 3_{0\ 1\ 3\ 9\ 5\ 4}$$
$$4_{0\ 1\ 3\ 9\ 5\ 4} \quad 5_{0\ 1\ 3\ 9\ 5\ 4},$$
$$(l_{13}) = 1_{0\ 1\ 3\ 9\ 5\ 4} \quad 2_{0\ 1\ 3\ 9\ 5\ 4} \quad 3_{0\ 1\ 3\ 9\ 5\ 4} \quad 4_{0\ 1\ 3\ 9\ 5\ 4}$$
$$5_{0\ 1\ 3\ 9\ 5\ 4} \quad 6_{0\ 1\ 3\ 9\ 5\ 4}.$$

We now show that $\mathscr{D}$ is not embeddable in a $2-(144, 66, 30)$ symmetric design. Let $1, 2, \ldots, 66$ be a set of 66 new points. Consider the blocks l_3, l_6 and l_{12} of $\mathscr{D}$. We have $[l_3, l_6] = 5$, $[l_{12}, l_3] = 10$, $[l_{12}, l_6] = 20$. Thus the extended blocks l_3^* and l_6^* have 25 new points in common. Let X be this common new point set of l_3^* and l_6^*. From the intersection number of l_{12}^* with l_3^* we see that an extended block l_{12}^* has precisely 20 common new points with l_3^*. This implies that l_{12}^* contains at least 15 new points from X, which is a contradiction, since we have $[l_{12}, l_6] = 20$.

Hence $\mathscr{D}$ is not embeddable. We have

THEOREM 4.7. *There exists a* $2 - (78, 143, 66, 36, 30)$ *nonembeddable quasi-residual design.*

4.8 Families of quasi-residual designs with parameters.

$$2 - (25^m, \frac{5}{3}(25^m - 1), \frac{2}{3}(25^m - 1), 10.25^{m-1}, \frac{2}{3}(10.25^{m-1} - 1)).$$

In this section we are dealing with the construction of nonembeddable quasi-residual designs with given parameters. In fact we construct the first member of the families, namely a 2-(25,40,16,10,6) design. Then applying Corollary 2.2 and 2.3 we get new infinite families. The designs in one of these families are of Bhattacharya type. In the following we construct two nonembeddable quasi-residual designs for $2-(25, 40, 16, 10, 6)$ admitting an automorphism group $H = F \times G$, where $F = \langle \rho, \zeta \mid \rho^5 = 1, \zeta^2 = 1, \zeta^{-1}\rho\zeta = \rho^{-1} \rangle$ is a Frobenius group of order 10 and $G = \langle \theta \mid \theta^2 = 1 \rangle$ is a cyclic group of order 2. Here ρ should act fixed-point-free on $\mathscr{D}$, ζ and θ each have 5 fixed points. We denote by $\infty, 1, 2, 3, 4, 5$ five point orbits of ρ. The action of ρ, ζ and θ on the points of $\mathscr{D}$ are as follows:

$$\rho = (I_0, I_1, I_2, I_3, I_4),$$
$$\zeta = (I_0)(I_1, I_4)(I_2, I_3),$$

where $I = 1, 2, 3, 4, 5$, and

$$\theta = (1_i, 2_i)(3_i, 4_i)(5_i),$$

where $i = 0, 1, 2, 3, 4$.

We present here two non-isomorphic $\mathscr{D}_1$ and $\mathscr{D}_2$. The 2 base blocks of $\mathscr{D}_i$, $i = 1, 2$, with respect to $\langle \rho \rangle$ are denoted by $l_1, l_6, l_{11}, l_{16}, l_{21}, l_{26}, l_{31}, l_{36}$:

$$\mathscr{D}_1$$

$$(l_1) = 1_{1\,4\,2\,3} \quad 2_0 \quad 3_0 \quad 4_0 \quad 5_{0\,2\,3},$$
$$(l_6) = 2_{1\,4\,2\,3} \quad 1_0 \quad 4_0 \quad 3_0 \quad 5_{0\,2\,3},$$
$$(l_{11}) = 1_0 \quad 2_0 \quad 3_{1\,4\,2\,3} \quad 4_0 \quad 5_{0\,1\,4},$$
$$(l_{16}) = 2_0 \quad 1_0 \quad 4_{1\,4\,2\,3} \quad 3_0 \quad 5_{0\,1\,4},$$
$$(l_{21}) = 1_{2\,3} \quad 2_{0\,2\,3} \quad 3_{1\,4} \quad 4_{2\,3} \quad 5_0,$$
$$(l_{26}) = 2_{2\,3} \quad 1_{0\,2\,3} \quad 4_{1\,4} \quad 3_{2\,3} \quad 5_0,$$
$$(l_{31}) = 1_{1\,4} \quad 2_{2\,3} \quad 3_{0\,1\,4} \quad 4_{1\,4} \quad 5_0,$$
$$(l_{36}) = 2_{1\,4} \quad 1_{2\,3} \quad 4_{0\,1\,4} \quad 3_{1\,4} \quad 5_0.$$

In order to show that $\mathscr{D}_1$ is nonembeddable we look at the blocks l_{21}, l_{26}, l_{31}, l_{36}, and see that $[l_i, l_j] = 5$, for $i, j = 21, 26, 31, 36$, $i \neq j$. Using the inequality $x(2\lambda + 1 - x) \leq 2(k + \lambda)$ for a necessary condition of an extendable quasi-residual design of Lemma 2.1 we see that $\mathscr{D}_1$ is not embeddable.

$$\mathscr{D}_2$$

$$(l_1) = 1_{0\,2\,3} \quad 2_0 \quad 3_0 \quad 4_{1\,4} \quad 5_{0\,1\,4},$$
$$(l_6) = 2_{0\,2\,3} \quad 1_0 \quad 4_0 \quad 3_{1\,4} \quad 5_{0\,1\,4},$$
$$(l_{11}) = 1_{2\,3} \quad 2_0 \quad 3_0 \quad 4_{0\,1\,4} \quad 5_{0\,2\,3},$$
$$(l_{16}) = 2_{2\,3} \quad 1_0 \quad 4_0 \quad 3_{0\,1\,4} \quad 5_{0\,2\,3},$$
$$(l_{21}) = 1_{1\,4\,2\,3} \quad 2_{2\,3} \quad 3_{2\,3} \quad 4_0 \quad 5_0,$$
$$(l_{26}) = 2_{1\,4\,2\,3} \quad 1_{2\,3} \quad 4_{2\,3} \quad 3_0 \quad 5_0,$$
$$(l_{31}) = 1_{1\,4} \quad 2_0 \quad 3_{1\,4\,2\,3} \quad 4_{1\,4} \quad 5_0,$$
$$(l_{36}) = 2_{1\,4} \quad 1_0 \quad 4_{1\,4\,2\,3} \quad 3_{1\,4} \quad 5_0.$$

We have that $[l_1, l_{11}] = [l_6, l_{16}] = 7$, so $\mathscr{D}_2$ is of Bhattacharya type and therefore nonembeddable.

Furthermore the designs $\mathscr{D}_1$ and $\mathscr{D}_2$ are not isomorphic with the design given in [21]. Hence applying Corollary 2.2 and 2.3 we have the following result:

THEOREM 4.8. *There exist at least three infinite families of quasi-residual designs with parameters*

$$2 - (25^m, \frac{5}{3}(25^m - 1), \frac{2}{3}(25^m - 1), 10.25^{m-1}, \frac{2}{3}(10.25^{m-1} - 1))$$

for $m \geq 1$; one of them is of Bhattacharya type and hence nonembeddable, the other two have the first member nonembeddable.

We conjecture that the designs in the families of non-Bhattacharya type are allnonembeddable as well.

4.9 A nonembeddable quasi-residual design for 2-(55, 99, 45, 25, 20). In this section we construct a nonembeddable quasi-residual design $\mathscr{D}$ with parameters $2 - (55, 99, 45, 25, 20)$ having an automorphism group $H = F \times G$, where $F = \langle \rho, \zeta \mid \rho^{11} = 1, \zeta^5 = 1, \zeta^{-1}\rho\zeta = \rho^3 \rangle$ is a Frobenius group of order 55 and $G = \langle \theta \mid \theta^2 = 1 \rangle$ is a cyclic group of order 2. Moreover ρ should act fixed-point-free on $\mathscr{D}$, ζ has 5 fixed points and θ has 11 fixed points.

Denoting by 1, 2, 3, 4, 5 five point orbits of ρ, we may put the action of ρ, ζ and θ on the points of $\mathscr{D}$ as follows:

$$\rho = (I_0, I_1, I_2, I_3, I_4, I_5, I_6, I_7, I_8, I_9, I_{10}),$$
$$\zeta = (I_0)(I, I_3, I_9, I_5, I_4)(I_2, I_6, I_7, I_{10}, I_8),$$

where $I = 1, 2, 3, 4, 5$, and

$$\theta = (1_i)(2_i, 5_i)(3_i, 4_i),$$

where $i = 0, 1, 2, 3, 4, 5, 6, 7, 8, 9, 10$.

Below are 9 base blocks of $\mathscr{D}$ with respect to $\langle \rho \rangle$:

$$(l_1) = 1_{2\,6\,7\,10\,8}\quad 2_{1\,3\,9\,5\,4}\quad 3_{2\,6\,7\,10\,8}\quad 4_{1\,3\,9\,5\,4}\quad 5_{1\,3\,9\,5\,4},$$
$$(l_2) = 1_{2\,6\,7\,10\,8}\quad 5_{1\,3\,9\,5\,4}\quad 4_{2\,6\,7\,10\,8}\quad 3_{1\,3\,9\,5\,4}\quad 2_{1\,3\,9\,5\,4},$$
$$(l_3) = 1_{2\,6\,7\,10\,8}\quad 2_{1\,3\,9\,5\,4}\quad 3_{1\,3\,9\,5\,4}\quad 4_{1\,3\,9\,5\,4}\quad 5_{2\,6\,7\,10\,8},$$
$$(l_4) = 1_{2\,6\,7\,10\,8}\quad 5_{1\,3\,9\,5\,4}\quad 4_{1\,3\,9\,5\,4}\quad 3_{1\,3\,9\,5\,4}\quad 2_{2\,6\,7\,10\,8},$$
$$(l_5) = 1_{0\,2\,6\,7\,10\,8}\quad 2_{0\,1\,3\,9\,5\,4}\quad 3_0\quad 4_{0\,1\,3\,9\,5\,4}\quad 5_{0\,2\,6\,7\,10\,8},$$
$$(l_6) = 1_{0\,2\,6\,7\,10\,8}\quad 5_{0\,1\,3\,9\,5\,4}\quad 4_0\quad 3_{0\,1\,3\,9\,5\,4}\quad 2_{0\,2\,6\,7\,10\,8},$$
$$(l_7) = 1_{0\,2\,6\,7\,10\,8}\quad 2_{0\,2\,6\,7\,10\,8}\quad 3_{0\,2\,6\,7\,10\,8}\quad 4_{0\,1\,3\,9\,5\,4}\quad 5_0,$$
$$(l_8) = 1_{0\,2\,6\,7\,10\,8}\quad 5_{0\,2\,6\,7\,10\,8}\quad 4_{0\,2\,6\,7\,10\,8}\quad 3_{0\,1\,3\,9\,5\,4}\quad 2_0,$$
$$(l_9) = 1_0\quad 2_{0\,1\,3\,9\,5\,4}\quad 5_{0\,1\,3\,9\,5\,4}\quad 3_{0\,2\,6\,7\,10\,8}\quad 4_{0\,2\,6\,7\,10\,8}.$$

We now show that $\mathscr{D}$ is not embeddable in a $2 - (100, 45, 20)$ symmetric design by looking at the blocks l_4, l_6 and l_7 of $\mathscr{D}$. We have $[l_4, l_6] = 20$, $[l_4, l_7] = 15$, $[l_6, l_7] = 15$. Let S be a set of 45 new points. As $[l_4, l_6] = 20$ we have $l_4^* = l_4 \cup X$ and $l_6^* = l_6 \cup Y$, where $X, Y \subset S$, $|X| = |Y| = 20$ and $X \cap Y = \varnothing$. Further $[l_4, l_7] = 15$ and $[l_6, l_7] = 15$ imply that an extended block l_7^* has 5 new points from X and 5 new points from Y. Therefore l_7^* must contain 10 new points from $S \setminus X \cup Y$, which is a contradiction since $|S \setminus X \cup Y| = 5$. Hence $\mathscr{D}$ is not embeddable and we have

THEOREM 4.9. *There exists a* $2 - (55, 99, 45, 25, 20)$ *nonembeddable quasi-residual design.*

4.10 Two families of quasi-residual designs of Bhattacharya type. In this section we first construct a series of quasi-residual designs of Bhattacharya type withparameters $2 - ((q + 1)^2, 2q(q + 1), q^2, q(q + 1)/2, q(q - 1)/2)$, where $q \equiv 3 \pmod 4$ is a prime power, using incidence matrices of Hadamard designs of Paley type. Each design in this family has q^2 pairs of blocks having $(q^2 - 1)/2$ common points. We should remark that the first member of this series has parameters $2 - (16, 24, 9, 6, 3)$, for which Bhattacharya constructed for the first time a nonembeddable quasi-residual design. Then applying Corollary 2.2 and 2.3 to some special values of q we obtain another new family of nonembeddable quasi-residual designs of Bhattacharya type.

Let $\mathscr{H}_S(q)$ (resp. $\mathscr{H}_{NS}(q)$) be the Hadamard design obtained by taking the set of nonzero squares (resp. the set of nonsquares) of the Galois field $GF(q)$ as a base block (we recall that a Hadamard design is a $2 - (x, (x-1)/2, (x-3)/4)$ symmetric design).

Let us describe some notations used in the following construction. ¡

$\quad H_S(q)$: The incidence matrix of $\mathscr{H}_S(q)$.

$\quad H_{NS}(q)$: The incidence matrix of $\mathscr{H}_{NS}(q)$.

$\quad CH_S(q)$: The incidence matrix of the complement of $\mathscr{H}_S(q)$.

$\quad CH_{NS}(q)$: The incidence matrix of the complement of $\mathscr{H}_{NS}(q)$.

$\quad R(q)$: The 1 by q matrix with all its entries 1.

$\quad R(q)^T$: The transpose of $R(q)$.

$\quad J(q)$: The q by q matrix with all its entries 1.

$\quad X \otimes Y$: The Kronecker product of two matrices $X = (x_{ij})$ and $Y = (y_{kl})$, which is defined as follows: $X \otimes Y = (x_{ij}Y)$.

Below is the incidence matrix of a $2 - ((q+1)^2, 2q(q+1), q^2, q(q+1)/2, q(q-1)/2)$ quasi-residual design of Bhattacharya type, where $q \equiv 3 \pmod 4$ is a prime power. It is fairly easy to check that the above matrix is an incidence matrix of a 2-design. Moreover, each $\mathscr{H}_S(q)$ or $\mathscr{H}_{NS}(q)$ (and hence $\mathscr{CH}_S(q)$ or $\mathscr{CH}_{NS}(q)$) has an automorphism group A, which is a semidirect product of an elementary abelian group of order q with a cyclic group of order $(q-1)/2$. Further the Kronecker product in Diagram 4.10.1 implies that the corresponding design also admits an elementary abelian group B of order q. Thus the obtained design has $A \times B$ as an automorphism group. It is interesting to determine the full automorphism group of this design. For more information about the full automorphism group of $\mathscr{H}_S(q)$ see [12].

We have

THEOREM 4.10.1. *Let* $q \equiv 3 \pmod 4$ *be a prime power, then there exists an infinite family of quasi-residual of Bhattacharya type with parameters*

$$2 - ((q+1)^2, 2q(q+1), q^2, q(q+1)/2, q(q-1)/2),$$

in which each design has q^2 *pairs of blocks meeting in* $\Lambda = (q^2-1)/2$ *points.*

For illustration we explicitly give this incidence matrix for $q = 3$. In this

$R(q) \otimes R(q)$	0	0	0
$R(q) \otimes H_{NS}(q)$	$R(q) \otimes CH_{NS}(q)$	0	0
$R(q) \otimes H_{NS}(q)$	$R(q) \otimes H_{NS}(q)$	$J(q)$	0
$H_S(q) \otimes CH_{NS}(q)$ $\cup$ $H_{NS}(q) \otimes H_{NS}(q)$	$H_S(q) \otimes CH_{NS}(q)$ $\cup$ $H_{NS}(q) \otimes H_{NS}(q)$	$R(q)^T$ $\otimes$ $H_S(q)$	Rq^T $\otimes$ $CH_{NS}(q)$

DIAGRAM 4.10.1

case the parameters become $2 - (16, 24, 9, 6, 3)$. We have

$$H_S(3) = \begin{pmatrix} 0 & 0 & 1 \\ 1 & 0 & 0 \\ 0 & 1 & 0 \end{pmatrix}, \qquad H_{NS}(3) = \begin{pmatrix} 0 & 1 & 0 \\ 0 & 0 & 1 \\ 1 & 0 & 0 \end{pmatrix},$$

$$CH_S(3) = \begin{pmatrix} 1 & 1 & 0 \\ 0 & 1 & 1 \\ 1 & 0 & 1 \end{pmatrix}, \qquad CH_{NS}(3) = \begin{pmatrix} 1 & 0 & 1 \\ 1 & 1 & 0 \\ 0 & 1 & 1 \end{pmatrix}$$

Below is the incidence matrix of a $2 - (16, 24, 9, 6, 3)$ quasi-residual design of Bhattacharya type from the above construction, which has 9 pairs of blocks meeting in 4 points.

1 1 1 1 1 1 1 1 1	0 0 0 0 0 0 0 0 0	0 0 0	0 0 0
0 1 0 0 1 0 0 1 0 0 0 1 0 0 1 0 0 1 1 0 0 1 0 0 1 0 0	1 0 1 1 0 1 1 0 1 1 1 0 1 1 0 1 1 0 0 1 1 0 1 1 0 1 1	0 0 0 0 0 0 0 0 0	0 0 0 0 0 0 0 0 0
0 1 0 0 1 0 0 1 0 0 0 1 0 0 1 0 0 1 1 0 0 1 0 0 1 0 0	0 1 0 0 1 0 0 1 0 0 0 1 0 0 1 0 0 1 1 0 0 1 0 0 1 0 0	1 1 1 1 1 1 1 1 1	0 0 0 0 0 0 0 0 0
0 0 0 0 1 0 1 0 1 0 0 0 0 0 1 1 1 0 0 0 0 1 0 0 0 1 1	0 0 0 0 1 0 1 0 1 0 0 0 0 0 1 1 1 0 0 0 0 1 0 0 0 1 1	0 0 1 1 0 0 0 1 0	1 0 1 1 1 0 0 1 1
1 0 1 0 0 0 0 1 0 1 1 0 0 0 0 0 0 1 0 1 1 0 0 0 1 0 0	1 0 1 0 0 0 0 1 0 1 1 0 0 0 0 0 0 1 0 1 1 0 0 0 1 0 0	0 0 1 1 0 0 0 1 0	1 0 1 1 1 0 0 1 1
0 1 0 1 0 1 0 0 0 0 0 1 1 1 0 0 0 0 1 0 0 0 1 1 0 0 0	0 1 0 1 0 1 0 0 0 0 0 1 1 1 0 0 0 0 1 0 0 0 1 1 0 0 0	0 0 1 1 0 0 0 1 0	1 0 1 1 1 0 0 1 1

DIAGRAM 4.10.2

The nine pairs of blocks corresponding to the column pairs i and $i + 9$, where $i = 1, 2, \ldots, 9$, in Diagram 4.10.2, have 4 common points.

Some further examples for small q of Theorem 4.10.1.

For $q = 7$ we have a $2 - (64, 112, 49, 28, 21)$ design with $\Lambda = 24$, and for $q = 11$, a $2 - (144, 264, 121, 66, 55)$ design with $\Lambda = 60$.

Now if q is a Mersenne prime then $(q + 1)^2$ is a two power, so we can use Corollary 2.2 and 2.3 to obtain a new family of nonembeddable quasi-residual designs of Bhattacharya type from Theorem 4.10.1.

We have:

THEOREM 4.10.2. *If q is a Mersenne prime, then there exists an infinite family of quasi-residual of Bhattacharya type with parameters*

$$2 - ((q + 1)^{2m}, 2(q + 1)(q + 1)^{2m} - 1)/(q + 2), q((q + 1)^{2m} - 1)/(q + 2),$$
$$(q + 1)^{2m-1}/2, q((q + 1)^{2m-2} - 1)/(q + 2) + q(q - 1)(q + 1)^{2m-2}/2),$$

in which each design has q^2 pairs of blocks meeting in $\Lambda = (q + 1)^{2m-1}$ $(q - 1)/2$ points.

For instance, if $q = 3$ and $m = 2$ then we have a $2 - (256, 408, 153, 96, 57)$ quasi-residual design of Bhattacharya type with $\lambda = 64$.

4.11 Some further families of quasi-residual designs. In this section we construct some series of quasi-residual designs with parameters $2 - ((q)^2, 2q(q - 1), (q - 1)^2, q(q - 1)/2, (q - 1)(q - 2)/2)$, where $q \equiv 1 \pmod{4}$ and $p = q - 2 \equiv 3 \pmod{4}$ are prime powers, using incidence matrices of Hadamard designs of Paley type and incidence matrices of other structures defined from the Galois field $GF(q)$.

Let $\mathcal{M}_S(q)$ (resp. $\mathcal{M}_{N}S(q)$) be the incidence structure obtained by taking the set of nonzero squares (resp. the set of nonsquares) of the Galois field $GF(q)$ as a base block for $q \equiv 1 \pmod{4}$.

The following notations are used for the construction. ¡

$M_S(q)$: The incidence matrix of $\mathcal{M}_S(q)$.
$M_{NS}(q)$: The incidence matrix of $\mathcal{M}_{NS}(q)$.
$CM_S(q)$: The incidence matrix of the complement of $\mathcal{M}_S(q)$.
$CM_{NS}(q)$: The incidence matrix of the complement of $\mathcal{M}_{NS}(q)$.
$I(q)$: The identity matrix of order q.

We give here the incidence matrix of a $2 - ((q)^2, 2q(q - 1), (q - 1)^2, q(q - 1)/2, (q - 1)(q - 2)/2)$ quasi-residual design, where $q \equiv 1 \pmod{4}$ and $p = q - 2 \equiv 3 \pmod{4}$ are both prime powers.

For example, if $q = 5$ and $p = q - 2 = 3$ we have a $2 - (25, 40, 16, 10, 6)$ quasi-residual design $\mathcal{B}_1$, the first member of this family; in this case we have

$$M_S(5) = \begin{pmatrix} 0 & 1 & 0 & 0 & 1 \\ 1 & 0 & 1 & 0 & 0 \\ 0 & 1 & 0 & 1 & 0 \\ 0 & 0 & 1 & 0 & 1 \\ 1 & 0 & 0 & 1 & 0 \end{pmatrix}, \quad M_{NS}(5) = \begin{pmatrix} 0 & 0 & 1 & 1 & 0 \\ 0 & 0 & 0 & 1 & 1 \\ 1 & 0 & 0 & 0 & 1 \\ 1 & 1 & 0 & 0 & 0 \\ 0 & 1 & 1 & 0 & 0 \end{pmatrix},$$

$R(p) \otimes M_S(q)$	$R(p) \otimes M_{NS}(q)$	0	$J(q)-$ $I(q)$
$R(p) \otimes CM_{NS}(q)$	$R(p) \otimes M_{NS}(q)$	$I(q)$	0
$H_S(p) \otimes CM_{NS}(q)$ $\cup$ $H_{NS}(p) \otimes M_{NS}(q)$	$I(p) \otimes I(q)$ $\cup$ $H_S(p) \otimes M_S(q)$ $\cup$ $H_{NS}(q) \otimes CM_{NS}(q)$	$R(p)^T$ $\otimes$ $CM_S(q)$	$R(p)^T$ $\otimes$ $M_{NS}(q)$

DIAGRAM 4.11.1

$$CM_S(5) = \begin{pmatrix} 1 & 0 & 1 & 1 & 0 \\ 0 & 1 & 0 & 1 & 1 \\ 1 & 0 & 1 & 0 & 1 \\ 1 & 1 & 0 & 1 & 0 \\ 0 & 1 & 1 & 0 & 1 \end{pmatrix}, \quad CM_{NS}(5) = \begin{pmatrix} 1 & 1 & 0 & 0 & 1 \\ 1 & 1 & 1 & 0 & 0 \\ 0 & 1 & 1 & 1 & 0 \\ 0 & 0 & 1 & 1 & 1 \\ 1 & 0 & 0 & 1 & 1 \end{pmatrix}.$$

Diagram 4.11.2 presents the incidence matrix of $\mathscr{B}_1$.

We do not know whether the designs in the family of Diagram 4.11.1 are embeddable or not, but we shall show that its first member, namely $\mathscr{B}_1$, is nonembeddable. We should remark that $\mathscr{B}_1$ is not isomorphic to one of the designs with the same parameters in Theorem 4.8 and [21].

We denote by (i) the block of $\mathscr{B}_1$ corresponding to the i-column in the Diagram 4.11.2 for $i = 1, 2, \ldots, 40$. Let $1, 2, \ldots, 16$ be a set of 16 new points. As before $(i)^*$ denotes an extended block of (i). We have $[(1), (6)] = [(1), (11)] = [(6), (11)] = 5$, so we can choose

$$(1)^* = (1) \cup 1\ 2\ 3\ 4\ 5\ 6,$$
$$(6)^* = (6) \cup 1\ 7\ 8\ 9\ 10\ 11.$$

Suppose $(11)^* = (11) \cup 1\ 12\ 13\ 14\ 15\ 16$. As $[(33), (1)] = [(33), (6)] = [(33), (11)] = 3$ we see that it is impossible to get an extended block $(33)^*$ compatible with $(1)^*$, $(6)^*$ and $(11)^*$. Hence we may put $(11)^* = (11) \cup 2\ 7\ 12\ 13\ 14\ 15$.

We observe that if a block (i) of $\mathscr{B}_1$ with $[(i), (1)] = [(i), (6)] = [(i), (11)] = 3$ then $(i)^*$ must contain the new points 1, 2 and 7. As $[(34), (1)] = [(34), (6)] = [(34), (11)] = [(34), (33)] = 3$, so from this observation we may put

$$(33)^* = (33) \cup 1\ 2\ 7\ 5\ 10\ 14,$$
$$(34)^* = (34) \cup 1\ 2\ 7\ 6\ 11\ 15.$$

Furthermore $[(i), (j)] = 3$ for i=37, 40 and j=1, 6, 11, 33, 34. Hence we may choose

$$(37)^* = (37) \cup 1\ 2\ 7\ 3\ 8\ 12,$$
$$(40)^* = (40) \cup 1\ 2\ 7\ 4\ 9\ 13.$$

01001	01001	01001	00110	00110	00110	00000	01111
10100	10100	10100	00011	00011	00011	00000	01111
01010	01010	01010	10001	10001	10001	00000	11011
00101	00101	00101	11000	11000	11000	00000	11101
10010	100010	10010	01100	01100	01100	00000	11110
11001	11001	11001	01001	01001	01001	10000	00000
11100	11100	11100	10100	10100	10100	01000	00000
01110	01110	01110	01010	01010	01010	00100	00000
00111	00111	00111	00101	00101	00101	00010	00000
10011	10011	10011	10010	10010	10010	00001	00000
00000	00110	01001	10000	10110	01001	11001	00110
00000	00011	10100	01000	01011	10100	11100	00011
00000	10001	01010	00100	10101	01010	01110	10001
00000	11000	00101	00010	11010	00101	00111	11000
00000	01100	10010	00001	01101	10010	10011	01100
01001	00000	00110	01001	10000	10110	11001	00110
10100	00000	00011	10100	01000	01011	11100	00011
01010	00000	10001	01010	00100	10101	01110	10001
00101	00000	11000	00101	00010	11010	00111	11000
10010	00000	01100	10010	00001	01101	10011	01100
00110	01001	00000	10110	01001	10000	11001	00110
00011	10100	00000	01011	10100	01000	11100	00011
10001	01010	00000	10101	01010	00100	01110	10001
11000	00101	00000	11010	00101	00010	00111	11000
01100	10010	00000	01101	10010	00001	10011	01100

DIAGRAM 4.11.2

We have $[(16), (1)] = 6$, $[(16), (6)] = [(16), (11)] = 3$, $[(21), (6)] = 6$, $[(21), (1)] = [(21), (11)] = 3$, $[(26), (11)] = 6$, $[(26), (1)] = [(26), (6)] = 3$, $[(16), (21)] = [(16), (26)] = [(21), (26)] = 5$. Further $[(i), (j)] = 4$ for $i = 16, 21, 26$ and $j = 33, 34, 37, 40$, implies that one of $1, 2, 7$ must be in $(16)^*$, $(21)^*$ and $(26)^*$. But comparing these with $(1)^*$, $(6)^*$, $(11)^*$ we see that $(16)^*$, $(21)^*$, $(26)^*$ must contain the new point 16. Hence by permuting the points in the sets $\{3, 4, 5, 6\}$, $\{8, 9, 10, 11\}$, $\{12, 13, 14, 15\}$, if necessary we may choose

$$(16)^* = (16) \cup 7\ 8\ 9\ 14\ 15\ 16,$$
$$(21)^* = (21) \cup 2\ 5\ 6\ 12\ 13\ 16,$$
$$(26)^* = (26) \cup 1\ 3\ 4\ 10\ 11\ 16.$$

Now look at the block (29). From $[(29), (16)] = [(29), (21)] = 3$ and $[(29), (26)] = 4$ it follows that $16 \in (29)^*$. Moreover $[(29), (i)] = 4$ for $i = 33, 34, 37, 40$, shows that $(29)^*$ contains exactly one of the point set $\{1, 2, 7\}$. We show that $1, 7 \notin (29)^*$.

Suppose $(29)^* = (29) \cup 1****16$. Since $[(29), (6)] = 5$, so $8, 9, 10, 11 \notin (29)^*$, and $[(29), (16)] = 3$ gives $14, 15 \in (29)^*$ and $12, 13 \notin (29)^*$. Further $[(29), (21)] = 3$ forces that $5, 6 \in (29)^*$, thus we have $(29)^* = (29) \cup 1\ 5\ 6\ 14\ 15\ 16$, which contradicts $[(29), (33)] = 4$.

Suppose $(29)^* = (29) \cup 7****16$, then $[(29), (6)] = 5$ shows that $8, 9, 10, 11 \notin (29)^*$; as $[(29), (26)] = 4$, at most one of $\{3, 4\}$ is in $(29)^*$. This fact together with $[(29), (1)] = 3$ imply that $5, 6 \in (29)^*$. On the other hand $[(29), (16)] = 3$ shows that 14 or 15 must appear in $(29)^*$, thus we have $(29)^* = (29) \cup 7\ a\ b\ 5\ 6\ 16$, where $a \in \{3, 4\}$ and $b \in \{14, 15\}$, which contradicts $[(29), (33)] = [(29), (34)] = 4$.

Hence we have $(29)^* = (29) \cup 2****16$. We see that both of 5 and 6 cannot occur in $(29)^*$ since $[(29), (21)] = 3$. If $5, 6 \notin (29)^*$ then $3, 4 \in (29)^*$ since $[(29), (1)] = 3$, which contradicts $[(29), (26)] = 4$. Thus we have shown that 5 or 6 must be in $(29)^*$. This fact together with $[(29), (1)] = 3$ yield that 3 or 4 must appear in $(29)^*$. Further $[(29), (26)] = 4$ and 3 or $4 \in (29)^*$ show that $10, 11 \notin (29)^*$. As $[(29), (6)] = 5$, either 8 or 9 must appear in $(29)^*$. Moreover $[(29), (21)] = 3$ gives $12, 13 \notin (29)^*$ and therefore 14 or $15 \in (29)^*$, as $[(29), (11)] = 4$. Hence we have

$$(29)^* = (29) \cup 2\ \heartsuit\ \diamondsuit\ \clubsuit\ \spadesuit 16,$$

where $\heartsuit \in \{3, 4\}$, $\diamondsuit \in \{5, 6\}$, $\clubsuit \in \{8, 9\}$ and $\spadesuit \in \{14, 15\}$.

Now consider the block (30). We have $[(30), (26)] = [(39), (29)] = 2$ and $[(39), (i)] = 4$ for $i = 1, 6, 11, 16, 21, 33, 34, 37, 40$. By the same argument for the block $(29)^*$ we see that $16 \in (30)^*$ and exactly one of $\{1, 2, 7\}$ must appear in $(30)^*$.

Suppose $(30)^* = (30) \cup 1****16$. As $[(30), (26)] = 2$, exactly two of $\{3, 4, 10, 11\}$ must be in $(30)^*$; we remark that both of 3 and 4 cannot occur in $(30)^*$ since $[(30), (1)] = 4$; if $3, 4 \notin (30)^*$ then $10, 11 \in (30)^*$, which is impossible since $[(30), (6)] = 2$. Thus 3 or 4 (resp. 10 or 11) must be in $(30)^*$. It follows that $5, 6 \notin (30)^*$. Further $[(30), (21)] = 4$ (resp. $[(30), (11)] = 4$) implies that 12 or 13 (resp. 14 or 15) must belong to $(30)^*$. Thus we have $(30)^* = (30) \cup 1\ a\ b\ c\ d\ 16$, where $a \in \{3, 4\}$, $b \in \{10, 11\}$, $c \in \{12, 13\}$, and $d \in \{14, 15\}$, which contradicts $[(30), (29)] = 2$.

Suppose $(30)^* = (30) \cup 2****16$. First we have $5, 6, 12, 13 \notin (30)^*$ since $[(30), (21)] = 4$. As $[(30), (1)] = 4$, so 3 or 4 is in $(30)^*$; this together with $[(30), (26)] = 2$ show $10, 11 \in (30)^*$. Further $[(30), (11)] = 4$ says that 14 or 15 is in $(30)^*$. Thus we have $(30)^* = (30) \cup 1\ a\ b\ 10\ 11\ 16$, where $a \in \{3, 4\}$, $b \in \{14, 15\}$, which contradicts $[(30), (33)] = [(30), (34)] = 4$. Hence we have $(30)^* = (30) \cup 7****16$. As $[(30), (26)] = 2$ we see that three of $\{3, 4, 10, 11\}$ must be in $(30)^*$. This fact together with $[(30), (6)] = 4$ mean that 10 or 11 belongs to $(30)^*$, and therefore $3, 4 \in (30)^*$ and $5, 6 \notin (30)^*$. Further 12 or 13 must be in $(30)^*$ since $[(30), (21)] = 4$. Thus we have $(30)^* = (30) \cup 7\ 3\ 4\ a\ b\ 16$, where

$a \in \{10, 11\}$, $b \in \{12, 13\}$, which is impossible since $[(30), (37)] = [(30), (40)] = 4$. This final contradiction shows that $\mathscr{B}_1$ is nonembeddable.

Now applying Corollary 2.2 and 2.3 to the family of quasi-residual designs given in Diagram 4.11.1 we get another new family. Hence we have

THEOREM 4.11.1. *If* $q \equiv 1 \pmod 4$ *and* $p = q - 2 \equiv 3 \pmod 4$ *are prime powers, then there is a quasi-residual design with the following parameters:*

(i) $2 - ((q)^2, 2q(q - 1), (q - 1)^2, q(q - 1)/2, (q - 1)(q - 2)/2,$

(ii)
$$2 - (q^{2m}, 2q(q^{2m} - 1)/(q + 1), (q^{2m} - 1)(q - 1)/(q + 1), q^{2m-1}(q - 1)/2,$$
$$q^{2m-1}(q - 1)^2/2(q + 1)).$$

Moreover the first member of (i) *is nonembeddable.*

We should remark here that the existence of symmetric designs corresponding to (i) and (ii) is still in doubt except for $q = 5$ in (i) (see [20]).

The Diagram 4.11.3 gives the incidence matrix of a quasi-residual design with parameters as in Theorem 4.11.1. (i). This design is not isomorphic to one in (i) since the former has more pairs of blocks meeting in $(q-1)(q-2)/2$ points than the latter. Hence there are at least two non-isomorphic families quasi-residual designs for each parameter set (i) and (ii) of Theorem 4.11.1.

$R(p) \otimes M_S(q)$	$R(p) \otimes M_{NS}(q)$	0	$J(q)-$ $I(q)$
$R(p) \otimes CM_{NS}(q)$	$R(p) \otimes M_{NS}(q)$	$I(q)$	0
$H_S(p) \otimes CM_{NS}(q)$ $\cup$ $H_{NS}(p) \otimes CM_{NS}(q)$	$I(p) \otimes I(q)$ $\cup$ $H_S(p) \otimes M_S(q)$ $\cup$ $H_{NS}(p) \otimes CM_{NS}(q)$	$R(p)^T$ $\otimes$ $CM_S(q)$	$R(p)^T$ $\otimes$ $M_{NS}(q)$

DIAGRAM 4.11.3

We conjecture that the designs in Theorem 4.11.1 are all nonembeddable.

Appendix A

The following table presents parameters $2 - (v, b, r, k, \lambda)$ of known nonembeddable quasi-residual designs with $k < \frac{1}{2}v$.

$2-(v,b,r,k,\lambda)$	REFS.
1. $2-(16,24,9,6,3)$	[2,4,13,14], Th. 4.10.1
2. $2-(21,35,15,9,6)$	Th. 4.2
3. $2-(25,40,16,10,6)$	[21]Th. 4.8
4. $2-(27,39,13,9,4)$	Th. 4.1
5. $2-(33,48,16,11,5)$	Th. 4.4
6. $2-(36,63,28,16,12)$	[21]
7. $2-(40,65,26,16,10)$	Th. 4.5
8. $2-(49,84,36,21,15)$	Th. 4.3
9. $2-(52,78,27,18,9)$	Th. 4.6
10. $2-(55,99,45,25,20)$	Th. 4.9
11. $2-(64,112,49,28,21)$	Th. 4.10.1
12. $2-(78,143,66,36,30)$	Th. 4.7

13. $2-(25^m, \frac{5}{3}(25^m-1), \frac{2}{3}(25^m-1),$
    $10.25^{m-1}, \frac{2}{3}(10.25^{m-1}-1)),$
    $m \geq 1$ Th. 4.8

14. $2-(49^m, \frac{7}{4}(49^m-1), \frac{3}{4}(49^m-1),$
    $21.49^{m-1}, \frac{3}{4}(21.49^{m-1}-1)),$
    $m \geq 1$ Th. 4.3

15. $2-((q+1)^2, 2q(q+1), q^2,$
    $q(q+1)/2, q(q-1)/2),$
    $q \equiv 3 \pmod 4$, a prime power   Th. 4.10.1

16. $2-((q+1)^{2m}, \frac{2(q+1)}{(q+2)}$
    $((q+1)^{2m}-1), \frac{q}{(q+2)}$
    $((q+1)^{2m}-1), \frac{q}{2}(q+1)^{2m-1},$
    $\frac{q}{(q+2)}((q+1)^{2m-2}-1)$
    $+\frac{q(q-1)}{2}(q+1)^{2m-2}),$
    q, a Mersenne prime, $m \geq 1$ Th.4.10.2

Appendix B

The following table lists parameters $2-(v,b,r,k,\lambda)$ of nonembeddable quasi-residual designs with $k \geq \frac{1}{2}v$ constructed in this paper.

$2-(v,b,r,k,\lambda)$	REFS.

1. $2-(q^3+1, q^2(q^2-q+1), q^2(q-1)^2,$
   $(q+1)(q-1)^2, (q-1)^2(q^2-q-1)),$
   $q=9$, or q, a Fermat prime, or $(q-1) \neq 2, 7$,
        a Mersenne prime   Th. 3.1

2. $2-(2^{2n-1}-2^{n-1}, (2^{2n}-1)(2^{n-1}-1)/3,$
   $(2^n+1)(2^{n-1}-1)(2^n-4)/3, 2^{n+1}(2^{n-2}-1),$
       $4(2^{3n-3}-2^{2n}+7.2^{n-2}+1)/3)),$
   $n \geq 3$, n even, or n odd and $(2^{n-2}-1)$,
       a nonsquare   Cor. 3.1

3. $2-(2^{2n-2}-2^n+2^{n-2}, (2^n+1)(2^n-3)$
   $(2^{n-2}-1)/5, (2^n+1)(2^{n-2}-1)(2^n-8)/5,$
      $2^{n+1}(2^{n-3}-1), (2^{n-2}-1)$
      $(2^{2n}-12.2^n-8)/5+2^n),$
       $n \equiv 2$ or $3 \pmod 4 \geq 3$, n even,
   or n odd and $(2^{n-3}-1)$, a nonsquare   Cor. 3.1

4. $2 - (31^m, 93.(31^m - 1)/32, 63.(31^m - 1)/32,$
   $21.31^{m-1}, 63.(31^{m-1} - 1)/32 + 42.31^{m-1}),$
   $m \equiv 1 \pmod 2$ Th. 3.4

5. $2 - (49^m, 147.(49^m - 1)/48, 99.(49^m - 1)/48,$
   $33.49^{m-1}, 99.(49^{m-1} - 1)/48 + 66.49^{m-1}),$
   $m \equiv 1 \pmod 2$ Th. 3.4

6. $2 - (37^m, 333.(37^m - 1)/36, 297.(37^m - 1)/36,$
   $33.37^{m-1}, 297.(37^{m-1} - 1)/36$
   $+264.37^{m-1}), m \equiv 1 \pmod 2$ Th. 3.4

7. $2 - (61^m, 915.(61^m - 1)/60, 855.(61^m - 1)/60,$
   $57.61^{m-1}, 855.(61^{m-1} - 1/60 + 798.61^{m-1}),$
   $m \equiv 1 \pmod 2$ Th. 3.4

8. $2 - (151^m, 3775.(151^m - 1)/150,$
   $3625.(151^m - 1)/150, 145.151^{m-1},$
   $3625.(151^{m-1} - 1)/150 + 3480.151^{m-1}),$
   $m \equiv 1 \pmod 2$ Th. 3.4

9. $2 - (109^m, 2943.(109^m - 1)/108,$
   $2835.(109^m - 1)/108, 105.109^{m-1},$
   $2835.(109^{m-1} - 1)/108 + 2730.109^{m-1}),$
   $m \equiv 1 \pmod 2$ Th.3.4

10. $2 - (739^m, 6651.(739^m - 1)/738,$
    $5913.(739^m - 1)/738, 657.739^{m-1},$
    $5913.(739^{m-1} - 1)/738 + 5256.739^{m-1}),$
    $m \equiv 1 \pmod 2$ Th.3.4

REFERENCES

1. Beth, Th., Jungnickel, D., Lenz, H., *Design Theory*, Bibliographisches Institut, Mannheim-Wien-Zürich, 1985.
2. Bhattacharya, K. N., *A new balanced incomplete block design*, Sci. Culture **9** (1944), 508.
3. Bose, R. C., Shrikhande, S. S. and Singhi, N. M., *Edge regular multigraphs and partial geometric designs with an application to the embedding of quasi-residual designs*, Atti dei convegni Lincei, Theorie combinatorie, Acad. Nazionale Lincei, Rome Tomo **1** (1976), 47–81.
4. Brown, R. B., *A new non-extensible* (16, 24, 9, 6, 3) *design*, J. Combin. Theory Ser. A **19** (1975),.
5. Dembowski, H. P., *Finite Geometries*, Springer, Berlin-Heidelberg-New York, 1968.
6. Denniston, R. H.F., *Some maximal arcs in finite projective planes*, J. Combin. Theory **6** (1969), 317–319.
7. Hall, M. and Connor, W. S., *An embedding theorem for balanced incomplete block design*, Canad. J. Math. **6** (1954), 35–41.
8. Hall, M., *Combinatorial Theory*, Blaisdel, Waltham, 1967.
9. Hanani, H., *Balanced incomplete block designs and related designs*, Discr. Math **11** (1975), 255–369.
10. Hughes, D. R. and Piper, F. C., *Design Theory*, Cambridge University Press, Cambridge, 1985.
11. Janko, Z. and Tran van Trung, *Construction of a new symmetric block design for* (78, 22, 6) *with the help of tactical decompositions*, J. Combin. Theory Ser. A **40** (1975), 451–455.
12. Kantor, W. M., *2-Transitive symmetric designs*, Trans. Amer. Math. Soc. **146** (1969), 1–28.
13. Lawless, J. F., *An investigation of Bhattacharya-type designs*, J. Combin. Theory **11** (1971), 139–147.
14. van Lint, J. H., *Non-embeddable quasi-residual designs*, Ned. Akad. Wetensch. Indag. Math. **40** (1978), 269–275.

15. van Lint, J. H. and Tonchev, V. D., *Nonembeddable quasi-residual designs with large k*, J. Combin. Theory Ser. A **37** (1984), 359–362.

16. Mathon, R. and Rosa, A., *Tables of Parameters of BIBDs with r ≤ 41 including Existence, Enumeration, and Resolvability Results*, Ann. Discr. Math. **26** (1985), 375–308.

17. Majumdar, K. N., *On some theorems in combinatorics relating to incomplete block designs*, Ann. Math. Statist. **24** (1953), 377–389.

18. Parker, E. T., *A result in balanced incomplete block designs*, J. Combin. Theory **3** (1967), 283-285.

19. Stanton, R. G. and Sprott, D. A., *Block intersections in balanced incomplete block designs*, Canad. Math. Bull. **7** (1964), 539–548.

20. Tran van Trung, *The existence of symmetric block designs with parameters* $(41, 16, 6)$ *and* $(66, 26, 10)$, J. Combin. Theory Ser. A **33** (1982), 201–204.

21. _____, *Non-embeddable quasi-residual designs with* $k < \frac{1}{2}v$, J. Combin. Theory Ser. A **40** (1986), 133-137.

MATHEMATISCHES INSTITUT, DER UNIVERSITÄT HEIDELBERG, IM NEUENHEIMER FELD 288, 6900 HEIDELBERG, WEST GERMANY

Contemporary Mathematics
Volume **111**, 1990

Finite Planes And Clique Partitions

W. D. WALLIS

ABSTRACT. One can interpret the set of lines of a finite projective plane as a way of decomposing a complete graph into cliques. This view is exploited in order to give bounds on the clique partition numbers of complements of some relatively sparse graphs.

1. Clique partitions. Throughout this paper graphs will be finite and undirected. They have no loops and no multiple edges. The graph G has $v(G)$ vertices and $e(G)$ edges. A complete graph K_n consists of n vertices and all $\frac{1}{2}n(n-1)$ possible edges; a subgraph of G which is a complete graph is called a *clique* of G.

By a *clique partition* of a graph G we mean a set of edge-disjoint cliques which between them contain every edge of G precisely once. The *clique partition number* $cp(G)$ is the smallest possible cardinality of a clique partition of G.

To show that $cp(G)$ is well-defined, it suffices to observe that one can always partition a graph into cliques: simply associate with each edge of G the corresponding K_2. This affords a clique partition of G into $e(G)$ cliques, so $cp(G) \leq e(G)$. But the partition is clearly not minimal unless G is triangle-free.

If G and H are any graphs we write $G - H$ for the graph obtained from G by deleting any edges which are in H. $G - H$ is called the *complement of H in G*; in particular, the *complement of H*, written $\bar{H}$, is defined to be $K_{v(H)} - H$, where the $K_{v(H)}$ has the same vertex-set as H.

The determination of clique partition numbers of complements has been of interest for some time (see [5, 6]). The most difficult cases involve the complements of relatively sparse graphs: for example, $cp(K_n - K_m)$ is completely determined when m is large relative to n. (roughly, $m \geq n/2$; see

1980 *Mathematics Subject Classification* (1985 *Revision*). Primary 05C70; Secondary 05B25.
This research was supported by NSF grant DMS 8601868.
This paper is in final form and no version of it will be submitted for publication elsewhere.

[6] for details), but for small m it is only completely determined for $m = 2$; in other cases only finitely many values have been discovered.

In this paper we give some asymptotic results: if $\{G_n\}$ is a family of graphs and G_n has n vertices, what can we say about the behavior of $cp(G_n)$ as n approaches infinity? Even here, very little is known.

We use the standard notations: $f(n) = o(g(n))$, $f(n) \sim g(n)$ and $f(n) = O(g(n))$ mean respectively that

$$\lim_{n \to \infty} \frac{f(n)}{g(n)} = 0,$$

$$\lim_{n \to \infty} \frac{f(n)}{g(n)} = 1,$$

and

$$\lim_{n \to \infty} \sup \frac{f(n)}{g(n)} \text{ is bounded.}$$

Clearly $cp(G_n) < \frac{1}{2}n^2$ is obvious for any n-vertex graph G_n; but there are few families for which we can even show that $cp(G) = o(n^2)$.

2. Finite planes and clique partitions. One can interpret a balanced incomplete block design with parameters $v, b, r, k, 1$ as a way of partitioning a complete graph into cliques: the graph has v vertices, corresponding to the treatments of the design, and a block B corresponds to a clique with vertex-set B; the partition will have every clique the same size. More generally, a finite linear space is simply a clique partition of a complete graph.

With this interpretation, the Theorem of de Bruijn and Erdös [1] can be stated as follows.

THEOREM 1. *Suppose there is a clique partition of K_n which does not consist of just one clique. Then the partition contains at least n cliques; moreover, if it contains exactly n cliques, then one of the following is true:*

(i) the partition consists of one clique K_{n-1} and $n-1$ cliques K_2: the K_{n-1} contains all vertices except one specified vertex x, and the K_2's correspond to the edges through x;

(ii) $n = t^2 + t + 1$ for some t; the cliques are all K_{t+1}'s and correspond to the lines of a finite projective plane $PG(2, t)$.

As an immediate corollary one sees that, when $n > 2$,

$$cp(K_n - K_2) = n - 1;$$

for a partition into k parts, where $1 \le k < n - 1$, together with the K_2, gives a clique partition of K_n into $k + 1$ parts, and $2 \le k + 1 \le n - 1$, which is impossible, but the value $n - 1$ can be obtained from part (i) of the Theorem. Similarly, part (ii) gives

$$cp(K_{t^2+t+1} - K_{t+1}) = t^2 + t,$$

whenever a plane exists. In this paper we shall exploit this second observation.

One problem is our lack of knowledge of finite planes when t is not a prime power. To get around this, the following (related) results are useful.

LEMMA 1 [4, Chapter 14]. *Suppose α is any real number greater than 0.6. Then there exists an integer S such that there exists a prime p in the range*

$$s \leq p \leq s + s^{\alpha},$$

whenever $s \geq S$.

LEMMA 2 [3]. *Suppose q and p are consecutive primes. Then there is a function $\delta(q)$ such that*

$$p \leq (1 + \delta(q))(q + q^{0.6}),$$

and $\delta(q)$ approaches 0 as p and q together approach infinity.

3. Complements of small graphs. Suppose H is a graph with h vertices. We write G_n for the complement of H in K_n. We shall assume h is relatively small with regard to n: either $h \leq \sqrt{n}$, or $h = o(\sqrt{n})$, or h is constant.

Suppose t is the smallest integer such that $t \geq \sqrt{n}$ and such that there is a $PG(2, t)$; by Lemma 1 we can assume that for sufficiently large n

$$\sqrt{n} \leq t \leq \sqrt{n} + n^{.31},$$

whence

$$n \leq t^2 \leq n + o(n).$$

Since G_n is an induced subgraph of $K_{t^2+h} - H$, clearly

$$cp(G_n) < cp(K_{t^2+h} - H).$$

Select a $PG(2, t)$, and construct a copy of K_{t^2+t+1} with a distinguished subgraph H by identifying the points of the plane with vertices in such a way that the vertices of H are identified with points of one line, L say. (This is possible because $h \leq \sqrt{n} = t$.) Next delete from the K_{t^2+t+1} all points of L other than those in H. The result is a K_{t^2+h} with a partition into t^2+t+1 cliques — one of size h, the others of size t or $t+1$ (according as the corresponding line did or did not meet one of the points of L which were not in H). Finally, replace the h-clique by a set of $cp(K_h - H)$ cliques which form a partition of $K_h - H$. We have a partition of $K_{t^2+h} - H$ into

$$t^2 + t + cp(G_h)$$

cliques, so

$$cp(G_n) \leq cp(K_{t^2+h} - H) \leq t^2 + t + cp(G_h).$$

It is clear that $cp(G_h) \leq \frac{1}{2}h(h-1) < t^2$, so

$$cp(G_n) \leq O(t^2) = O(n).$$

If in fact h is $o(\sqrt{n})$, so that $\frac{1}{2}h(h-1)$ is $o(n)$, we have

$$cp\,(G_n) \leq n + o(n).$$

It is clear from part (ii) of Theorem 1 that $cp\,(G_n)$ cannot be significantly less than n, so we have:

THEOREM 2. *If H has $\sqrt{n}$ or less vertices, then*

$$cp\,(K_n - H) = O(n).$$

If H has $o(\sqrt{n})$ vertices, then

$$cp\,(K_n - H) \sim n.$$

(Part (ii) comes from [7]; part (i) is implicit in that paper.)

4. The complement of a one-factor. A one-factor F on $2n$ points is a perfect matching—a set of n disjoint edges which precisely cover the $2n$ points. The complement $T_{2n} = K_{2n} - F$ is sometimes called the *cocktail party graph* on $2n$ vertices. Orlin [5] studied $cp\,(T_{2n})$; estimating it seems to be particularly difficult.

We shall show that $cp\,(T_{2n})$ is of order at worst $n\log\log n$. However we first need a preliminary result, and quickly prove that $cp\,(T_{2n})$ is $o(n^2)$. This can be done as follows (see [8]). Let q be the smallest power of 2 such that

$$2n < q^2 + q.$$

Clearly $q^2 < 8n$, for otherwise $q/2$ could be used. Since $cp\,(T_{2n})$ is clearly monotone (T_{2n} is a spanning subgraph of T_{2n+2}),

$$cp\,(T_{2n}) \leq cp\,(T_{q^2+q}).$$

To get a bound on $cp\,(T_{q^2+q})$ we consider a $PG\,(2,q)$, and select one point x. Delete x from the plane; replace every line L which contains x by a copy of T_q with vertex-set $L\backslash\{x\}$, and replace every other line by a K_{q+1}. We obtain a decomposition into edge-disjoint subgraphs of the form

$$T_{q^2+q} = q^2 K_{q+1} + (q+1)\,T_q,$$

whence

$$cp\,(T_{2n}) \leq q^2 + (q+1)\,cp\,(T_q)$$
$$< 8n + (2\sqrt{2n} + 1)\,cp\,(T_{\sqrt{8n}}).$$

Now $cp\,(T_{\sqrt{8n}}) < \frac{1}{2}(\sqrt{8n})^2 = 4n$, so

$$cp\,(T_{2n}) < 8n + (2\sqrt{2n} + 1)\cdot 4n,$$

whence $cp\,(T_{2n}) \leq O(n^{3/2})$. We have:

LEMMA 3. $cp(T_{2n}) = o(n^2)$.

It is shown in [8] that similar reasoning leads to the result that $cp(T_{2n}) = o(n^{1+\epsilon})$ for any positive ϵ, but we do not need that result.

THEOREM 3 [2]. *For each* $\epsilon > 0$, $cp(T_v) < (1 + \epsilon)\, v \log\log v$ *for all sufficiently large even integers* v.

PROOF. It is convenient to work with the function $c(x)$:

$$c(x) = cp(T_{2n}),$$

where $2n$ is the smallest even integer not less than x. Then $c(x)$ is monotone increasing.

Given x, say d is the largest even integer such that $d \le \sqrt{2n}$; let e be the smallest integer such that $de \ge 2n$. Clearly $e \le d + 5$. Let t be the smallest prime power such that $t \ge e$. In an affine plane of order t, identify lines with cliques as before. Select one parallel class; delete all but d of the t lines in that parallel class. Now select another parallel class, replace e of the t lines (which now have only d points each) with copies of the graph T_d, and delete the points on the remaining $t - e$ lines. We now have de points remaining; the lines in the remaining $t - 1$ parallel classes can be interpreted as cliques, as can those in the first class. So we have a decomposition of T_{de} into $t^2 - t + d$ cliques and e copies of T_d. Since $cp(T_{2n})$ is clearly monotone in n (T_{2n} is an induced subgraph of T_{2n+2}),

$$cp(T_{2n}) \le cp(T_{de}) \le t^2 - t + d + e\,cp(T_d) < t^2 + (d + 5)cp(T_d).$$

In terms of x, we can deduce

(1) $$c(x) < t^2 + (\sqrt{x} + 5)\, c\,(\sqrt{x}).$$

Say q is the largest prime less than t. From Lemma 2,

$$t \le (1 + \delta(q))(q + q^{0.6}),$$

where $\delta(q)$ approaches 0 as t approaches ∞. (In fact, the condition "t is prime" would have been sufficient for this, rather than "t is a prime power".) And by definition

$$q \le e - 1 \le \sqrt{x} + 4.$$

This means that $(q + q^{0.6})^2 = x + o(x)$, so $t^2 = (1 + 2\delta(q) + \delta(q)^2)(x + o(x))$. Since Lemma 3 implies that $c(\sqrt{x})$ is $o(x)$, we have from (1) that for a given $\epsilon > 0$,

$$c(x) < (1 + \tfrac{1}{4}\epsilon)x + \sqrt{x}\, c\,(\sqrt{x}) + o\,(x),$$

—one merely takes x (and consequently t and q) large enough so that $2\delta(q) + q^2$ is less than $\tfrac{1}{4}\epsilon$. So

$$\frac{c(x)}{x} < 1 + \frac{1}{2}\epsilon + \frac{c\sqrt{x}}{(\sqrt{x})},$$

for sufficiently large x (the "$o(x)/x$" terms will eventually be smaller than $\frac{1}{4}\epsilon$).

We now write $x = 2^{2^z}$, or $z = \log\log x$, and define

$$h(z) = c(x)/x.$$

Then we have

$$h(z) < 1 + \frac{1}{2}\epsilon + h(z-1),$$

for all sufficiently large z. It follows that there is a constant M such that

$$h(z) < (1 + \frac{1}{2}\epsilon)z + M,$$

for all z. So, given ϵ,

$$c(x) < (1 + \frac{1}{2}\epsilon)x \log\log x + Mx,$$

and if x is large enough then M will be less than $\frac{1}{2}\epsilon \log\log x$, so

$$c(x) < (1 + \epsilon)x \log\log x.$$

Taking x an even integer, the result follows.

5. Complements of paths and cycles. Suppose S_n denotes the graph $K_n \backslash P_n$, where P_n is a path on n vertices. It is easy to show, similarly to Lemma 3, that $cp(S_n)$ is $o(n^2)$. We then proceed as follows. Let d be the integer part of $\sqrt{n}$, and choose e to be the smallest even integer such that $de \geq n$: at worst, $e \leq d+3$. If t is the smallest prime power not less than $\sqrt{n}$, we embed a path P_n in a copy of the affine plane of parameter t. One parallel class P_1 is chosen, and all but d of its lines are deleted. Now select another parallel class P_2, delete $t-e$ of its lines, and replace the e remaining lines (which now are of size d) by copies of S_d. Say the point of intersection of line i of P_1 with line j of P_2 is a_{ij}. Then we can assume that instead of the graph K_d, line j of P_2 is represented by the complement of the path

$$(a_{1j}, a_{2j}, \dots, a_{dj}).$$

Finally, the first and d-th lines of the first parallel class are replaced—line 1 by the complement of the set of edges

$$a_{11}a_{12}, a_{13}a_{14}, \dots$$

and line d by the complement of the set

$$a_{12}a_{13}, a_{14}a_{15}, \dots.$$

The first of these is a copy of T_e; the second is derived from a T_e by adding the edge $a_{1e}a_{11}$. So their clique partition numbers are $cp(T_e)$ and (at most) $cp(T_e) + 1$ respectively. So we can deduce

$$cp(S_n) \leq (t^2 - t + d - 2) + e \, cp(S_d) + 2cp(T_e) + 1.$$

Again, writing $c(x) = cp(S_{[x]})$ we obtain

$$c(x) \leq t^2 + (\sqrt{x} + 4)\, c(\sqrt{x}) + o(x),$$

which is similar to (1), and the result of Theorem 3 can be obtained.

In conclusion, we observe that the complement of a cycle can be handled in the same sort of way. In fact, there is no difficulty in showing that the result of Theorem 4 holds for the complement of any graph with all degrees 1 or 2. However, the application of finite planes does not seem likely to yield a linear asymptotic bound on the clique partition number of the complement of a one-factor, cycle or path, although we believe that such bounds exist in all three cases.

References

1. N. G. de Bruijn and P. Erdös, *On a combinatorial problem*, Indag. Math. **10** (1948), 421–423.
2. D. A. Gregory, S. McGuiness and W. D. Wallis, *Clique partitions of the cocktail party graph*, Discrete Math. **59** (1986), 267–273.
3. D. R. Heath-Brown and H. Iwaniek, *On the difference between consecutive primes*, Inventiones Math. **55** (1979), 49–69.
4. H. L. Montgomery, *Topics In Multiplicative Number Theory*, Lecture Notes in Mathematics **227**, Springer-Verlag, Berlin, 1971.
5. J. Orlin, *Contentment in graph theory: covering graphs with cliques*, Indag. Math. **39** (1977), 406–424.
6. N. J. Pullman and A. Donald, *Clique coverings of graphs* II: *complements of cliques*, Utilitas Math **19** (1981), 207–213.
7. W. D. Wallis, *Asymptotic values of clique partition numbers*, Combinatorica **2** (1982), 99–101.
8. ____, *Clique partitions of the complement of a one-factor*, Congressus Num. **46** (1985), 317–319.

DEPARTMENT OF MATHEMATICS, SOUTHERN ILLINOIS UNIVERSITY, CARBONDALE, ILLINOIS 62901-4408

Contemporary Mathematics
Volume 111, 1990

Oval Designs in Quadrics

MICHAEL A. WERTHEIMER

ABSTRACT. The points (x, y) of the Desarguesian affine plane $AG(2, 2^m)$ may be partitioned into the complementary sets: trace $(\alpha x^2 + xy + \alpha y^2) = 0$ and trace $(\alpha x^2 + xy + \alpha y^2) = 1$. Identifying the points of $AG(2, 2^m)$ with the vector space $V = [GF(2)]^{2m}$ and assuming trace $(\alpha) = 1$ the former set describes an elliptic quadric in V and the latter set an affine translate of a hyperbolic quadric in V. The classical symplectic group $Sp(2m, 2)$ is known to have a representation which acts 2-transitively on both the elliptic quadric and its (hyperbolic) complement. By examining the orbits under the rank 2^{m-1} subgroup $PGL(2, 2^m)$ of $Sp(2m, 2)$ of certain ovals and arcs of $AG(2, 2^m)$ which lie in the elliptic quadric we construct three infinite families of balanced incomplete block designs. These designs have parameters (v, k, λ) equal to $(\frac{1}{2}q(q-1), q+1, q)$, $(\frac{1}{2}q(q-1), q+2, q+2)$, and $(\frac{1}{2}q(q-1), q-1, q)$ where $q = 2^m$, $m > 1$. The first of these families is minimal (in the sense of attaining the smallest possible value of λ for the given v and k), the last resolvable, all are simple (i.e., have no repeated blocks). In particular, setting $m = 3$ we get the first example of a resolvable $(28, 7, 8)$ block design.

1. Introduction. The use of groups acting on sets to construct (v, k, λ) balanced incomplete block designs (BIBDs) is a standard technique made all the simpler when the group action is 2-transitive. In this case the orbit of an arbitrary k-subset is always a BIBD. However, such designs often have the undesirable feature of possessing a large value of 'λ'. One way around this problem is to find k-subsets with large block stabilizers in the group G; another is to look at orbits of a k-subset under suitable subgroups of G which may not act 2-transitively but may still yield block designs.

In this paper we take as our starting point the symplectic group $Sp(2m, 2)$ acting 2-transitively on the points of an elliptic quadric in $V = [GF(2)]^{2m}$. We choose as our subgroup $PGL(2, 2^m)$ which we show acts with rank 2^{m-1} on the quadric. For k-subsets we choose certain *geometric* entities, namely

1980 *Mathematics Subject Classification.* Primary 05B05, 05B25, 51E05, 51E20.

This paper is in final form and no version of it will be submitted for publication elsewhere.

ovals and arcs of $AG(2, 2^m)$. The resulting orbits yield block designs for all $m > 1$.

2. Group action. For notational consistency elements of the Galois field $GF(q)$ will be denoted throughout by lower case γ, β, x, y, etc. Elements of $AG(2, q)$ will be denoted by upper case X, Y, etc. If σ is an element of a group G acting on a set S then the image of $s \in S$ under σ is written s^σ.

Representations of $Sp(2m, 2)$ acting 2-transitively on sets of sizes $\frac{1}{2}q(q-1)$ and $\frac{1}{2}q(q+1)$, $q = 2^m$, are well known (see, e.g., [5] or [12]). The inherited action of the subgroup $PGL(2, q) \subset Sp(2m, 2)$ can be made quite explicit as follows.

Begin by considering the elliptic quadratic form

$$Q(X) = Q(x, y) = \alpha x^2 + xy + \alpha y^2,$$

defined on the points of $AG(2, q)$ where α is chosen so that $\text{trace}(\alpha) = 1$, trace going from $GF(q)$ to $GF(2)$, i.e.,

$$\text{trace}(x) = x + x^2 + x^4 + \cdots + x^{2^{m-1}}.$$

For the set S take the $\frac{1}{2}q(q-1)$ solutions to $\text{trace}(Q(X)) = 0$. It is known that $\text{trace}(Q(X))$ determines an elliptic quadratic form in the $2m$-dimensional vector space $V = [GF(2)]^{2m}$ (with corresponding elliptic quadric S) identifying the points of V and $AG(2, q)$ in the natural way (see [17]).

If $X = (x_1, x_2)$ and $Y = (y_1, y_2)$ are two points of $AG(2, q)$ the *bilinear form* B associated with Q is defined to be

$$B(X, Y) = Q(X + Y) + Q(X) + Q(Y) = x_1 y_2 + x_2 y_1.$$

The set of *linear* transformations of $AG(2, q)$ which preserve this bilinear form is the symplectic group $Sp(2, q)$ which is easily seen to be $SL(2, q) \simeq PGL(2, q)$. The subgroup of $PGL(2, q)$ which preserves the quadratic form Q is the orthogonal group $0^-(2, q)$ which is dihedral of order $2(q + 1)$. (Note: if α is chosen so that $\text{trace}(\alpha) = 0$ the quadratic form is called *hyperbolic* and is fixed by $0^+(2, q)$ which is dihedral of order $2(q - 1)$).

Pick $\sigma \in Sp(2, q) \simeq PGL(2, q)$. Note that we have the following relations for $X, Y \in AG(2, q)$:

(1) $B(X, Y) = Q(X + Y) + Q(X) + Q(Y)$,
(2) $B(X^\sigma, Y^\sigma) = B(X, Y) = Q(X^\sigma + Y^\sigma) + Q(X^\sigma) + Q(Y^\sigma)$,
     which together imply
(3) $Q(X^\sigma + Y^\sigma) + Q(X + Y) = Q(X) + Q(Y) + Q(X^\sigma) + Q(Y^\sigma)$.

If we define the function $H(X) = Q(X) + Q(X^\sigma)$, then from (3) we see that $H(X + Y) = H(X) + H(Y)$ and $H(\lambda X) = \lambda^2 H(X)$ for $\lambda \in GF(q)$. This implies that $H(X)$ must be the *square of a linear function*. Since the bilinear form B is nondegenerate we may write

(4) $H(X) = Q(X^\sigma) + Q(X) = B(X, C_\sigma)^2$ for some $C_\sigma \in AG(2, q)$.

If we apply the trace from $\mathrm{GF}(q)$ to $\mathrm{GF}(2)$ to both sides of (4) we get

$$\mathrm{trace}(Q(X^\sigma)) + \mathrm{trace}(Q(X)) = \mathrm{trace}(B(X, C_\sigma)),$$

which simplifies to

$$\mathrm{trace}\big[Q([X + C_\sigma]^\sigma)\big] = \mathrm{trace}(Q(X)) + \mathrm{trace}(Q(C_\sigma^\sigma)).$$

Since the nonsingular affine transformation $X \to (X + C_\sigma)^\sigma$ permutes the points of $\mathrm{AG}(2, q)$, the number of zeroes on both sides of this last equation must be the same. Since $\mathrm{trace}\,(Q(X)) = 0$ has $\frac{1}{2}q(q-1) < \frac{1}{2}q^2$ solutions we conclude that $\mathrm{trace}\,(Q(C_\sigma^\sigma))$ must be zero. Finally, we write

$$\mathrm{trace}\big[Q([X + C_\sigma]^\sigma)\big] = \mathrm{trace}(Q(X)).$$

Thus, each σ, C_σ pair gives rise to a permutation of the points of the quadric: $\{X \in \mathrm{AG}(2, q) \mid \mathrm{trace}(Q(X)) = 0\}$ (as well as its complement: $\{X \in \mathrm{AG}(2, q) \mid \mathrm{trace}(Q(X)) = 1\}$. The set of all such maps is easily seen to form a group G isomorphic to $\mathrm{PGL}(2, q)$ [17]. The stabilizer in G of $(0, 0) \in \mathrm{AG}(2, q)$ is clearly $0^-(2, q)$ (which implies transitivity on the quadric since $|G|/\big|0^-(2, q)\big| = \frac{1}{2}q(q-1)$) and by Witt's theorem [18] $0^-(2, q)$ partitions the non-zero points of the quadric into $q/2 - 1$ suborbits according to $Q(X) = i$, $\mathrm{trace}\,(i) = 0$, $i \neq 0$, each suborbit of size $q+1$. The stabilizer in G of $(\sqrt{\alpha}, \sqrt{\alpha}) \in \mathrm{AG}(2, q)$ is $0^+(2, q)$ (which implies transitivity on the complement of the quadric since $|G|/\big|0^+(2, q)\big| = \frac{1}{2}q(q+1)$) and again by Witt's theorem $0^+(2, q)$ partitions the non-zero points of the complement into $q/2$ suborbits according to $Q(X) = i$, $\mathrm{trace}\,(i) = 1$, one suborbit of size $2(q-1)$ the remaining suborbits of size $q-1$. We summarize these results in

PROPOSITION 1. *Let* $Q(X) = Q(x, y) = \alpha x^2 + xy + \alpha y^2$ *be an elliptic quadratic form defined over* $\mathrm{AG}(2, q)$, $q = 2^m$, $m > 1$, *with associated bilinear form* B. *Let* G *be the set of affine transformation of the form*

$$X \to (X + C_\sigma)^\sigma,$$

where $\sigma \in \mathrm{Sp}(2, q) \simeq \mathrm{PGL}(2, q)$ *and* C_σ *is given by the unique solution to*

$$Q(X^\sigma) = Q(X) + B(X, C_\sigma)^2.$$

Then,

(1) $G \simeq \mathrm{PGL}(2, q)$.

(2) *The stabilizer in* G *of* $(0, 0) \in \mathrm{AG}(2, q)$ *is the orthogonal group* $0^-(2, q)$ *which has suborbits*

$$O_i = \{X \in \mathrm{AG}(2, q) \mid Q(X) = i\}, \ \mathrm{trace}(i) = 0, \ i \neq 0.$$

(3) G *acts with rank* 2^{m-1} *on the elliptic quadric*:

$$\{X \in \mathrm{AG}(2, q) \mid \mathrm{trace}(Q(X)) = 0\}.$$

(4) *The stabilizer in* G *of* $(\sqrt{\alpha}, \sqrt{\alpha}) \in AG(2, q)$ *is the orthogonal group* $O^+(2, q)$ *which has suborbits*

$$O_i = \{X \in AG(2, q) \mid Q(X) = i\}, \quad \text{trace}(i) = 1.$$

(5) G *acts with rank* $1 + 2^{m-1}$ *on the complementary set*:

$$\{X \in AG(2, q) \mid \text{trace}(Q(X)) = 1\}. \quad \square$$

For $\sigma \in PGL(2, q)$ computation of the parameter C_σ is straightforward and yields the following explicit group action of $PGL(2, q)$ on the quadric (and its complement):

For $\sigma_{a,b,c,d} \in PGL(2, q)$, $ad + bc = 1$, $a, b, c, d, \in GF(q)$, and for (x, y) in the quadric (or its complement) set

$$(x, y)^{\sigma_{a,b,c,d}} = \left[ax + cy + \sqrt{Q(a, c)} + \sqrt{\alpha}, \, bx + dy + \sqrt{Q(b, d)} + \sqrt{\alpha}\right]$$

where, as always, $Q(x, y) = \alpha x^2 + xy + \alpha y^2$.

Now, $PGL(2, q)$ acting on the 2-*subsets* of the quadric forms $2^{m-1} - 1$ equally sized orbits indexed, via Proposition 1, by the non-zero elements of $GF(q)$ with trace equal to zero. Specifically, if X and Y are two points in the quadric, $X \neq Y$, pick any $\sigma \in PGL(2, q)$ such that $C_\sigma = X$. Then $\{X, Y\}$ is in orbit O_i where

$$i = Q((Y + X)^\sigma) = Q(X + Y) + B(X, Y)^2.$$

3. Lines, arcs, ovals in the quadric. The representation of an elliptic quadric in $[GF(2)]^{2m}$ as a subset of $AG(2, q)$ allows us to view the geometry of the affine plane which is inherited by the quadric S. It is known (see [6]) that the lines of $AG(2, q)$ intersect S in only two cardinalities, specifically in $q/2$ or 0 points. We make this precise in

PROPOSITION 2. *Let* $Q(X) = Q(x, y) = ax^2 + xy + ay^2$ *be an elliptic quadratic form* (trace$(\alpha) = 1$) *defined on* $AG(2, q)$, $q = 2^m$ *with corresponding bilinear form* B. *Let* $S = \{X \in AG(2, q) \mid \text{trace}(Q(X)) = 0\}$. *Then,*

(1) *The lines of* $AG(2, q)$ *meet* S *in either* $q/2$ *or* 0 *points (hence meet its complement in* $q/2$ *or* q *points, respectively).*

(2) *If* ℓ *is the line of* $AG(2, q)$ *given by*

$$B(X, P) = c,$$

then ℓ *meets* S *in* 0 *points (hence meets its complement in* q *points) if and only if* $Q(P) = c^2$.

PROOF. Let ℓ be a line of $AG(2, q)$, then ℓ is of the form

$$p_1 x + p_2 y = c,$$

for $p_1, p_2, c \in \mathrm{GF}(q)$, i.e.,

$$B((x, y), (p_2, p_1)) = c.$$

Set $X = (x, y)$ and $P = (p_2, p_1)$. We wish to count the number of solutions X to

$$\mathrm{trace}(Q(X)) = 0,$$

subject to the constraint $B(X, P) = c$. Our claim is that there are $q/2$ solutions except when $Q(P) = c^2$ when there are *no* solutions. Assume $p_1 \neq 0$. Then

$$
\begin{aligned}
Q(x, y) &= [1/p_1^2] Q(p_1 x, p_1 y) \\
        &= [1/p_1^2] Q(c + p_2 y, p_1 y) \\
        &= [1/p_1^2] \{ Q(c, 0) + Q(p_2 y, p_1 y) + B((c, 0), (p_2 y, p_1 y)) \}.
\end{aligned}
$$

Using the fact that trace $(x^2) = $ trace (x) we simplify this last expression to

$$\mathrm{trace}(Q(x, y)) = \mathrm{trace} \left\{ [y^2/p_1^2](Q(P) + c^2) + \alpha c^2/p_1^2 \right\}.$$

Now if $Q(P) = c^2$ one sees that trace $(\alpha c^2/p_1^2) = 1$ (since trace $(\alpha) = 1$) hence under this assumption there are no solutions to trace $(Q(X)) = 0$. If $Q(P) \neq c^2$ we are reduced to solving trace $(k_1 y) = \mathrm{trace}(k_2)$ where $k_1, k_2 \in \mathrm{GF}(q)$, $k_1 \neq 0$ for which there are precisely $q/2$ solutions y since the trace map is rank 1 GF(2) linear.

If $p_1 = 0$ then $p_2 \neq 0$ and the previous argument applies substituting p_2 for p_1 and x for y, concluding the proof. $\square$

Proposition 2 implies that exactly one line from each parallel class of $\mathrm{AG}(2, q)$ is disjoint from the quadric S. Thus, the intersections of the lines of $\mathrm{AG}(2, q)$ with S form a *resolvable* block design with parameters

$$(v, k, \lambda) = (\tfrac{1}{2}q(q-1), \tfrac{1}{2}q, 1) \quad q = 2^m, \qquad m > 1.$$

These designs have been discovered independently by a number of investigators (see, e.g., [3]) in the more general setting as duals of complements of complete ovals in $\mathrm{PG}(2, q)$. The correspondence between descriptions is made by noting that the $q + 1$ lines of the affine plane disjoint from the quadric form a *line oval* in $\mathrm{AG}(2, q)$ which can be extended uniquely to a complete (line) oval in $\mathrm{PG}(2, q)$ with the addition of the "line at infinity". We shall have occasion in the following section to refer back to these designs.

In $\mathrm{PG}(2, q)$ a set of μ points no three collinear is said to be a μ-arc. It is known that $\mu \leq q + 1$ when q is odd and $\mu \leq q + 2$ otherwise. A line of the plane is called a secant, a tangent, or an external line if it meets the arc in two, one, or zero points. When q is even, which is the case we are considering, it is known [15] that all of the tangents of a $q + 1$-arc are concurrent. The point at which the tangents meet is called the *nucleus* or *knot* of the arc. A $q + 1$-arc together with its knot forms a $q + 2$-arc without tangents. Conversely, any

$q + 2$-arc in $\mathrm{PG}(2, q)$ (or $\mathrm{AG}(2, q)$) has no tangents. A $q + 2$-arc is called an *oval* .

In $\mathrm{PG}(2, q)$ an irreducible conic is a $q + 1$-arc. The converse is true for q odd. For $q = 2^m$, $q + 2$-arcs exist which do not contain any set of $q + 1$ points forming a conic provided $m = 4$, 5 or $m \geq 7$. When $m = 1$, 2, or 3 every $q + 2$-arc can be obtained by adjoining the knot to the $q + 1$ points of an irreducible conic [**16**].

4. The designs. As our point set S we continue to take the elliptic quadric

$$S = \{X \in \mathrm{AG}(2, q) \mid \mathrm{trace}(Q(X)) = 0\},$$

where $q = 2^m$ and $Q(X) = Q(x, y) = \alpha x^2 + xy + \alpha y^2$, $\mathrm{trace}\,(\alpha) = 1$. We recall $|S| = \frac{1}{2}q(q - 1)$. We may think of S as the union of the origin $(0, 0) \in \mathrm{AG}(2, q)$ with the $\frac{1}{2}q - 1$ sets

$$\mathscr{S}_i = \{X \in \mathrm{AG}(2, q) \mid Q(X) = i\},$$

$\mathrm{trace}\,(i) = 0$, $i \neq 0$, $|\mathscr{S}_i| = q + 1$.

Notice that the $\mathscr{S}_i$ are precisely the suborbits $\mathscr{O}_i$ of Proposition 1. Moreover, each $\mathscr{S}_i$ is an irreducible conic of $\mathrm{AG}(2, q)$, hence a $q + 1$-arc of $\mathrm{AG}(2, q)$. The knot of each $\mathscr{S}_i$ in $\mathrm{AG}(2, q)$ is easily verified to be $(0, 0)$. This partitioning of the quadric into conics leads to our first design family.

We recall that a (v, k, λ) design is *minimal* provided any design with parameters (v, k, λ') necessarily has $\lambda' \geq \lambda$.

THEOREM 1. *Let* $Q(X) = Q(x, y) = \alpha x^2 + xy + \alpha y^2$ *be an elliptic quadratic form defined on* $\mathrm{AG}(2, q)$, $q = 2^m$, $m > 1$. *Then the union of orbits under* $\mathrm{PGL}(2, q)$ *of the sets*

$$\mathscr{S}_i = \{X \in \mathrm{AG}(2, q) \mid Q(X)) = i\},$$

$i \neq 0$, $\mathrm{trace}\,(i) = 0$ *forms a minimal* BIBD *with parameters*

$$(v, k, \lambda, r, b) = \left[\frac{1}{2}q(q - 1),\ q + 1,\ q,\ \frac{1}{2}q(q - 1) - 1,\ \frac{1}{4}q(q - 1)(q - 2)\right]$$

Each block of this design is a $q + 1$-*arc (in fact a conic) of* $\mathrm{AG}(2, q)$ *with block stabilizer* $0^-(2, q)$.

PROOF. From the previous discussion we recall that the $\mathscr{S}_i$ are the suborbits of the quadric S under the stabilizer $(0^-(2, q))$ of $(0, 0)$ in $\mathrm{PGL}(2, q)$. If $\sigma \in \mathrm{PGL}(2, q)$ then $\mathscr{S}_i^\sigma$ is a conic with knot $(0, 0)^\sigma = P$ since σ is a nonsingular affine transformation of $\mathrm{AG}(2, q)$, hence an automorphism mapping conics to conics. The image of $\mathscr{S}_i$ is the set

$$\mathscr{S}_i^\sigma = \{X \in \mathrm{AG}(2, q) \mid Q(X + P) + B(X, P)^2 = i\}.$$

Now pick $X, Y \in S$ arbitrarily. We wish to prove that there are precisely 'q' conics of the form $\mathscr{S}_i^\sigma$ which contain both X and Y. By transitivity we may assume without loss of generality that $X = 0 = (0, 0)$ (this makes the computation simpler). Now $X = 0$ and Y lie in a set $\mathscr{S}_i^\sigma$ if and only if there exists a $P \in S$, $P \neq 0$, such that

$$Q(0 + P) + B(0, P)^2 = Q(Y + P) + B(Y, P)^2,$$

i.e.,

$$Q(P) = Q(Y + P) + B(Y, P)^2,$$

which reduces to

$$B(Y, P)^2 + B(Y, P) + Q(Y) = 0.$$

Since trace $(Q(Y)) = 0$, as a polynomial in $B(Y, P)$ this has solutions

$$B(Y, P) = c \quad \text{and} \quad B(Y, P) = 1 + c$$

for some $c \in \mathrm{GF}(q)$. From Proposition 2 we will get $q/2$ solutions $P \in S$ to each of these equations provided $Q(Y) \neq c^2$ and $Q(Y) \neq 1 + c^2$. If we assume $Q(Y) = \beta + c^2$ where $\beta \in \{0, 1\}$ the resulting equations

$$Q(Y) = \beta + c^2,$$
$$Q(Y) = c + c^2,$$

imply $c = 0$ or $c = 1$. But then $Q(Y) = 0$ forcing $Y = 0$ which is a contradiction. Thus, X and Y appear in precisely q "blocks" of the form $\mathscr{S}_i^\sigma$, which proves the resulting configuration is a BIBD. Minimality follows by noting that for arbitrary designs $(v - 1)\lambda = r(k - 1)$ which in this case requires λ to be a multiple of q. $\square$

If we complete each of the conics (blocks) of Theorem 1 to an oval by including its knot the block size (k) increases by one. Moreover, each pair of points X and Y will now by covered an additional *two* times: once when X is the distinguished knot and once when Y is the distinguished knot (a knot being distinguished when the remaining $q + 1$ points of the oval form a conic). This configuration then determines a BIBD with $k = q + 2$ and $\lambda = q + 2$. We summarize this in

THEOREM 2. *Let* $Q(X) = Q(x, y) = \alpha x^2 + xy + \alpha y^2$ *be an elliptic quadratic form defined on* $\mathrm{AG}(2, q)$, $q = 2^m$, $m > 1$. *Then the union of orbits under* $\mathrm{PGL}(2, q)$ *of the sets*

$$\mathscr{S}_i = \{0, 0\} \cup \{X \in \mathrm{AG}(2, q) \mid Q(X) = i\},$$

$i \neq 0$, $\mathrm{trace}(i) = 0$, *forms a BIBD with parameters*

$$(v, k, \lambda, r, b) = \left[\frac{1}{2}q(q - 1), \; q + 2, \; q + 2, \; \frac{1}{2}(q^2 - 4), \; \frac{1}{4}q(q - 1)(q - 2) \right]$$

Each block of this design is an oval (in fact a conic + knot) of $\mathrm{AG}(2, q)$ *with block stabilizer* $0^-(2, q)$. $\square$

Setting $q = 8$ in Theorems 1 and 2 results in block designs with parameters $(28, 9, 8, 27, 84)$ and $(28, 10, 10, 30, 84)$, respectively. These two designs are, in fact, isomorphic to those constructed in [14] and are listed in the design tables of [13]. They can also be readily obtained from the pseudocyclic association schemes appearing in [8].

In [1] the concepts of arc and oval are extended to arbitrary block designs. A μ-arc of a block design is a set of μ points no three of which lie on any block of the design. For a BIBD with parameters (v, k, λ, r, b) the size of a μ-arc is bounded as follows:

(1) If $r - \lambda$ is odd or $r - \lambda$ is even but $\lambda \nmid r$ then

$$\mu \leq \frac{r + \lambda - 1}{\lambda}.$$

(2) If $r - \lambda$ is even and $\lambda \mid r$ then

$$\mu \leq \frac{r + \lambda}{\lambda}.$$

An *oval* is an arc which meets the appropriate bound (1) or (2).

With respect to ovals the designs D of Theorem 2 fall into category (2) above. For these D $(r + \lambda)/\lambda$ equals $q/2$. Recall now that the lines of $AG(2, q)$ which intersect the quadric non-trivially do so in $q/2$ points. Any such line intersects a block of D in 0 or 2 points since each block of D is itself an oval of $AG(2, q)$. Revisiting our discussion of Section 3 we conclude

COROLLARY. *Fix* $q = 2^m$. *The lines of* $AG(2, q)$ *which intersect the quadric* $\{X \in AG(2, q) \mid \text{trace}(Q(X)) = 0\}$ *nontrivially are ovals for the design of Theorem 2 with* $q = 2^m$. *Moreover, these ovals themselves form a resolvable* BIBD *with parameters*

$$(v, k, \lambda, r, b) = \left[\frac{1}{2} q (q - 1), \ \frac{1}{2} q, \ 1, \ q + 1, \ q^2 - 1 \right]. \quad \square$$

The existence of ovals for the designs of Theorem 2 provides information regarding the binary codes generated by the incidence matrices of these designs. Specifically, for a design D we consider the GF(2) span of the rows of the $b \times v$ matrix $A = (a_{ij})$ where $a_{ij} = 1$ if point v_j lies on block B_i and $a_{ij} = 0$ otherwise. The linear space spanned by the rows of A is called the binary code $\mathcal{C}$ associated with the design. The dual code $\mathcal{C}^{\perp}$ is the GF(2) space spanned by the vectors perpendicular (under the ordinary dot product) to all the vectors in $\mathcal{C}$. Each vector in a code is called a *codeword* and the *weight* of a codeword is the number of its nonzero components. The *weight enumerator* of a code is a listing of the number of codewords of each weight in the code.

In [1] it is shown that for a design D with code $\mathcal{C}$, $r - \lambda$ even, and $\lambda \mid r$ the minimum nonzero weight of $\mathcal{C}^{\perp}$ is *at least* the size of an oval. Moreover, codewords of this weight in $\mathcal{C}^{\perp}$ are precisely the ovals of D.

Interpreting this information in light of the corollary, the minimum nonzero weight in the dual codes to the corresponding designs is $q/2$. Moreover, since the ovals themselves contain a design which satisfies the conditions $r-\lambda$ is even and $\lambda \mid r$ we see that the minimum nonzero weight in the code itself must be $q + 2$. In Table 1 we explicitly compute the weight enumerators of $\mathcal{C}$ and $\mathcal{C}^{\perp}$ for the case $q = 8$.

Weight	Code $\mathcal{C}$	Dual Code $\mathcal{C}^{\perp}$
	(dim.=9)	(dim.=19)
0	1	1
2	0	0
4	0	63
6	0	1512
8	0	12285
10	84	50400
12	63	120771
14	216	154224
16	63	120771
18	84	50400
20	0	12285
22	0	1512
24	0	63
26	0	0
28	1	1

TABLE 1
Codes for (28, 10, 10) BIBD

This code is well known (see [4]). It has $P\Gamma L(2, 8)$ as its automorphism group (the small Ree group) acting 2-transitively on the 28 points of the quadric. Thus, the words of each fixed weight in both $\mathcal{C}$ and $\mathcal{C}^{\perp}$ form block designs. In this case the (28, 4, 1, 9, 63) oval design (comprised of the 63 codewords of weight 4 in $\mathcal{C}^{\perp}$) is the *Ree Unital* whose ovals, in turn, form the (28, 10, 10, 30, 84) design of Theorem 2.

In moving to our final construction we recall that solutions X to $Q(X + P) + B(X, P)^2 = c$, $c \neq 0$, trace $(c) = 0$ form a $q + 1$-arc in $AG(2, q)$ with knot P. In Theorem 1 we chose P to be in the elliptic quadric $S = \{X \in AG(2, q) \mid \text{trace}(Q(X)) = 0\}$. In this case the entire $q + 1$-arc as well as the knot P lay in S. By considering such arcs at each point of S we constructed a family of block designs. Now we examine the effect of choosing $P \in AG(2, q)/S$. Since

$$c = Q(X + P) + B(X, P)^2 = Q(X) + Q(P) + B(X, P) + B(X, P)^2$$

implies $\text{trace}(c) = \text{trace}(Q(X)) + \text{trace}(Q(P)) = 0 + 1 = 1$ we only consider conics of the form

$$Q(X + P) + B(X, P)^2 = c \ \text{trace}(c) = 1.$$

Setting $P = (\sqrt{\alpha}, \sqrt{\alpha})$ and $X = (x_1, x_2)$ this last equation reduces to

$$[x_1 + \sqrt{\alpha}][x_2 + \sqrt{\alpha}] = c,$$

which clearly has $q - 1$ solutions. Two additional points of this conic lie on the "line at infinity" and the knot is at $(\sqrt{\alpha}, \sqrt{\alpha})$. The transitivity of PGL$(2, q)$ on both S and AG$(2, q)/S$ implies that the orbit of this conic consists of conics with *each point* P of AG$(2, q)/S$ serving as the knot for the image set

$$\{X \in \text{AG}(2, q) \mid Q(X + P) + B^2(X, P) = c\}.$$

Notice that for a fixed P as c runs over the $q/2$ elements with trace one we get a spread of the quadric by conics. Thus the resulting configuration is resolvable; in fact it is a design.

THEOREM 4. *Let* $Q(X) = Q(x, y) = \alpha x^2 + xy + \alpha y^2$ *be an elliptic quadratic form defined on* AG$(2, q)$, $q = 2^m$, $m > 1$. *Then the union of orbits under* PGL$(2, q)$ *of the sets*

$$\mathscr{S}_i = \{X \in \text{AG}(2, q) \mid Q(X + P) + B(X, P)^2 = i\},$$

$P = (\sqrt{\alpha}, \sqrt{\alpha})$, trace$(i) = 1$ *forms a resolvable* BIBD *with parameters*

$$(v, k, \lambda, r, b) = \left[\frac{1}{2}q(q - 1), \ q - 1, \ q, \ \frac{1}{2}q(q + 1), \ \frac{1}{4}q^2(q + 1)\right].$$

Each block of this design is a $q - 1$-*arc (in fact a conic) of* AG$(2, q)$ *with block stabilizer* $0^+(2, q)$.

PROOF. Proceeding exactly as in the proof of Theorem 2, let X and Y be elements of the quadric. By transitivity assume $X = 0$. Arguing as before we seek the number of solutions P to

$$Q(Y + P) + B(Y, P)^2 = Q(0 + P) + B(0, P)^2 = Q(P).$$

But this was computed to be 'q' in the proof of Theorem 2, which concludes this proof. □

Setting $q = 8$ in Theorem 4 we get a resolvable

$$(28, 7, 8, 36, 144) - \text{BIBD}.$$

In [13] 5432 designs are listed with these parameters none of which is known to be resolvable [see 11]. Thus, Theorem 4 provides the first example of a resolvable $(28, 7, 8)$ design.

5. Subdesigns. In all of the designs mentioned above the automorphism group PGL$(2, q)$ may be extended to PΓL$(2, q)$ in the natural way as a subgroup of Sp$(2m, 2)$. When $q = 8$, PΓL$(2, 8)$ acts 2-transitively (as the small Ree group) on the 28 points of the quadric. Thus we get designs by virtue of the group action alone. Interestingly, this means that the $(28, 7, 8)$ design of Theorem 4 (setting $q = 8$) actually contains a $(28, 7, 2)$ design (set $i = 1$ and take the orbit of $\mathscr{S}_1$ under $P\Gamma L(2,8)$) which is, in fact, a residual of the non-self-dual $(37, 9, 2)$ biplane constructed by Hussain [10]. Continuing in this vein we note that setting $q = 8$ in Theorem 2 results in a $(28, 10, 10)$ design. Although the existence of a $(28, 10, 5)$ design has recently been demonstrated, it is worth noting that the constructed design *does not* split into two $(28, 10, 5)$ designs. However, examining the orbits under $P\Gamma L(2,8)$ of the 50,400 words of weight 10 in the dual code of Table 1 a second $(28, 10, 10)$ design was discovered. This second design also fails to split but it does generate the full 19 dimensional dual code implying *it* has no ovals. These constructions provide an example of two nonisomorphic designs with the same parameters and the further property that all the blocks of one design meet all the blocks of the other in an even number of points.

BIBLIOGRAPHY

1. Andriamanalimanana, B., *Ovals, unitals, and codes*, Ph. D. thesis, Lehigh University (1979).
2. Beth, T., Jungknickel, D., and Lenz H., *Design Theory*, Bibliographisches Institut, Mannheim/Wien/Zürich, 1985.
3. Bose, R. C., *On the construction of balanced incomplete block designs*, Ann. Eugenics **9** (1939), 353–399.
4. Brouwer, A. E., *Some unitals on 28 points and their embeddings in projective planes of order 9*, Geometries and Groups, Lecture Notes in Mathematics **893** (1981), 183–188.
5. Cameron, P. J., *Finite permutation groups and finite simple groups*, Bull. London Math. Soc. **13** (1981), 1–22.
6. Denniston, R. H. F., *Some maximal arcs in finite projective planes*, J.Comb. Theory **6** (1969), 317–319.
7. Hanani, H., *Balanced incomplete block designs and related designs*, Discrete Math. **11** (1975), 255–369.
8. Hollman, H. D. L., *Pseudocyclic 3-class association schemes on 28 points*, Discrete Math. **52** (1984), 209–224.
9. Hughes, D. R. and Piper F. C., *Design Theory*, Cambridge University Press, Cambridge/London/New York, 1985.
10. Hussain, Q. M., *Symmetrical incomplete block designs with $\lambda = 2$, $k = 8$ or 9*, Bull. Calcutta Math. Soc. **37** (1945), 115–123.
11. Jungnickel, D., *Quasimultiples of biplanes and residual biplanes*, Ars Comb. **19** (1985), 179–186.
12. Kantor, W. M., *Symplectic groups, symmetric designs, and line ovals*, J. of Algebra **33** (1975), 45–58.
13. Mathon, R. and Rosa, A., *Tables of BIBDs with $r \leq 41$ including existence, enumeration, and resolvability results*, Ann. Disc. Math. **26** (1985), 275–308.
14. _____, *Some results on the existence and enumeration of BIBD's*, Mathematical Report 125, McMaster University (1985).
15. Qvist, B., *Some remarks concerning curves of the second degree in a finite plane*, Ann. Acad. Sci. Fenn. **134** (1952), 1–27.
16. Segre, B., *Ovals in a finite projective plane*, Canadian J. Math, **7** (1955), 414–416.
17. Wertheimer, M. A., *Designs in quadrics*, Ph. D. thesis, University of Pennsylvania (1986).
18. Witt, E., *Theorie der quadratischen Formen in beliebigen Körpern*, J. Reine angew. Math. **176** (1937), 31–44.

DEPARTMENT OF DEFENSE, 9800 SAVAGE ROAD, FORT GEORGE G. MEADE, MARYLAND 20755

Contemporary Mathematics
Volume **111**, 1990

On the Order of a Finite Projective Plane and its Collineation Group

CHAT YIN HO

In this article some of the recent results given in the talk by the author will be reported. First on projective planes of odd composite order pq, we report the following result from [3].

THEOREM 1. *Any collineation group of a projective plane of order* 15 *is solvable and its order divides* 2^6; $2^3 \cdot 3$; $2 \cdot 5$; $3 \cdot 5$; $2^3 \cdot 3 \cdot 7$; *or* $2^6 \cdot 7$.

Regarding a collineation group as a permutation group on the points of the projective plane we obtained the following result which is also known to Prof. A. Prince:

THEOREM 2. *The order of a finite projective plane admitting a group G as a collineation group is bounded by a function of $|G|$ and the number of the regular G-orbits of points.*

The proof of the above theorem uses the permutation character of G on the points of the projective plane and a crude estimation of the number of the fixed points of a collineation. For details see [**4**, Theorem A].

There are two extreme cases. The first is when each G-orbit of points is irregular (i.e. the number of regular G-orbit is 0). In this case we call G *totally irregular*. Finite Desarguesian planes are very close to having totally irregular collineation groups. In fact these planes have been classified by various totally irregular collineation groups, e.g., a doubly transitive collineation group (Ostrom-Wagner); a collineation group such that either each point is a center or each line is an axis of a perspectivity in this group (Cofman-Piper). In the same spirit, explicit calculations, case analyses and improving the bound of Theorem 2 in the case when $G \cong \mathrm{PSL}(2, 4)$ acts totally irregularly on the points of the projective plane yield the following result in [**4**].

1980 *Mathematics Subject Classification* (1985 *Revision*). Primary 51O17; Secondary 20O51.
The final version of this paper has been submitted for publication elsewhere.

THEOREM 3. *A finite projective plane admitting* $\mathrm{PSL}(2, 4) \cong A_5$ *as a totally irregular collineation group is Desarguesian of order* 4, 5, 9, *or* 11.

Each finite Desarguesian plane admits a totally irregular collineation group isomorphic to $\mathrm{PSL}(2, q)$ for some prime power q. It will be interesting to see whether the converse of this is also true. Theorem 3 is a step in this direction. In general it is not true that the total irregularity of the collineation group G implies that G has no regular orbit of lines. For example, in the projective plane of order 3 let G be the collineation group generated by the two collineations corresponding to the following two matrices over the field of three elements:

$$\begin{pmatrix} 1 & 0 & 0 \\ 1 & 1 & 0 \\ 0 & 0 & 1 \end{pmatrix}, \qquad \begin{pmatrix} 1 & 0 & 0 \\ 0 & -1 & 0 \\ 0 & 0 & -1 \end{pmatrix}$$

Then $G \cong S_3$ with point orbits sizes : 1,1,2,3,3,3 ; but line orbits sizes are 1,1,1,1,3,6.

The other extreme case occurs when each G-orbit of points is regular. Then $|G|$ is odd by Baer's involution theorem. So G is solvable by Feit-Thompson's theorem. Besides these remarks, this case seems to be very difficult. The special situation, when G is cyclic and has only one regular orbit of points, (G is called a Singer group when these conditions are satisfied,) studies finite cyclic projective planes. The outstanding conjecture in this area is that every finite cyclic plane is Desarguesian. However even the conjecture that the order of a finite cyclic plane is a prime power has not been verified. Besides group theory and geometry, it seems that some non-trivial number theory is needed. Let Π be a finite cyclic plane with a Singer group G. Let the full collineation group of Π be F. Let X be a point of Π. Then $N := N_F(G) = G \cdot N_X$. The group $M := N/S \cong N_X$ is independent of G and X. We call M the multiplier group of Π. If G is not normal in F, then Π is Desarguesian by Ott's result [7]. We now mention some known facts about M.

(A) Hall [6, p.269–271] proves that three divides $|M|$.

(B) The order of the multiplier group of a Desarguesian plane of order p^k for some prime p is $3k$.

(C) The element in $\mathrm{Aut}(G)$, which inverts every element of G, is not in M [1].

Now (C) implies that $|M| \leq |\mathrm{Aut}(G)|/2$. Using Galois theory of cyclotomic fields, we classify cyclic planes satisfying $|M| = |\mathrm{Aut}(G)|/2$ with the help of the Gaussian quadratic sum. More precisely, the following result is obtained in [5].

THEOREM 4. *Let* Π *be a finite cyclic plane of order* n *with the multiplier group* M. *Then the following holds.*

(1) *The Sylow 2-subgroup* T *of* M *is cyclic. If* $|T| = 2^a$ *for some integer* $a \geq 0$, *then* $n = m^{2^a}$ *for some integer* $m \geq 2$.

(2) *For $k = 3$, 5 or 2^a for some integer $a \geq 0$, we have $|M| = 3k$ if and only if $n = p^k$ for some prime p.*

(3) *We have $|M| = |\mathrm{Aut}(G)|/2$ if and only if $n = 2$ or 4. Furthermore, $n = 2$ occurs if and only if $|M| = $ odd in this case.*

(4) *We have $|M| = |\mathrm{Aut}(G)|/4$ if and only if $n = 3$.*

Some more remarks are in order. The result of Hall shows that F is totally irregular even though G is regular. Also in Theorem 4, statement (1) improves a result of Ostrom and Wagner on planar 2-subgroups for cyclic planes [2]. Statement (2) and (3) treat the extreme values of the order of M. In particular, (2) gives a characterization of the order of a finite cyclic plane being a prime if and only if its multiplier group has order 3. Finally, it would be very interesting if one could prove the following: The multiplier group of a finite cyclic plane of order n has order n if and only if $n = 3$.

REFERENCES

1. L. D. Baumert, "Cyclic difference sets," Springer Lecture notes 182, New York, 1971.
2. P. Dembowski, "Finite Geometries," Springer, New York, 1968.
3. C. Y. Ho, *Projective planes of order 15 and other odd composite orders*, Geom. Dedicata (to appear).
4. _____, *On the order of a projective plane and planes with a totally irregular collineation group*, Proceedings of Symposia in Pure Mathematics **47**, Amer. Math. Soc., 423–429, 1987.
5. _____, *On multiplier groups of finite cyclic planes*, J. of Algebra (to appear).
6. D. Hughes and F. Piper, *Projective planes*, Springer, New York, 1973.
7. U. Ott, *Endliche Zyklische Ebenen*, Math. Z. **144**, 195–215.

DEPARTMENT OF MATHEMATICS, UNIVERSITY OF FLORIDA, 201 WALKER HALL, GAINESVILLE, FLORIDA 32611

Contemporary Mathematics
Volume **111**, 1990

Some Geometric Aspects of Root Finding in $GF(q^m)$

P. C. VAN OORSCHOT AND S. A. VANSTONE

EXTENDED ABSTRACT

Finding roots of polynomials over finite fields is a fundamental problem in algebraic computation and computational number theory. The problem has received considerable attention, with much of the work pioneered by Berlekamp [1]. Lidl and Niederreiter [2] survey a number of efficient techniques. Here we are concerned with extension fields $\mathbf{F} = GF(q^m)$, q a prime or prime power and $m \geq 1$. We consider some combinatorial and geometric aspects of root finding.

Let $f(x) \in \mathbf{F}[x]$ and suppose $\gcd(f(x), x^{q^m} - x) = f(x)$, i.e. $f(x)$ has all of its roots distinct and in $\mathbf{F}$. Since $\mathbf{F}$ is finite, a naive approach for finding the roots of $f(x)$ is to simply check all $\alpha \in \mathbf{F}$ to see which satisfy $f(\alpha) = 0$. Of course, for q and m of even moderate size this technique becomes infeasible, and a more efficient approach is required. We begin our discussion with a simple fundamental observation.

Let $\mathcal{P}$ be an arbitrary partition of the elements of $\mathbf{F}$ into at most n parts. Denote these parts by $B_1, B_2, \ldots, B_t$ where $t \leq n$. Associate with each part B_i a polynomial

$$\Phi_{B_i}(x) = \prod_{\delta \in B_i} (x - \delta), \qquad 1 \leq i \leq t.$$

Let R be the root set of $f(x)$, and let

$$d_i(x) = \gcd(f(x), \Phi_{B_i}(x)), \qquad 1 \leq i \leq t.$$

OBSERVATION. $d_i(x)$ is a non-trivial factor of $f(x)$ if and only if $0 < |R \cap B_i| < |R|$.

If a particular partition does not lead to the discovery of non-trivial factors of $f(x)$ via this observation, then another partition could be selected. This

---

1980 *Mathematics Subject Classification* (1985 *Revision*). Primary 68Q40; Secondary 11T06, 05B25.

The final detailed version of this paper has been submitted for publication elsewhere.

leads to the consideration of a set of partitions which will guarantee the complete factorization of any such polynomial $f(x)$. To this end we define a *pair-splitting set* to be a set of partitions having the property that there is no pair of distinct field elements which is contained in some part(block) of each partition.

It is a simple matter to determine a lower bound on the number of partitions in a pair-splitting set. Let $P_1, P_2, \ldots, P_r$ be the r partitions in a pair-splitting set. Label the parts in each partition arbitrarily with the integers 1 through n. Form an $r \times q^m$ matrix A where the rows are labelled by the partitions and the columns by the elements of $\mathbf{F}$. In cell $A[i, \alpha]$ place the label of the part of P_i which contains field element α. Then A is a matrix defined on the set $U = \{1, 2, \ldots, n\}$, and the set of partitions is pair-splitting if and only if A does not have two identical columns. Since there are n^r r-tuples over U, if the set is pair-splitting then we must have $q^m \leq n^r$, implying $r \geq \log_n q^m$. A pair-splitting set with r partitions meeting the lower bound $\lceil \log_n q^m \rceil$ can thus be constructed from any matrix A over U which has no repeated columns.

We see that in general pair-splitting sets are easily found. However, in practice most turn out to be of little use for finding the roots of a polynomial $f(x)$. This is due to the nature of the polynomial representation $\Phi_B(x)$ of a block B in a partition. In general, $\Phi_B(x)$ will have about $|B|$ non-zero terms. Now since it may be required to check each part in a partition, it is desirable that n be small. Hence since some block in a partition must have at least q^m/n elements, for q^m large the polynomial representation of parts of a partition will be unmanageably large. To overcome this problem, we consider special types of partitions.

We first consider a partition of use in the case when q^m is odd. Let B_0 be the set of quadratic residues in $\mathbf{F}$, and let B_1 be the quadratic non-residues along with the field element 0. Let P_0 be the partition with blocks B_0 and B_1. The polynomial representation of these blocks is then simply

$$\Phi_{B_0}(x) = x^{(q^m-1)/2} - 1, \qquad \Phi_{B_1}(x) = x(x^{(q^m-1)/2} + 1).$$

Define P_α to be the partition consisting of blocks $B_0 + \alpha$ and $B_1 + \alpha$, where $B_i + \alpha = \{b + \alpha : b \in B_i\}$. The polynomial representation of these blocks is then $(x - \alpha)^{(q^m-1)/2} - 1$ and $(x - \alpha)((x - \alpha)^{(q^m-1)/2} + 1)$ respectively. If we limit our choice to partitions of this particular type, then polynomial representations will be compact, but finding a pair-splitting set with a minimum number of blocks becomes a much more difficult problem. In fact, pair-splitting sets having $\lceil \log_n q^m \rceil$ partitions may not even always exist.

EXAMPLE. Consider $\mathbf{F} = GF(31)$ and the partition $P_0 = (B_0, B_1)$ where

$$B_0 = \{1, 2, 4, 5, 7, 8, 9, 10, 14, 16, 18, 19, 20, 25, 28\},$$
$$B_1 = \{0, 3, 6, 11, 12, 13, 15, 17, 21, 22, 23, 24, 26, 27, 29, 30\}.$$

Note here that $\lceil \log_2 31 \rceil = 5$. Exhaustive search reveals that among the partitions $P_\alpha = (B_0 + \alpha, B_1 + \alpha)$, $\alpha \in \mathbf{F}$, no pair-splitting set of cardinality 5 exists. However, many pair-splitting sets of cardinality 6 can be found; one such pair-splitting set is $\{P_0, P_1, P_2, P_3, P_5, P_{28}\}$.

Although splitting sets meeting the lower bound cannot always be found, it has been shown [3] ,[4] that splitting sets having cardinality at most $2 \log_2 q^m$ always exist among such partitions. Unfortunately, no efficient (i.e. with time polynomial in $m \log(q)$) method is yet known for finding such sets. In fact, we do not even know whether this problem is in the class NP.

Inability to efficiently find pair-splitting sets when using partitions of this type rules out easy construction of a deterministic algorithm. Nonetheless, this partitioning is the basis for a very effective probabilistic (Las Vegas) root finding algorithm [1] , [5] .

For the case when q is small (including the case of even q) and $m > 1$ is large, a more effective partitioning is obtained by considering the affine geometry of dimension m and order q, $AG(m, q)$. Denote the set of all k-flats in $AG(m, q)$ by $AG_k(m, q)$. As any two m-dimensional vector spaces over $GF(q)$ are isomorphic, we may identify the underlying vector space of $AG(m, q)$ with the field $\mathbf{F} = GF(q^m)$. Now let P be a partition of $\mathbf{F}$ consisting of a parallel class of k-flats in $AG_k(m, q)$. If $B \in P$ then

$$\text{(1)} \qquad \Phi_B(x) = \lambda + \sum_{i=0}^{k} \lambda_i x^{q^i},$$

where $\lambda, \lambda_i \in \mathbf{F}$. Here note that $|B| = q^k$ but $\Phi_B(x)$ has at most $k + 2$ non-zero terms.

It is easily seen from the lower bound discussed above that using partitions composed of k-flats, a lower bound on the cardinality of a pair-splitting set here is $\lceil m/(m-k) \rceil$. For $k = m - 1$, it was observed by Berlekamp [1] that this lower bound can be realized, leading to an efficient deterministic root finding algorithm; a complete characterization of such splitting sets is given in [6]. It can further be shown (see discussion below) that the bound $\lceil m/(m - k) \rceil$ can be met for arbitrary k. Before proceeding here to characterize a special type of such pair-splitting sets, we shall require additional notation.

Again identifying $AG(m, q)$ with $\mathbf{F}$, each $\alpha \in \mathbf{F}^* = \mathbf{F} \setminus \{0\}$ induces a collineation $\bar{\alpha}$ of $AG(m, q)$ by defining

$$\bar{\alpha}(\beta) = \alpha\beta \quad \text{for all } \beta \in \mathbf{F}.$$

$\bar{\alpha}$ is a multiplicative automorphism, usually denoted simply α. The set of all multiplicative automorphisms forms a group G isomorphic to $\mathbf{F}^*$, and thus isomorphic to the cyclic group of order $q^m - 1$. We call an orbit of parallel classes of k-flats under the action of the group G a (multiplicative) $k - orbit$ (of parallel classes). The polynomial representations of k-flats in a common orbit are closely related. In addition to (1), if P^* is any parallel class of k-flats in the orbit of P, then for some fixed $\alpha \in \mathbf{F}$ each k-flat $B^* \in P^*$ has polynomial representation

$$\Phi_{B^*}(x) = \Phi_B(\alpha x) \quad \text{for some } B \in P.$$

We call a pair-splitting set contained in a k-orbit a *multiplicative pair-splitting set*. The following characterization of multiplicative pair-splitting sets can then be established. For $V \subseteq \mathbf{F}$, denote the set $\{v\alpha : v \in V\}$ by $V\alpha$.

THEOREM. *Let B be the k-orbit determined by a k-dimensional linear subspace V of $\mathbf{F} = GF(q^m)$ over $GF(q)$, and let $\alpha_1, \alpha_2, \ldots, \alpha_s$ be distinct elements in $\mathbf{F}^*$. Then the parallel classes determined by $V\alpha_1, V\alpha_2, \ldots, V\alpha_s$ form a pair-splitting set of $AG_k(m, q)$ if and only if the subspace U spanned by the set $\{\alpha_1^{-1}, \alpha_2^{-1}, \ldots, \alpha_s^{-1}\}$ is not contained in any k-dimensional linear subspace of B.*

PROOF. Assume that $V\alpha_1, V\alpha_2, \ldots, V\alpha_s$ determine a pair-splitting set. Then

$$(2) \qquad\qquad \bigcap_{i=1}^{s} V\alpha_i = \{0\}.$$

Suppose that U is contained in $V\delta$ for some $\delta \in \mathbf{F}^*$. Then $U\delta^{-1}$ is a subspace of V. Thus $\alpha_i^{-1}\delta^{-1} \in V$ for $1 \le i \le s$, implying $\delta^{-1} \in V\alpha_i$ for $1 \le i \le s$, contradicting (2). We conclude U is not contained in any k-space of B.

Conversely, assume that U is not contained in any k-dimensional linear subspace of B. If $V\alpha_1, V\alpha_2, \ldots, V\alpha_s$ do not determine a pair-splitting set, then there exist distinct elements $a, b \in \mathbf{F}$ contained in the same coset of $V\alpha_i$ for all $1 \le i \le s$. Let $c = a - b \ne 0$. Then $c \in \bigcap_{i=1}^{s} V\alpha_i$, implying $c\alpha_i^{-1} \in V$ for $1 \le i \le s$. But then Uc, the span of $\{c\alpha_1^{-1}, c\alpha_2^{-1}, \ldots, c\alpha_s^{-1}\}$, is contained in V, yielding the contradiction $U \subset Vc^{-1}$. $\square$

This result tells us that the size of a smallest multiplicative pair-splitting set is equal to the dimension d of a smallest subspace not contained in any k-flat of that k-orbit. Our lower bound given earlier implies $d \ge \lceil m/(m-k) \rceil$. In [7] it is shown that in any k-orbit, regardless of its size, a subspace of dimension $d = \lceil m/(m-k) \rceil$ exists which is not contained in any k-flat of this k-orbit. It follows that a pair-splitting set of minimum size can be found among the parallel classes of any k-orbit.

The proof of this last statement is one of existence, and in general does not give an effective method for finding such multiplicative pair-splitting sets. In the case when $m - k$ divides m, an efficient constructive technique is available, and is described in [7]. If $m - k$ does not divide m, then the technique can still be used to find a pair-splitting set of minimum size, but the splitting set is now not guaranteed to be contained within a k-orbit. However, multiplicative pair-splitting sets of minimum size appear to be abundant in any k-orbit. Thus probabilistic (Las Vegas) methods for determining pair-splitting set are feasible.

We note that given a set of r parallel classes constructed from k-flats, here, unlike in the general case, it can be efficiently checked whether or not this set is pair-splitting. This is possible due to the special polynomial representation of blocks. A set of partitions constructed from k-flats is pair-splitting if and only if (2) is satisfied. Now the intersection of blocks corresponds to the greatest common divisor of their polynomial representations, and here, the blocks being k-flats have affine polynomial representations. Furthermore, the greatest common divisor of two affine polynomials is again an affine polynomial, and moreover division of affine polynomials may be expedited by using exponentiation to align high order terms where multiplication would be used in standard long division of polynomials.

We summarize with the statement that for the purposes of a deterministic root finding algorithm, it is useful to exploit the rich group structure available in the affine geometry.

References

1. E. R. Berlekamp, *Factoring polynomials over large finite fields*, Math. Comp. **24** (1970), 713–735.
2. R. Lidl and H. Niederreiter, " Finite Fields," Addison-Wesley, 1983.
3. P. Camion, *A deterministic algorithm for factorizing polynomials of $F_q|x|$*, Annals of Discrete Mathematics **17** (1983), 149–157.
4. P. C. van Oorschot and S. A. Vanstone, *On splitting sets in block designs and finding roots of polynomials,* Discrete Math. **35**, No. 2 (1989).
5. M. O. Rabin, *Probabilistic algorithms in finite fields*, SIAM J. Comput. **9**(1980), 273–280.
6. P.C. van Oorschot and S. A. Vanstone, *A geometric approach to root finding in $GF(q^m)$*, IEEE Transactions on Information Theory **35**, No. 2 (1989).
7. A . Beutelspacher, D. Jungnickel, P. C. van Oorschot and S.A. Vanstone, *Pair-splitting sets in $AG(m, q)$*, Research Report CORR **87-41**, Department of Combinatorics and Optimization, University of Waterloo, December 1987.

DEPARTMENT OF COMPUTER SCIENCE AND DEPARTMENT OF COMBINATORICS AND OPTIMIZATION, UNIVERSITY OF WATERLOO, WATERLOO, ONTARIO, N2L 3G1 CANADA
Current address, P. C. van Oorschot: Bell-Northern Research Ltd., P.O. Box 3511, Station C, Ottawa, Ontario K1Y 4H7 Canada

Contemporary Mathematics
Volume 111, 1990

Automorphism Groups as Linear Groups

JOHANNES SIEMONS

ABSTRACT. A permutation group acting on a set Ω can always be viewed as a linear group acting on $F\Omega$. This is the vector space over a chosen field F with basis indexed by the elements of Ω. In this note we examine automorphism groups acting on the point set and on the block set of a finite incidence structure. We report about results obtained in [1, 3] which concern the relationship between such permutation actions when viewed as linear representations.

EXTENDED ABSTRACT

I. Introduction. Incidence structures for our purpose are very general objects: they consist of two finite, disjoint sets P and B with some incidence relation. This relation can be viewed simply as some subset I of $P \times B$. Let F be some field and let FP and FB be the vector spaces with basis indexed by the elements of P and B respectively. The incidence relation then gives rise to incidence maps between FP and FB which in turn lead to a standard decomposition of these modules. This is described in the first section.

Automorphisms of $(P, B; I)$ consist of pairs of permutations which preserve the incidence relation. Consequently, when viewed as maps of FP and FB, an automorphism commutes with the incidence maps. This fact is the reason for an intertwining relationship between the linear representations on points and blocks. In Section 2 more detail and a number of corollaries are given.

In this note we can only attempt to produce the principal ideas involved. Both theorems with most corollaries are proved in a forthcoming paper [1].

II. Spectral decompositions. Let P be a finite set of points, B a finite set of blocks, disjoint for P, and I some subset of $P \times B$. We say that p is *incident* with b if (p, b) belongs to I. The triple (P, B, I) then an

---

1980 *Mathematics Subject Classification* (1985 *Revision*). Primary 05B25; Secondary 20B25.
This paper is in final form and no version of it will be submitted for publication elsewhere.

incidence structure.

For a field F we construct the point module FP as usual by taking points as basis vectors. In the same way the block module FB is defined. Now two incidence maps $\delta^+ : FP \to FB$ and $\delta^- : FB \to FP$ arise by putting

$$\delta^+(p) = \sum_{(p,b)\in I} b \quad \text{and} \quad \delta^-(b) = \sum_{(p,b)\in I} p.$$

Denote the standard inner products on FP and FB by $\langle \, , \rangle$ in both cases. Then $\langle p, \delta^-(b) \rangle = \langle \delta^+(p), b \rangle$ so that δ^- and δ^+ are adjoint to each other. This means that $\delta^-\delta^+ : FP \to FP$ and $\delta^+\delta^- : FB \to FB$ are both self-joint (or symmetric) maps. We say that F is a *suitable* field if the characteristic polynomials of these maps split into linear factors where furthermore the multiplicity of each eigenvalue is the same as the dimension of the corresponding eigenspace. So the field of real numbers, for instance, is suitable. Then the spectrum of $(P, B; I)$, denoted by spec, are the eigenvalues of $\delta^-\delta^+ : FP \to FP$ counted with multiplicities. For λ in spec the corresponding eigenspaces of $\delta^-\delta^+ : FP \to FP$ and of $\delta^+\delta^- : FB \to FB$ are denoted by E_λ and by E_λ^* respectively.

THEOREM A. *Let $(P, B; I)$ be an incidence structure and let F be a suitable field. Then*

$$FP = \bigoplus_{\lambda\in\text{spec}} E_\lambda \quad \text{and} \quad FB = \bigoplus_{0\neq\lambda\in\text{spec}} E_\lambda^* \oplus E_0^*.$$

Furthermore, if $\lambda \neq 0$ then $\delta^+ : E_\lambda \to E_\lambda^*$ is an isomorphism.

Of general interest are the incidence structures whose elements are the k- and ℓ-element subsets of an n-set, $k < \ell$, and $\ell + k < n$, with containment as incidence relation. The spectrum of such a structure consists of integer values $\lambda_0 > \lambda_1 \cdots > \lambda_k > 0$ where λ_i has multiplicity $\binom{n}{i} - \binom{n}{i-1}$. The situation is very similar when k- and ℓ-dimensional subspaces of an n-dimensional vector space over a Galois field are considered. There again the spectrum consists of $k + l$ integer values $\lambda_0 > \lambda_1 > \cdots > \lambda_k > 0$ with multiplicities given in terms of q-binomial coefficients. For details of these results see [3].

III. Intertwining. An automorphism of $(P, B; I)$ is a pair of permutations (g_P, g_B) of P and B so that (p, b) belongs to I precisely when $(g_P(p), g_B(b))$ belongs to I. Let π and β denote the permutation characters on the points and the blocks of the incidence structure.

Let F be a field and let FP, FB be the associated modules as above. The permutations of P and B extend to linear actions on FP and FB in the usual way. Now observe that if (g_P, g_B) is an automorphism, then

$$(*) \qquad\qquad \delta^+ g_P = g_B\delta^- \quad \text{and} \quad \delta^- g_B = g_P\delta^+.$$

These relations can be viewed as an intertwining between the linear actions on points and blocks. It follows from $(*)$ that the eigenspaces E_λ and E_λ^*

of the previous section are G-invariant. Hence let π_λ and β_λ denote the characters of G acting on E_λ and E_λ^*. As a corollary to theorem A we have

COROLLARY A1. *Over a suitable field F we have*

$$\pi = \sum_{\lambda \in \text{ spec}} \pi_\lambda \quad and \quad \beta = \sum_{0 \neq \lambda \in \text{ spec}} \beta_\lambda + \beta_0,$$

where furthermore $\pi_\lambda = \beta_\lambda$ for $\lambda \neq 0$.

Now we turn to the general situation of arbitrary fields where spectra may not be defined.

THEOREM B. *Let $(P, B; I)$ be an incidence structure, G a group of automorphism and F a field. Suppose that either $\delta^- \delta^+ : FP \to FP$ is a bijection or that $\delta^- : FB \to FP$ is surjective and $|G| \neq 0$ in F. Then $\beta = \pi + \psi$ where ψ is the character of the F-representation on the kernel of $\delta^- : FB \to FP$.*

Note that the hypotheses of the theorem are equivalent when F has characteristic zero. In this case we say that an incidence structure satisfying the hypothesis is of *maximal linear rank*. We state a number of consequences of Theorem B. The first is rather well known.

COROLLARY B1 (BLOCK'S LEMMA). *Let $(P, B; I)$ be an incidence structure of maximal linear rank and let G be an automorphism group with n_P and n_B orbits on points and blocks. Then $0 \leq n_B - n_P$ is the multiplicity of the principal character in $\psi = \beta - \pi$ where π and β are as in theorem B.*

The permutation rank of G on points, denoted $r_P(G)$, is the number of G-orbits on $P \times P$. The permutation rank on blocks $r_B(G)$ is defined equivalently. Let $f(G)$ denote the number of G-orbits on flags (incident point-block pairs) and let $a(G)$ denote the number of G-orbits on anti-flags (non-incident point-block pairs).

COROLLARY B2. *Let $(P, B; I)$ be an incidence structure of maximal linear rank with automorphism group G. Then $r_P(G) \leq a(G) + f(G) \leq r_B(G)$ and the following are equivalent:*

    (1) $|P| = |B|$,
    (2) $a(G) + f(G) = r_B(G)$,
    (3) $r_P(G) = r_B(G)$.

Furthermore $a(G) + f(G) = r_P(G)$ if and only if π and $\beta - \pi$ have no character of G in common.

A set S of permutations on B is *sharply transitive* if for every pair (b, b') from B there is precisely one $g \in S$ with $g(b) = b'$. The following is based on O'Nan's work in [2].

COROLLARY B3. *Let $(P, B; I)$ be of maximal linear rank with $|B|$ not divisible by $|P|$. Then no set of automorphisms can act sharply transitively on B.*

References

1. A.R. Camina & I. J. Siemons, *Intertwining automorphisms in finite incidence structures*, Linear Algebra Appl. **117** (1989), 25–34.
2. M. O'Nan, *Sharply 2-transitive sets of permutations*, Proceedings of the Rutgers group theory year, 1983-1984, Cambridge University Press, Cambridge-New York, 1985.
3. I. J. Siemons, *On a class of partially ordered sets and their linear invariants*, Geom. Dedicata (to appear).

SCHOOL OF MATHEMATICS, UNIVERSITY OF EAST ANGLIA, NORWICH NR4 7TJ, UNITED KINGDOM